Intelligent Systems Reference Library

Volume 188

The aim of this series is to publish a Reference Library, including novel advances and developments in all aspects of Intelligent Systems in an easily accessible and well structured form. The series includes reference works, handbooks, compendia, textbooks, well-structured monographs, dictionaries, and encyclopedias. It contains well integrated knowledge and current information in the field of Intelligent Systems. The series covers the theory, applications, and design methods of Intelligent Systems. Virtually all disciplines such as engineering, computer science, avionics, business, e-commerce, environment, healthcare, physics and life science are included. The list of topics spans all the areas of modern intelligent systems such as: Ambient intelligence, Computational intelligence, Social intelligence, Computational neuroscience, Artificial life, Virtual society, Cognitive systems, DNA and immunity-based systems, e-Learning and teaching, Human-centred computing and Machine ethics, Intelligent control, Intelligent data analysis, Knowledge-based paradigms, Knowledge management, Intelligent agents, Intelligent decision making, Intelligent network security, Interactive entertainment, Learning paradigms, Recommender systems, Robotics and Mechatronics including human-machine teaming, Self-organizing and adaptive systems, Soft computing including Neural systems, Fuzzy systems, Evolutionary computing and the Fusion of these paradigms, Perception and Vision, Web intelligence and Multimedia.

** Indexing: The books of this series are submitted to ISI Web of Science, SCOPUS, DBLP and Springerlink.

More information about this series at http://www.springer.com/series/8578

Alfred Zimmermann · Rainer Schmidt ·
Lakhmi C. Jain
Editors

Architecting the Digital Transformation

Digital Business, Technology, Decision Support, Management

Springer

Editors
Alfred Zimmermann
Faculty of Informatics
Reutlingen University
Reutlingen, Baden-Württemberg, Germany

Rainer Schmidt
Faculty of Computer Science
Munich University of Applied Sciences
Munich, Bayern, Germany

Lakhmi C. Jain
University of Technology Sydney
Sydney, NSW, Australia

ISSN 1868-4394 ISSN 1868-4408 (electronic)
Intelligent Systems Reference Library
ISBN 978-3-030-49642-5 ISBN 978-3-030-49640-1 (eBook)
https://doi.org/10.1007/978-3-030-49640-1

This Springer imprint is published by the registered company Springer Nature Switzerland AG
The registered company address is: Gewerbestrasse 11, 6330 Cham, Switzerland

Foreword

Digital transformation is considered the most complex phase of digitalization because it has the potential to change markets, create new technological applications, and further advance products and services. This leads to changes in economic structures, requires new skill and competence profiles for employees, as well as, in broad terms, influences work and business practices. The digital transformation that we are currently experiencing creates many benefits for the companies that engage in it. For the society, the effects can be seen as disruptive with far-reaching social and economic consequences.

This book *Architecting the Digital Transformation* offers a comprehensible account of the state of the art in the area of enterprise architecture and digital transformation. In doing so, it also offers ideas and guidance for the paths ahead. It sets the fundamental principles of digital transformation within the context of enterprise architecture management. The book offers a wide range of methodological and technological contributions from the areas of business and enterprise modeling, service design, capability management, conceptual modeling, business process modeling and management, data-intensive systems and analytics, risk analysis, as well as adaptability and resilience. The topics are discussed from a holistic view and explored from multiple dimensions.

The material included in the book stems from research works, but it is strongly based on the practice in the field, which denotes scientific quality and relevance. The content of the book also has a significant educational value—it presents the background, cutting-edge contributions as well as future challenges. All this should be as relevant to those just starting their journeys in the fields of enterprise architecture and digital transformation as to established researchers and practitioners.

This book comes out at the time of the Covid-19 pandemic. At the time of writing this foreword, we are still in its beginning, but it is becoming increasingly apparent that the crisis will profoundly affect the future of business and society globally. The preliminary analysis indicates that we are to expect heavy economic impact and, possibly, sluggish recovery. This crisis shows us that nothing can be taken as granted and the need to adapt. At the same time, as Arundhati Roy writes in Financial Times (April 3, 2020), historically, pandemics have forced humans to break with the past and imagine their world anew. The ethos of this book is that of supporting digital transformation with agility, resilience, and systematic way of solving organizational and IT management and design problems. It provides knowledge on the basis of which many solutions can be created rapidly and innovatively as companies seek new application areas for their products and services or adapt their business profiles. It will be needed for the times ahead.

Janis Stirna
Stockholm University
Stockholm, Sweden

Preface

This research oriented book presents essential trends for architecting the digital transformation. New methods, models, and architectures for digital products and services are integrating novel information technology trends of our days. The potential of the Internet and related digital technologies, such as the Internet of Things, cognition and artificial intelligence, data analytics, services computing, cloud computing, mobile systems, collaboration networks, cyber-physical systems, and robotics are strategic drivers and both enablers of digital platforms with fast-evolving ecosystems of intelligent services exploiting a service-dominant logic.

Data, information, and knowledge are fundamental core concepts of our everyday activities and are driving the digital transformation of today's global society. New services and smart connected products expand physical components by adding information and connectivity services using the Internet. Influenced by the transition to digitalization, many enterprises are presently transforming their strategy, culture, processes, and their information systems to become more digital and adopt systems and services with artificial intelligence.

Digital technologies are strategic drivers for highly transformative digital business models that enable new disruptive businesses. In this way, digital technologies inspire current and future digital strategies and create new business opportunities for the digital transformation enabling next digital products and services. Digitalization fosters the development of IT systems with many, globally available and diverse, rather small and distributed structures. This has a strong impact for top-down architecting intelligent digital services and products integrating high distributed intelligent systems additionally in a bottom-up manner.

This book provides essential novel approaches to address challenges of architecting and managing the digital transformation to modern businesses. The chapters included in this book contribute to the understanding of relevant trends of current research and practice. The chapters of the book are aligned according following main parts: digital transformation, digital business, digital architecture, decision support, and digital applications. The book is directed to researchers, postgraduate and doctoral students, graduate and undergraduate students, professors and practitioners interested in the digital transformation and the current orientation of main businesses to digital business models and architectures. We are grateful to the contributors and peer reviewers for their valuable expertise and contributions, without them this book would not have existed. We wish to show our appreciation to Springer-Verlag for their support right from the concept development to the final phase of this book. The unconditional support provided by our universities is acknowledged.

Reutlingen, Germany — Alfred Zimmermann
Munich, Germany — Rainer Schmidt
Sydney, Australia — Lakhmi C. Jain

Contents

Part I Digital Transformation

About the Editors

Alfred Zimmermann is Professor of Computer Science for Digital Enterprise Architecture at Reutlingen University, Germany. He is director of research and speaker of the doctoral program services computing at the Herman Hollerith Center, Böblingen, Germany. His research is focused on digital transformation and digital enterprise architecture with decision analytics in close relationship with digital strategy and governance, software architecture and engineering, artificial intelligence, data analytics, Internet of Things, services computing, and cloud computing. He graduated in medical informatics at the University of Heidelberg, Germany, and obtained his Ph.D. in informatics from the University of Stuttgart, Germany. Besides his academic experience, he has a strong practical background as a technology manager and leading consultant at Daimler AG, Germany. Professor Zimmermann keeps academic relations of his home university to the German Computer Science Society (GI), the Association for Computing Machinery (ACM), and IEEE.

e-mail: alfred.zimmermann@reutlingen-university.de

Rainer Schmidt is Professor of Business Information Systems at Munich University of Applied Sciences. He holds a Ph.D. (KIT Karlsruhe) and an engineering degree in Computer Science. His current research areas include artificial intelligence, social information systems, business process management, and the integration of these themes. Dr. Schmidt has successfully directed several successful projects in the field of artificial intelligence. They include the detection of exceptional situations and their management in data centers, the automatic configuration of services according to objectives such as reliability and energy efficiency, and the structural analysis and optimization of IT systems. Rainer Schmidt is co-organizer of the BPMDS working conference at CAISE, the BPMS2 workshop series at BPM'08 to BPM'19, the SoEA4EE workshop series at EDOC since 2009, the IDEA workshop series at BIS, associate editor at WI 2017, and a member of the program committee of several workshops and conferences. Rainer Schmidt is serving on the editorial boards of International Journal of Information Systems in the Service Sector, and International Journal on Advances in Internet Technology. He has industrial experience as a management consultant and researcher. In 2018, he was awarded the Oskar von Miller Prize as the best researcher at Munich University of Applied Sciences. Rainer Schmidt applies his research in a number of projects and cooperation with industry.

e-mail: rainer.schmidt@hm.edu

Lakhmi C. Jain is a visiting professor at Liverpool Hope University, UK, and adjunct professor at University of Technology Sydney, Australia.

Dr. Jain founded the KES International for providing a professional community the opportunities for publications, knowledge exchange, cooperation, and teaming. Involving around 5000 researchers drawn from universities and companies worldwide, KES facilitates international cooperation and generate synergy in teaching and research. KES regularly provides networking opportunities for professional community through one of the largest conferences of its kind in the area of KES.

www.kesinternational.org

His interests focus on the artificial intelligence paradigms and their applications in complex systems, security, e-education, e-healthcare, unmanned air vehicles, and intelligent systems.

Part I
Digital Transformation

Chapter 1
Architecting the Digital Transformation: An Introduction

Alfred Zimmermann, Rainer Schmidt, and Lakhmi C. Jain

Abstract This chapter presents an introduction to the emerging trends for architecting the digital transformation having a strong focus on digital products, intelligent services, and related systems together with methods, models and architectures. The primary aim of this book is to highlight some of the most recent research results in the field. We are providing a focused set of brief descriptions of the chapters included in the book.

Keywords Digitalization · Digital transformation · Digital technologies · Digital business · Digital architecture · Digital applications

1.1 Introduction

Current companies and organizations are adapting their strategy, business models, products and services as well as business processes and information systems in order to increase their level of digitalization [1] through smart services and digitally enhanced products. The potential of the Internet and related digital technologies, such as the Internet of Things, cognition and artificial intelligence, data analytics, services computing, cloud computing, mobile systems, collaboration networks and cyber-physical systems, are both strategic drivers and enablers of digital platforms with fast-evolving ecosystems [2] of intelligent services for digital products.

Digitalization [3] fosters the development of IT systems with many globally available and diverse, rather small and distributed structures, like Internet of Things or

A. Zimmermann (✉)
Informatics, Reutlingen University, Herman Hollerith Center, Reutlingen, Germany
e-mail: alfred.zimmermann@reutlingen-university.de

R. Schmidt
Informatics and Mathematics, Munich University, Munich, Germany
e-mail: rainer.schmidt@hm.edu

L. C. Jain
University of Technology Sydney, Sydney, NSW, Australia
e-mail: jainlakhmi@gmail.com

A. Zimmermann et al. (eds.), *Architecting the Digital Transformation*,
Intelligent Systems Reference Library 188,
https://doi.org/10.1007/978-3-030-49640-1_1

mobile systems. This has a substantial impact on architecting intelligent digital services and products while bottom-up integrating highly distributed intelligent systems. Data, information and knowledge are fundamental core concepts of our everyday activities and are driving the digital transformation [4] of today's global society [3]. New services and smart connected products expand physical components by adding information and connectivity services using the Internet.

Digital transformation is the current dominant type of business transformation, having IT both as a technology enabler and as a strategic driver [5]. Digital technologies are the main drivers for digitalization because digital technologies are changing the way, how business is conducted and have the potential to disrupt existing businesses. In [1] SMACIT defines the strategic core of digital technologies, with abbreviations for social, mobile, analytics, cloud, and the Internet of Things. From today's view, we have to add more technologies and, therefore, enlarge this technological core by artificial intelligence and cognition, biometrics, robotics, blockchain, 3-D printing and edge computing. Therefore, digital technologies deliver three core capabilities for a changing business [1]: ubiquitous data, unlimited connectivity, and massive processing power.

When we use the term digitalization [6], we mean more than just digital technologies. Digitalization [3] bundles the mature phase of a digital transformation [4] from analog over digital to fully digital. Through digital substitution initially, only analog media are replaced by digital media considering the same existing business values, while digital augmentation functionally enriches related transformed analog media. In a further step of the digital transformation, new processing patterns or processes are made possible by a digitally supported modification of the basic terms or concepts. Finally, completely new forms of value propositions for disruptive businesses, services, products, processes and systems are enabled by digital redefinition of the processes, services and systems. Digitalization [6] is, therefore, more about shifting processes to attractive high automated digital business operations and not just communication by using the Internet. Suitable digital redefinition usually causes disruptive effects on business. Going beyond the value-oriented perspective of digitalization, digital business requires a careful adoption of human, ethical and social principles.

Considering close related concepts of digitization, digitalization and digital transformation [4, 6] we conclude: Digitization and digitalization is about digital technology, while digital transformation is about the changing role of digital customers and the digital change process based on new value propositions. We digitize information, we digitalize processes and roles for extended platform-based business operations, and we digitally transform the business by promoting a digital strategy, customer-centric and value-oriented digital business models, and an architecture-driven digital change.

We will proceed as follows. First, we are setting the fundamental architectural context of digitalization and digital transformation. Then we present essential perspectives of the digital business. With the context of digital transformation and digital business, we give insights to a new style of digital architecture. We are focusing on architectural decision support methods and mechanisms. Finally, we present exemplarily digital applications and architectures from selected domains.

Digital transformation and digital innovation both call for a methodical and technical integration of approaches from different areas of information systems research facilitating the required changes, such as digital business model management (BMM), capability management (CM) and enterprise architecture management (EAM). The book chapter on Digital Innovation and Transformation addresses this methodical and technical integration by proposing the Digital Innovations and Transformation Process (DITP), which has a focus on method support, and the Digital Business Architect (DBA), as a modern qualification profile for professionals in an enterprise working in innovation and transformation initiatives.

The chapter on Perpetual Evolution introduces a new approach that allows changing out elements of enterprise architecture quickly, adding new parts in no time, and incorporating the latest and greatest functionality. We also explore what is required to shift to this more modern approach. Lastly, we demonstrate the model with examples and explain how companies can streamline their legacy business processes and mindsets.

The chapter Enabling Rapid Digital Transformation introduces several methodologies to rapid digital transformation, including Segregated Business Services Design, based on pre-emptive identification and definition of eventually possible digital interfaces throughout the enterprise's business ecosystem. Adopted as a pattern-based modeling style this does not make the EA development more difficult, but it makes the enterprise intrinsically ready for rapid changes, such as client-facing service digitalization, digital outsourcing, digital consolidation of internal services, digital interaction with suppliers, and digital-oriented business model change.

The chapter on Digitally Transforming Live Communication presents results from a field study on the use of digital services for live communication. We formulate performance criteria and indicators to study the effects of digital services on event management. A framework for Event Resource Management provides insights on services that support managing related assets. We discuss how digital services act as enablers for event management and how their use drives firm capabilities for continued digital transformations in event management.

The chapter Digital Transformation of Public Administration analyzes essential factors of public administration business and propose the way to the application of the idea of digital transformation in public administration based on the concept of process-driven management as a representative of the essence of digital transformation.

The chapter Requirements for IT-Support of Personal Services in the Digital Age shows mechanisms to empirically elicit requirements for improved IT- support in the domain of personal services by combining workshops, interviews, and shadowing techniques in a triangulation approach. With our consolidated catalog of requirements, we promote the design of information systems that provide improved IT support for people working in personal service contexts.

The digital transformation is today's dominant business transformation having a strong influence on how digital services and products are designed in a service-dominant way. The chapter on Service-Dominant Design (SDD) gives insides for a comprehensive ideation method based on S-D logic. SDD is aimed at supporting teams in the transition to a service- and value-oriented perspective for designing digital products and services.

In the chapter on Semantically Integrated Conceptual Modeling (SICM) method we show how the method can be applied for business process re-engineering. A well-known case study is analyzed to demonstrate how service interactions among enterprise organizational components can be represented to detect the essential similarities and differences of business processes.

Enterprises are currently transforming their strategy, processes, and information systems to extend their level of digitalization. The main result of the book chapter on Architecting Digital Products and Services connects methods for integral digital strategies with value-oriented architectural models for digital products and services, which are defined in the framework of a multi-perspective digital enterprise architecture reference model.

With the increasing complexity of business models and information technologies, the usage of enterprise architectures becomes inevitable, so that the alignment of business IT is maintained. In the chapter on Adaptive and Resilient Business Architecture for the Digital Age we describe a method how to develop a resilient and adaptive minimal business architecture so that business requirements can be clearly defined, their support by the functionality of applications evaluated and the missing support resolved. The method is illustrated with the generally accepted EA standards TOGAF/ArchiMate, UML and BPMN and its benefits are discussed. We also provide the reader with a meta-model that specifies how these standards match together.

The chapter on Adaptive Integrated Digital Architecture Framework: Risk Management Case presents the unique and new solution of a resiliency architecture model together with a risk management model as well as the architecture assessment model, knowledge sharing model and social collaboration model, which are related to the "Adaptive Integrated Digital Architecture Framework—AIDAF," applied for crucial problems in digital transformation.

The chapter on Automated Enterprise Architecture Model Maintenance via Real-time IT Discovery presents a novel approach for an EA model creation that leverages runtime service instrumentation of the existing IT architecture to automatically create, update, and enhance static EA models with runtime information. An implemented prototype allows different stakeholders to explore information from both perspectives (static and runtime) based on a linked enterprise knowledge graph, which supports new use cases and analysis capabilities.

The chapter Digital Enterprise Architecture for Global Organizations introduces and briefs enterprise architecture (EA) frameworks such as TOGAF, FEAF, DoDAF, MODAF, Adaptive EA, etc. and analyses the selected EA frameworks from standpoints of Cloud/Mobile IT integration. The author shows the Strategic Architecture framework aligned with Digital IT strategies based on the above solutions, as the framework of Digital Enterprise Architecture.

To support organizations in developing their EA, the chapter on An Integrative Method for Decision-Making in EA Management introduces a novel integrative method that systematically integrates stakeholder interests into decision-making activities. By using the method, collaboration between stakeholders involved is improved by identifying points of contact. Furthermore, standardized activities make decision-making more transparent and comparable without limiting creativity.

The chapter Transformation and Enactment of Data-Intensive Business Process Using Advanced Architectural Styles proposes a method for specifying requirements towards data-intensive activities and uses these requirements to select appropriate implementation and enactment technologies. An example of business process redesign is discussed.

The chapter on A Tool Supporting Architectural Principles and Guidelines in Large-Scale Agile Development presents a prototypical web application called "Architecture Belt" that supports the establishment of architecture principles. It uses social design principles and the analogy of belts in martial arts to enforce the application of architecture principles by exerting institutional pressures on agile teams.

The collaboration between enterprise architects and agile teams is not always frictionless as both exhibit antithetic mindsets. Against this backdrop, we present in the chapter about Improving the Collaboration Between Enterprise Architects and Agile Teams an embedded multiple-case study of five leading German companies that aims to shed light on this field of tension. Based on our results from 68 semi-structured interviews with 25 industry experts, we present a set of tactics for improving the collaboration between enterprise architects and agile teams.

The chapter on Digitalization of Retail Banking discusses possible business models, applications, business risks and new opportunities for retail banks and their customers enabled by the upcoming digital transformation.

There are few visual modeling methods to design innovations for e-Healthcare services. In the final chapter on e-Healthcare Service Design using Model Based Jobs Theory (MBJT), we propose an environment which can visually model the Jobs Theory of Christensen. Moreover, we examine the applicability of MBJT by designing an e-Healthcare service.

References

1. Ross, J.W., Beath, C.M., Mocker, M.. Designed for Digital. How to Architect Your Business for Sustained Success. The MIT Press (2019)
2. Zimmermann, A., Gonen, B., Schmidt, R., El-Sheikh, E., Bagui, S., Wilde, N.: Adaptable enterprise architectures for software evolution of SmartLife ecosystems. In: 2014 IEEE 18th International Enterprise Distributed Object Computing Conference Workshops and Demonstrations, pp. 316–323. IEEE (2014)
3. McAfee, A., Brynjolfsson, E.: Machine, Platform, Crowd. Harnessing Our Digital Future. W. W. Norton & Company (2017)
4. Rogers, D. L.: The Digital Transformation Playbook. Columbia University Press (2016)

5. Zimmermann, A., Jugel, D., Sandkuhl, K., Schmidt, R., Schweda, C., Möhring, M.: Architectural decision management for digital transformation of products and services. J. Complex Syst. Inf. Model. Quart. **6**, 31–53 (2016)
6. Hamilton, E. R., Rosenberg, J. M., Akcaoglu, M.: Examining the substitution augmentation modification redefinition (SAMR) model for technology integration. Tech. Trends **60**, 433–441 (2016) (Springer)

Chapter 2
Digital Innovation and Transformation: Approach and Experiences

Matthias Wißotzki, Kurt Sandkuhl, and Johannes Wichmann

Abstract The transition from traditional to digital business models, i.e. digital transformation, requires a high degree of agility in enterprises and a way to systematically establish and control the required organizational and structural changes. Furthermore, digital business models often are triggered by technological innovations and lead to entirely new products and services. Consequently, new operational procedures or substantial redefinitions of the existing ones are required as part of digital innovation. Digital transformation and digital innovation both call for a methodical and technical integration of approaches from different areas of information systems research facilitating the required changes, such as digital business model management (BMM), capability management (CM) and enterprise architecture management (EAM). The book chapter addresses this methodical and technical integration by proposing the Digital Innovations and Transformation Process (DITP), which has a focus on method support, and the „Digital Business Architect" (DBA) as a modern qualification profile for professionals in an enterprise working in innovation and transformation initiatives. The DITP enables the DBA to moderate between business management and technology experts as part of the business executive team in an enterprise in order to support implementing their requirements under consideration of integrative Business-IT-Alignment concepts. DITP and DBA integrate and connect selected techniques from BMM, CM and EAM. The chapter presents the DITP and the DBA with a focus on their core features and uses application cases from digital transformation for illustrating the use of DITP in practice.

Keywords Digital Innovation · Digital Transformation · Business Modelling · Enterprise Modelling · Enterprise Architecture Management · Capability Management · Business-IT-Alignment

M. Wißotzki (✉) · J. Wichmann
Wismar University, Philipp-Müller-Str. 14, 23966 Wismar, Germany
e-mail: matthias.wissotzki@hs-wismar.de

K. Sandkuhl
University of Rostock, Albert-Einstein-Str. 22, 18059 Rostock, Germany
e-mail: kurt.sandkuhl@uni-rostock.de

A. Zimmermann et al. (eds.), *Architecting the Digital Transformation*,
Intelligent Systems Reference Library 188,
https://doi.org/10.1007/978-3-030-49640-1_2

2.1 Introduction

As part of the digitization efforts, corresponding effects on various industrial and service sectors can be recognized. The end-to-end digitalization of operational processes and the ongoing collection and evaluation of data on the use of products have resulted in new kinds of products and services.

Examples are new digital signage solutions changing the displayed content based on who is looking at them [1], smart garden appliance also hosting security functions [2], or tourist guides integrated into vehicles. In most of these examples, the changes and innovations on product and service level are an indicator for substantial changes in the business models, i.e. the value propositions and value creation predominantly based on digital technologies and transactions over information networks.

The transition from analogue to digital business models, i.e. digital transformation, requires a high degree of agility in enterprises and a way to systematically establish and control the required organizational and structural changes. Furthermore, digital business models are often triggered by technological innovations and lead to entirely new products and services. Consequently, new operational procedures or substantial redefinitions of the existing ones are required as part of digital innovation (see Sect. 2.2). Digital transformation and digital innovation both call for a methodical and technical integration of approaches from different areas of information systems research (see Sect. 2.3) facilitating the required changes, such as digital *business model management* (*BMM*), *capability management* (*CM*) and *enterprise architecture management* (*EAM*).

The book chapter addresses this methodical and technical integration by proposing the *Digital Innovations and Transformation Process* (*DITP*), which has a focus on method support, and the *Digital Business Architect* (*DBA*) as a modern qualification profile for professionals in an enterprise working in innovation and transformation initiatives. The new profile should not only enable personnel and students to develop digital business models but also to carry out the planning and implementation or providing significant support to do so with the help of suitable methods. It could be used as a kind of basic training profile for existing roles such as enterprise architect, business analyst, business architects or change manager in order to improve collaboration and understanding around the demands of the business units [3].

The *DITP* enables the *DBA* to moderate between the business executive team of an enterprise and technology experts. The aim is to combine the corresponding requirements of both aforementioned groups, taking into account integrative Business-IT-Alignment concepts. Therefore, the *DITP* as well as the *DBA* combine and integrate selected, established BMM, CM and EAM techniques.

The chapter presents the *DITP* and the *DBA* with a focus on their core features and uses application cases from digital transformation for illustrating the use of *DITP* in practice. The need for *DITP* emerged from the results of a dissertation [4]. The first *DITP* version was evaluated in the context of a master thesis at the University of Rostock [5]; the second version was applied in an industrial case [6]. Lessons learned from the evaluation were used for two additional industrial cases (Sects. 2.5 and 2.6) to apply parts of the third version which is subject of in this chapter.

The chapter starts by discussing the phenomenon of digital transformation in order to position the focus of the *DITP*'s intended contribution (Sect. 2.2). Afterwards, the theoretical background for *DITP* is given in Sect. 2.3, which includes BMM, CM and EAM. Section 2.4 presents the core activities of *DITP*; Sects. 2.5 and 2.6 illustrate the *DITP* usage in different industrial cases. Finally, the results are summarized and an outlook on further research activities is given in the last section.

2.2 Digital Transformation

Digitization and digital transformation are topics frequently discussed in research, industry and society with the expectation to have substantial effects on markets, companies and their operations. Digitalization can be understood, in simplified terms, as a generic term for efforts to convert information, processes, products or services into a form that can be processed or supported by information technology. In the scientific literature there are different approaches to subdivide digitization historically into phases which, depending on the point of view of the observer, are more technologically oriented [7], consider the socio-economic change triggered by digitization [8] or examine specific industries [9]—to name just a few examples. The current phase of digitization is often referred to as the "third" or "fourth industrial revolution" [10]. The focus here is on the disruptive social and economic consequences which, due to the potential of digital technologies to substantially change markets, lead to new technological application potentials and the resulting changes in economic structures, qualification requirements for employees and working life in general [8].

In the practice of many enterprises, digitization is associated with historically "earlier" phases of digitization and activities such as the optimization of internal business processes, sales channels or products; the automation of internal workflows, improvement of business processes, data exchange with customers and suppliers; and the use of possibilities of electronic business transactions, customer communication via social media, etc. Each of these topics is considered well researched in its own right, which is why these digitization topics are not the primary object of this investigation. Rather, the focus is on the transformation of enterprises with changing business models or markets, i.e. when emerging or changing customer needs, competitive situations, potentials for new service offerings or new partner structures affect both the range of services and service creation processes of a company as well as the organizational structures and requirements for employees and leadership. This is often triggered by substantial changes in the markets and accompanied technologically by the systematic collection and use of data.

Figure 2.1 illustrates the area of digital transformation and the focus of the paper (upper right quadrant). The two axes show on the one hand the operational processes of a company (value creation and service provision) and on the other hand the services

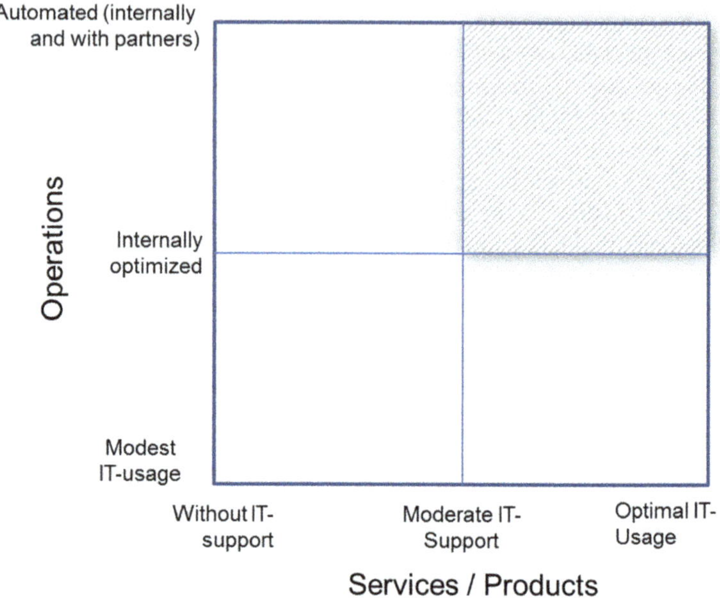

Fig. 2.1 Digital transformation dimensions, based on [7]

or products offered by the company (service offerings). The focus is on the transition from conventional to innovative business models.

Several case studies and surveys in industry indicate that companies feel a need to invest in digital transformation due to customer demand and market pressure, but at the same time also experience many challenges in planning and implementing digital transformation processes. An examples for such surveys is a study among 12,000 German companies [11] where 41% of all enterprises see a "strong push" towards digitization. Furthermore, [12, 13] conclude based on empirical material that there is a clear need for better support of enterprises in digital transformation.

2.3 Theoretical and Methodical Background

Profound knowledge of the anatomy of business models, necessary capabilities for adaptation or reorganization and the structures affected thereby within or outside the enterprise (e.g. processes, IT systems, roles) is a prerequisite for successful digital transformation projects. As part of new management approaches, this knowledge complements established disciplines, such as innovation management and organizational change. This also implies that enterprises have to rethink and upgrade the established management approaches to be able to change quickly within a digital environment.

Our proposal is to use an integrative approach of three disciplines: business model management (Sect. 2.3.1), capability management (Sect. 2.3.2) and enterprise architecture management (Sect. 2.3.3). These disciplines are topic of this section and described in the following. In the *DITP* (see Sect. 2.4), all three of them are connected through their respective outputs in the form of models in order to support enterprises with:

- Analysis and documentation of the current *company situation* comprising business model and the necessary architectural objects such as processes, structures and products,
- *Analysis of digital potentials* for additional and new value chains,
- Development and *concretization of the business strategy* for the implementation of digital objectives as well as derivation of the required business capabilities,
- *Determining the need for change* and planning of the architectures required for the digital transformation.

The approach presented in this work is intended to provide a methodological basis for the role and training profile of Digital Business Architect.

2.3.1 Business Model Management (BMM)

Business models have been an essential element of economic behavior since decades, but received significantly growing attention in research with the advent of the Internet [14] and expanding industries dependent on post-industrial technologies [15]. In general, the business model of an enterprise describes the essential elements that create and deliver a value proposition for the customers, including the economic model and underlying logic, the key resources and key processes.

Zott and Amit identified three major lines of work in their analysis of recent academic work in business model developments [16]:

- Business models for e-business scenarios and the use of IT in organizations
- The strategic role of business models in competitive advantage, value creation and organizational performance
- Business models in innovation and technology management.

For analyzing the business innovation potential of digital transformation, we consider both value creation based business models for the service industry and approaches from e-business as promising [17, 18]. For the purpose of this work, the proposal by [18] seems to be most suitable due to the explicit identification of six partial models: *capital model, procurement model, manufacturing model, market model, service offer model*, and *distribution model*. In this way the essential parts of value creation are covered. The *capital model* is subdivided into financing model and revenue model. The *financing model* describes the sources of the capital that is necessary for business activity. The revenue model provides means to generally systemize business models by four dimensions: direct or indirect generation

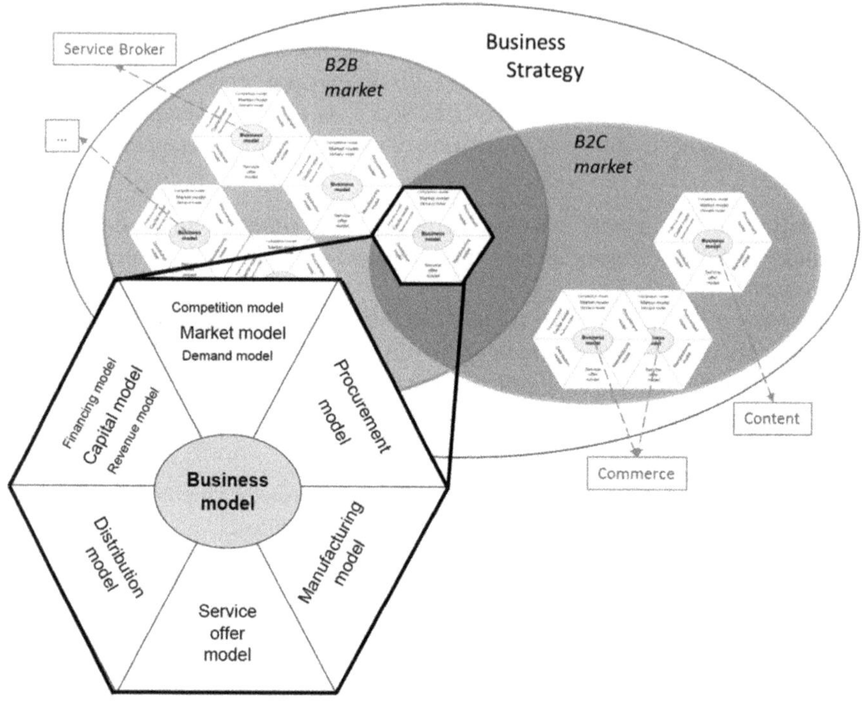

Fig. 2.2 Partial models of the integrated business model [18]

of revenue, as well as the transaction-dependent and the transaction-independent generation of revenue. The *procurement model* describes production factors and their sources. Here the distribution of power between suppliers and demanders is an important aspect. The *manufacturing model* covers the combination of input factors to new goods and services. Demand structures as well as the competitive situation are described by the respective sub-models of the *market model*. The *service offer model* defines which IT services are provided to the customers, while the *distribution model* focuses on the channels that are used to make the IT services available to the specific customer groups. Wirtz has proposed a categorization of e-business models by the kind of IT service offered (see Fig. 2.2) into the Business-to-Business (B2B) and the Business-to-Consumer (B2C) market.

2.3.2 Capability Management (CM)

Different areas of business information systems use the term "capability". Although in the literature there seems to be an agreement about the characteristics of the capability, there is no generally accepted definition of the term. The definitions mainly put the focus on "combination of resources" [19], "capacity to execute an activity" [20], "perform better than competitors" [21] and "possessed ability" [22].

Capabilities are enablers of competitive advantage and should help companies to continuously deliver a specific business value in dynamically changing circumstances [23]. They can be perceived from different organizational levels and thus are utilized for different purposes. According to [24], an enterprise performances best, when the enterprise maps its capabilities to IT applications. Capabilities as such are directly related to business processes that are affected from the changes in context, such as, regulations, customer preferences and system performance. As companies in rapidly changing environments need to anticipate these variations and respond to them [25], the affected processes/services need to be adjusted quickly. In other words, adaptations to changes in context can be realized promptly if the required variations to the standard processes have been anticipated and defined in advance and can be instantiated.

In this work, capability is defined as the ability and capacity that enable an enterprise to achieve a business goal in a certain context [26]. Ability refers to the level of available competence, where competence is understood as talent intelligence and disposition, of a subject or enterprise to accomplish a goal; capacity means availability of resources, e.g. money, time, personnel, tools. This definition utilizes the notion of context, thus stresses the need to take variations of the standard processes into consideration. To summarize, an enterprise capability represents the ability of an organization to join information and roles able to execute a specific activity with available resources in order to support strategy goals under consideration of its context. Thus, in terms of its organizational value capabilities support:

- high-level representation of organizational activities,
- strategic decisions like mergers & acquisition, out- and insourcing or budgeting,
- transparency,
- a common language between business and IT responsible,
- identification of new competitive advantages,
- the identification of organizational requirements for a successful strategy, implementation,
- scenario planning,
- relating IT perspective to business value.

In order to facilitate digital transformation, we propose an integrated design of capabilities, enterprise architectures and business services explicitly considering business model innovations.

But how do you find capabilities in a company? For this purpose, we used an approach that's supports identifying and managing capabilities required for an efficient operationalization of a business model innovation. The capabilities are derived systematically through a process, gathered and controlled in a discipline called *capability management*.

The used process is called *Capability Management Guide* (*CMG*) consists of four *Building Blocks* (*BB*'s) each focusing on distinct contents and having distinct outputs [4]. In short, the first building block sets preparation conditions like problem, scope, and stakeholder definition. The second building block designs the capability catalog structure, whereas the third block develops the detailed capability content. The governance building block covers catalog evaluation and maintenance issues. Every *BB* consists of several *working steps* (WS). These are shortly summarized by the following central goals:

- *Identification of involved parties* and definition of terms and preconditions
- *Identification of capability types* and corresponding capabilities for operationalizing of strategic goals
- *Systematic derivation of capabilities*, gathered and maintained in a repository called capability catalog.

The *CMG* encapsulates practices to provide a process that supports capability management activities. Thus, the approach can be used to derive capabilities through a structured process and gather them in an enterprise-specific catalog for different capability types. In particular, the approach provides a building block covering the continuous evaluation and maintenance in order to maintain capability and catalog quality.

2.3.3 Enterprise Architecture Management (EAM)

In general, an EA captures and structures all relevant components for describing an enterprise, including the processes used for development of the EA such as [27]. Research activities in EAM are manifold. The literature analysis included in [28] shows that elements of EAM [29], process and principles [30], and implementation drivers and strategies [31] are among the frequently researched subjects. Furthermore, work on architecture analysis [32], decision making based on architectures [33] and IT governance [34] exist. However, there are no investigations concerning a specific focus on the EAM combined with digital transformation procedures. Of relevance for digital transformation are EAM frameworks identifying recurring structures in EA. In this context, TOGAF [35] is considered by many researchers as industry standard and defines three different architectural levels that are visible in many other frameworks: *The Business Architecture* defines the business strategy, governance, organization and key business processes. *The Information Architecture* is often divided into two sub-layers: Data Architecture and Application Architecture. The Data Architecture describes the structure of an organization's logical and physical data assets and data management resources. The Application Architecture provides a blueprint for the individual application systems to be deployed, for their interactions and their relationships to the core business processes of an organization. *The Technology Architecture* describes the physical realization of an architectural solution. In addition to EAM frameworks there are also different modeling languages to support different EAM activities. One such language is ArchiMate, which is widely used for these purposes.

Ahlemann et al. [27] define EAM as a management practice that establishes, maintains and uses a coherent set of guidelines, architecture principles and governance styles to achieve an enterprise's vision and strategy. Facing opportunities and challenges derived from digital transformation, business leaders need new ways to conduct effective strategic decisions related to the increased digital enterprise [36]. With the huge diversity of IoT technologies and products, enterprises have to leverage and extend previous EA efforts to enable business value by integrating the concept of digitalization into their business environment [37]. On the way towards IoT-inclusive EAM, [10, 38] consider integrating the growing IoT architectural descriptions into a consistent enterprise architecture as a significant challenge.

The increased adoption of digitization through IoT, data analytics (big data), and cloud computing has opened new ways of thinking in many dimensions: customer involvement, optimized processes, and business models. In terms of business models, Chan [39] has presented a new business model that can be more suitable for organizing business in IoT age. This and other new business models emerging in the digitization age will have impact on the EAM practice.

2.4 Digital Innovation and Transformation Process (DITP)

This section is dedicated to introduce the digital innovation and transformation process (*DITP*) and the competences of the digital business architect (*DBA*). The process overview (Sect. 2.4.1) is followed by a detailed description of the process steps (Sect. 2.4.2).

2.4.1 Process Overview and Usage Scenarios

This section introduces the interaction of the disciplines (BMM, CM, EAM) and presents indications on how to use discipline-specific methods, in particular scenarios, of digital transformation and thus denote one of the foundations of actions of the *DBA*. A special feature of the approach presented in this work is the focus on the interaction of available methods of the respective disciplines considering a choice that should be preferably needs-based.

The following scenarios exemplify the procedure and the integrative use of the different disciplines and thus outline a concept for the different possible fields of *DITP* use:

1. The first scenario represents the case of an existing enterprise, which is facing new customer-, product- and/or market-related challenges due to digitalization and has to address them promptly.
2. The second scenario demonstrates the process of implementing a business concept for a start-up.

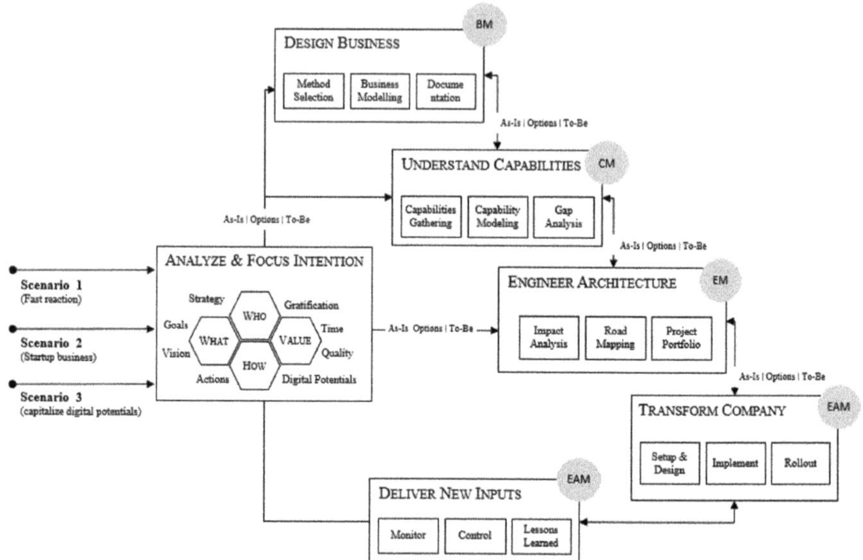

Fig. 2.3 Digital innovation and transformation process (DITP) including required disciplines

3. The third scenario represents a company that is not in an urgent situation concerning the digital transformation. Nevertheless, it is interested in completing or expanding its own corporate activities by leveraging its digital potentials.

In all three scenarios, the proposed process is not only an efficient means to tackle business-model and architecture-related problems but to foster communicative improvement as well as the knowledge about the enterprise.

Figure 2.3 shows the entire digital innovation and transformation process (*DITP*) as well as the separate process steps required by the *DBA*. The different steps are discussed in the next section.

2.4.2 *DITP Procedural Model*

This section outlines the *DITP* procedure as well as the core competences of the *DBA* necessary for the development and the implementation of digital business models. The initial idea for creating the *DITP* is based on involving business people who have not been trained in business or enterprise modeling in a participatory way through a business-related process in such a way that they actually take over a large part of the modeling themselves. It is important that the *DBA*, as the moderator of the participatory workshops, has the appropriate skills to implement the digital innovation ideas. As a basic education, a course of study that combines technical and economic points of view is advantageous for the *DBA*, for example business informatics or business engineering. The *DBA* should be familiar with the methods of business model- and enterprise architecture management, as these methods are the foundation of the *DITP*. The *DBA* is therefore a hybrid of business and enterprise architect. Regarding the achievement of objectives, it is important that the *DBA* is not part of the company in the sense of an employee but in terms of a fee relationship. The hierarchically unbound position makes it easier for the *DBA* to mediate neutrally in sense of the *DITP* purpose between the members of the organization and deliver feedback from various project stakeholders into the participative meetings. In addition, it is unlikely that a small to medium-sized company can provide the resources to employ a business or enterprise architect on a permanent basis.

As the starting point for the *DITP*, three typical enterprise situations were presented in Sect. 2.4.1. In the following, we will only focus on aspects of the second scenario since this is required for the case study companies (Sects. 2.5 and 2.6). More information about the first scenario is given in [6]. The first case study carried out was to accompany the realization of a new digital business, i.e. the founders of the new enterprise were supported by the *DBA* and the *DITP* procedure during their start-up process. The second case study was to identify digital potentials in zoo gardening facilities [40, 41].

The investigations begin with the first phase of the *DITP* "***Analyze & Focus Intention***". The aim of this step is to elaborate existing ideas, goals, visions and strategies. For further discussion of the subsequent *DITP* steps, it is necessary that the output of the first phase contains a clear business purpose. The determination follows a participatory approach that helps those involved to identify internal/external factors, requirements and implementation approaches [42].

In order to ensure that a common methodical starting point exists, the activities that are necessary for the development of visions and goals are derived from the conceptual development of a comprehensively documented business concept. That concept is the basis for a business model that will later be operationalized using a enterprise architecture [43]. Unfortunately, the components required for the design of a business model are not always obvious and manageable for an inexperienced company. Because of this, the declared goal of this process step is to determine the basic intentions for a company's digitalization projects. The procedure required for the analysis is based on the following four process steps: *DITP1.1 What?*, *DITP1.2 Who?*, *DITP1.3 How?* and *DITP1.4 Value?*

DITP1.1 What? aims to determine the basic motivation for a digitization project. Vision, mission goals and strategy are important elements of the corporate management. Questions such as: *What goals should be achieved in order to fulfill the intention? How does the desired result look like? What is the intention? What is currently available or should be offered to the customer group? Which performance promises should the customer group be informed of?* should help to describe the project in more detail [44–46]. A starting point could, for example, be an existing idea, which is specified with the aid of creative methods [5]. The mentioned case study of the transformation process for zoos follows an idea process, with the help of which internal/external factors, requirements and advantages can be identified and formulated. To answer the questions of *DITP1.1*, three mechanisms can support the procedure: (1) methods/processes for identifying ideas within a company, (2) activation methods for motivating the stakeholder management and (3) procedures for evaluating and idea portfolio. The activities to develop the basic idea are carried out by all founders in order to offer a common, methodical starting point. In the meantime, the corresponding competence of the *DBA* is required to systematically derive a concept from the ideas that meets the aforementioned requirements. The *DBA* uses methods from a pool of over 60 creativity techniques [5]. Ultimately, *DITP1.1* crates a fully documented business concept. *DITP1.2 Who?* deals with the customer groups for the aforementioned goals. Questions like: *Who is the recipient of the goal? What added value are generated? Which benefits should be communicated? What is the life situation and willingness to pay of the customer group? What kind of stakeholders exist?* are useful [47–49]. *How should the value proposition of the idea be operationalized as part of the value creation activities? What kind of resources and partners are needed for this?* are examples of questions answered by *DITP1.3* [47, 50]. In addition, the *DITP1.3* focuses on identifying digital trends, inspirations and best practices. In addition, the *DITP1.3* helps to identify possible digital approaches that could be important for defining the business concept [51]. This phase could be divided into activities like *Research of digital trends/building blocks and Evaluation of digital trends/building blocks.* Based on the character of the business concept, the first activity allows the research of different data sources in order to find digital trends/building blocks such as *Gartner Hype Cycle* [52], *Technology Radar* [53], *TechTrends* [54] and *Producthunt* [55]. Examples for digital technology building blocks are: *Automatization* (robotics, adaptive manufacture, autonomous vehicles), *Digital accesses* (social networks, mobility, apps, market places, gesture and voice control) *interconnectivity and exchange* (broadband, mobile Internet, IoT, Industrie 4.0, Cloud), *artificial intelligence* (smart cities & homes, agents & bots, machine learning), *Data* (big data, blockchain, predictive maintenance), *User Experiences* (Wearables, Virtual Reality, context-related individualization). In this phase, the *DBA* supports the research of digital trends/building blocks and gives advice on *HOW* to use digital potentials in possible application scenarios. The outcome of this phase is a technology map with suitable digital potentials that will be considered in the context of the business model development. The accuracy of this map will be specified more precisely later.

Regardless of the project and motivation, each digitization project should be able to add value to the company. Accordingly, *DITP1.4* determines the question of the degree of fulfillment that occurs for internal (e.g. employees) or external stakeholders (e.g. customers). In practical implementation, this means that the newly created digital functions (Fig. 2.1), e.g. the improvement of the quality of products and working conditions, the networking and/or automation of work processes and cost savings are realized. In this context, the goal is that the customer can buy the corresponding product cheaper, faster and/or with new functions. This results in the central question of *DITP1.4*: What kind of value is provided to the company by the revenue model in connection with the cost- and revenue structure? [47].

The second phase includes the modelling of the business model „***Design Business***". Different approaches can be used for modelling according to personal preferences or the requirements of the project [56]. Depending on the output of the first phase, a new business model is developed or the existing one is adapted, if necessary, taking options into account. If a business model does not yet exist, a future (To-Be) model must be developed. An example for the To-Be model is the implementation of a business idea as part of a start-up. However, the more frequent solution is that business activities already exist, which are to be supplemented by an idea. In this case, the existing business activities are mapped in an (As-Is) business model, which is supplemented by the components of the idea in order to create a future business model. In this context, it can happen that there are several possibilities/options when combining the existing As-Is model and idea components. Some of them have to be evaluated immediately in this step and/or at a later point in the DITP and selected for implementation. For this purpose of combining, the following activities have to be accomplished: *DITP2.1 Choice of business modelling approach, DITP2.2 Modelling of the business model, DITP2.3 Visualization of the business model & documentation.* For scenario 2, we selected the business model innovation method according to [56] which is one of the most comprehensive and most detailed approach for the detailed development of business models. The *DBA* plays the implementing role in this process step and supports the choice of the business model approach, the business model development and the choice of suitable documentation software. The outcome of this process is a comprehensive digital business model that serves as basis for the following operationalization activities.

In the process step „***Understand Capabilities***", we will resolve the following question: "*What capabilities are needed by our company or our future company in order to implement the developed business model?*" As a starting point, the discipline of capability management gaining growing attention in theory and practice will help to answer this question [57]. At present, capabilities are considered as missing links between business units and IT and therefore support the business IT alignment [4]. For the present phase, we suggest an efficient method (Capability Management Guide) that supports the identification, the structuring and the management of capabilities [4]. The following activities were carried out within this process step: *DITP3.1 Collection of required and existing capabilities, DITP3.2 Capability modeling, DITP3.3 Gap analysis and adaptations.* The CMG is based on an integrated capability approach that was developed from different scientific studies. The approach is integrated in a process comprising four building blocks (Sect. 2.3.2). They propose suitable procedures, concepts and supporting tools derived from specific theoretical and practical applications in order to detect and structure capabilities within the company [4, 58].

The *DBA* takes over the executing role during this process step and conducts the four building blocks. In doing so, the *DBA* integrates the founder and the disciplines in the elaboration of the capabilities catalogue. The outcome of this process is a catalogue describing the required capabilities for the implementation of the business model.

In the phase „**Engineer Architectures**", the future enterprise architecture model is derived and defined by means of the detected capabilities and possibly considering the present architecture. The aim of this phase is to draft an enterprise architecture model (EA) including the required architecture objects and their dependencies in order to obtain a transparent picture about the aspects necessary for the implementation of the business model. The defined capabilities catalogue serves as a basis for the retrieval of required enterprise architecture objects [4]. In this context, we propose the following activities: *DITP4.1 Choice of an approach for the impact analysis, DITP4.2 Development of the enterprise architecture including roadmap for integration, DITP4.3 Project portfolio for the implementation.* For *DITP4.1*, it is necessary to select a suitable modelling approach and tool. This can be done primarily based on the detected requirements and/or based on existing experiences. A method that meets these requirements is the 4EM-Method [42]. It describes an approach for a systematic and controlled procedure for the analysis, development and documentation of a company or an organizational unit. The aim is to represent structures and processes incrementally in the form of an EA, whereas the single elements can be already derived from the capabilities of the catalogue. For instance, to convey the mentioned EA activities known from the classic corporate enterprise modeling for e.g. zoo garden facilities (see case study 2 in Sect. 2.6), a simple, EA reference structure was used as the starting point, based on the elements illustrated in Fig. 2.4.

Concerning the elements of Fig. 2.4 it can be stated that each element represents a typical structural element of zoological facilities, which are defined by different

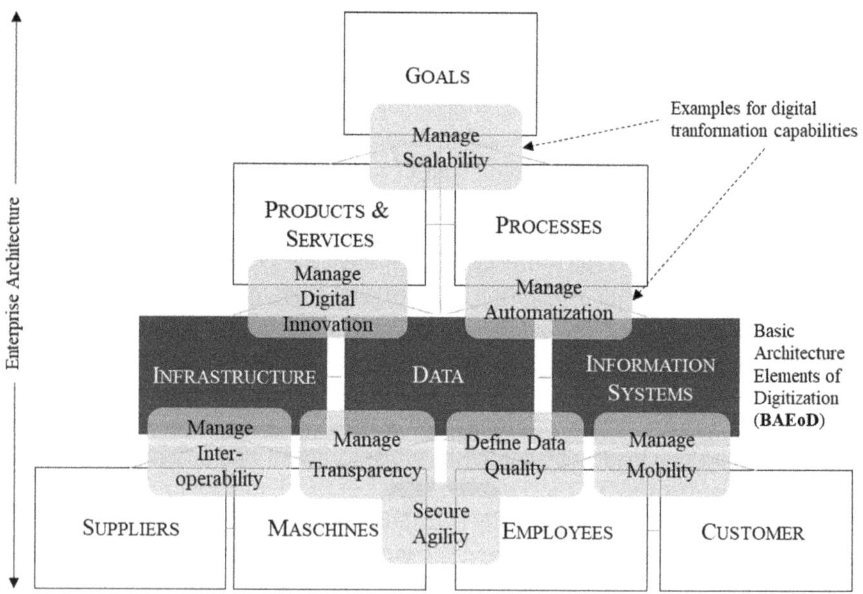

Fig. 2.4 Enterprise model approach including different digital transformation capabilities for zoo gardening facilities, based on [40, 41]

properties and relationships. It is fundamental that, starting from individual resources of the company (suppliers, machines, employees and customers) and their combination, an economic activity arises. Coupled with the *Basic Architecture Elements of Digitization* (*BAEoD*) Infrastructure, Data and Information Systems, new application scenarios arise (e.g. secure agility, define data quality—gray boxes between the elements). These new application scenarios are examples of digital transformation capabilities and are aimed at improving existing business areas or building new business areas. Since digital progress can take place in various areas of a company, the *BAEoD* must be combined with the other elements through horizontal and vertical connections. In summary, the development of EA requires analysis and consideration of various interest groups regardless of the type of company. This in turn is a fundamental part of managing an enterprise architecture [59].

To complete the elaboration of the enterprise architecture, the specific models have to be documented appropriately. For this purpose, we recommend software tools ranging from simple text programs, spreadsheet applications or graphic programs to comprehensive enterprise architecture modelling and management tools that need to be selected in accordance with the profile of requirements and maturity [60].

The *DBA* plays an implementing role in this process step and supports the choice of the modelling approach according to the requirements as well as the establishment and documentation of the architecture by means of suitable software tools. The outcome of this phase is the enterprise architecture required for the operationalization of the business model.

The way in which the elaborated EA is implemented and the change measures evolving from it are crucial for the success of the enterprise. The phase „***Transform Company***" focusses the activities *DITP5.1 Project set up, DITP5.2 Implementation and DITP5.3 Roll out* of the EA in the enterprise. The planning of strategic initiatives and/or single projects provides the framework for the following change measures and the implementation of the planned EA. Therefore, EA management (EAM) provides an approach for a systematic development of an EA in line with its goals by performing planning, transforming, and monitoring functions. There are many reasons for implementing an EA using EAM. On the one hand, it shows the connections between business and IT by means of a detailed description. On the other hand, it enables and supports the adjustments of IT according to the identification of problems, the alignment based on the company goals and the support in overcoming challenges. Furthermore, it comprises the definition of typical project management activities such as, for instance, the definitions of milestones, supply of information, personnel, escalation handling, implementation and roll out reporting.

Here, the recurrence of single parts of the *DITP* is possible. Therefore, the successful implementation of an EA depends on the planning already mentioned, but also on the degree of maturity of the EA models [27]—any insufficient degree will become evident during the planning phase of the project and must be increased in an iterative process. On the other hand, the users (e.g. executive board, business developers, line managers) must be convinced of the emerging structures [42]. In order to obtain a successful roll out as well as an extensive use, we recommend the following procedure: The roll out process should start with a specific change measure within the frame of a pilot project in a selected organizational unit. The project has to prove whether the expected advantages actually occur, or rather unexpected challenges emerge. After successful completion of the pilot project, the implemented change measures can be rolled out on additional organizational units or on the entire organization. This procedure can be repeated for further changes and organizational units whereas different organizational unit could be in charge of this. Based on the EA and by means of graphical models and simulations, different implementation approaches can be tested. The *DBA* supports the project teams(s) with the implementation. The result of this phase is an operationalized business model based on an EA.

Regarding the fact that enterprises are permanently facing new digital challenges in increasingly shorter time intervals, monitor and control activities must be set up in the enterprise. They control the new digital components of the enterprise architecture and can detect signals for digital changes at an early stage [38]. In this context, one of the most important functions of the phase „***Deliver New Inputs***" is to safeguard the digitalization objectives. In order to monitor the quality of short, medium and long-term digitalization objectives within the entire enterprise and the changes related to it, enterprises determine suitable indicators to be sustained by a corresponding data base [61]. Especially usage-oriented, trend and context-related data of introduced digital changes form the basis for new inputs of the *DITP*. Measuring tests, risk assessments and compliance tests are activities based on indicators that could support the implementation and the evaluation of goals as well as the derivation of new inputs for the procedure based on the collected data. The *DBA* supports the information needs analysis [62] of the management regarding the definition of appropriate indicators and compiles the required elements of the enterprise architecture.

2.5 Case Study 1—Start-Up Enterprise in E-commerce

The first case study illustrates the use of *DITP* for a scenario where a start-up company is supported to implement a new business concept which implements a digital business model. The section starts with an introduction of the case study company and continues with the steps performed in *DITP* and the results achieved.

2.5.1 Case Study Company

The case study was carried out with a start-up company located in the North of Germany. The company "myScreen" consists of four founders with complementing competences in different areas of management, marketing, IT platforms and software development. The start-up is positioned the area of e-commerce solutions with a specific focus on the "second screen" segment.

The target groups addressed by the company are end consumers when they are watching TV. More and more digital natives are using devices such as tablets and smartphones to obtain additional information about television program or other media content. For some years this type of interaction has been summarized under the name Second Screen and opens a whole range of new application areas. Just as Shazam already provides an automatic search for music titles, myScreen provides a much-simplified search for products and services visible in the current TV program or videos on platforms such as YouTube. The products shown in the videos or TV movies are identified by the app and sales offers for these products or similar versions are displayed for the end consumer. The prototype of the myScreen platform includes an app, back-office solution and connection to e-commerce platforms. Workflows for various roles and usage scenarios were defined. Tests performed with pilot users were carried out and will lead to new cycles of software development with the derived requirements.[1]

2.5.2 DITP Usage in Case Study Company

Starting point for the work with the start-up company was the „***Analyze & Focus Intention***" phase in which the MyScreen ideas were elaborated with the help of creative methods with focus on a clear business purpose. It followed an idea process that helped identifying and formulating, among others, internal/external factors, requirements and implementation approaches. Moreover, this process step was focused on the analysis of digital trends and building blocks and helped finding possible digital approaches that could be relevant for the business concept. Thus, the activities *Research of digital trends/building blocks* and *Evaluation of digital trends/building blocks* were performed.

Based on the documented business concept, the first activity recommends the research of different data sources in order to find digital trends/building blocks. The founders involved preselected potential digital trends/building blocks for the business model development of the next phase by means of feature or priority evaluations. This phase had also to be carried out by all founders. It is recommendable but not imperative to involve team members with a high affinity to technology.

The elaboration of the business model, i.e. „***Design Business Model***" *was the next step.* The founding team was convinced by this approach because of its comprehensive methodological support (meta model, questionnaires, examples) during the different phases of the business model development. The graphical representation of the business model using visualization software is a possibility of documenting the results. Another possibility of documentation is based on the proposed meta model by [56] that can be represented by means of modelling tools. The executive manager was

[1] myScreen v2.0 Project—Combining Video and Shopping Experience: https://www.fww.hs-wismar.de/fakultaet/personen-bereiche/professoren/wirtschaftsinformatik/matthias-wissotzki/projekte/.

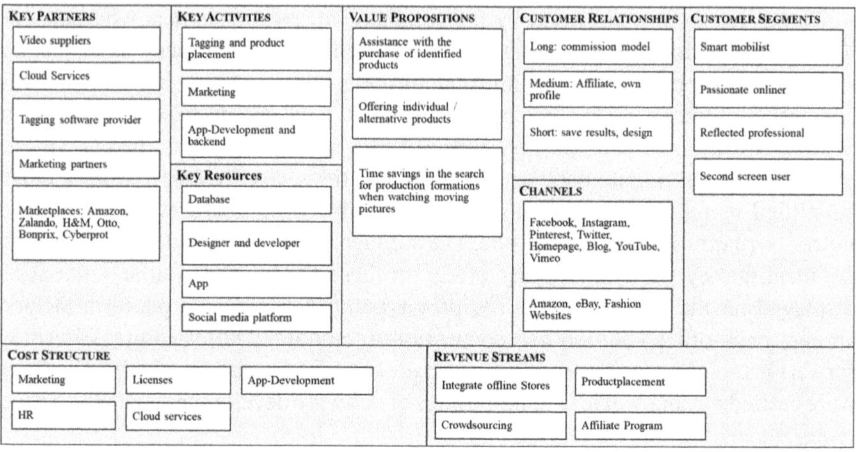

Fig. 2.5 MyScreen business model canvas

responsible for this process step and was an active member in the specific activities. Figure 2.5 exemplary illustrates the result of this phase for the myScreen project.

In the process step „***Understand Capabilities***", the required capabilities were in focus. The Capability Management Guide was applied which supports the identification, the structuring and the management of capabilities. The process was performed by the management team of the start-up. Figure 2.6 shows the final myScreen capability map.

In the phase „**Engineer Architectures**", the future enterprise architecture model was derived and defined. Since *DITP* scenario 2 is about an initial EA and with the resources of the founders are limited, they needed an uncomplicated and easy-to-learn method for the modelling. This method, however, has to be able to represent the entire architecture (e.g., processes, IT, roles, resources, objectives). We used the 4EM-Method [42]. To complete the elaboration of the enterprise architecture, the specific models had to be documented in ArchiMate, i.e. we used a graphical modelling tool in order to visualize the required dependencies (Fig. 2.7).

The phase „***Transform Company***" did not have to be performed since the start-up created a completely new company. This was done based on the results of the previous phases.

2.6 Cases Study 2—Zoo Gardening Facilities

The second case study described in this section was used for *DITP* validation purposes of the lessons learned of Case Study 1. The case study is presented with basic information about the use case and the *DITP* usage.

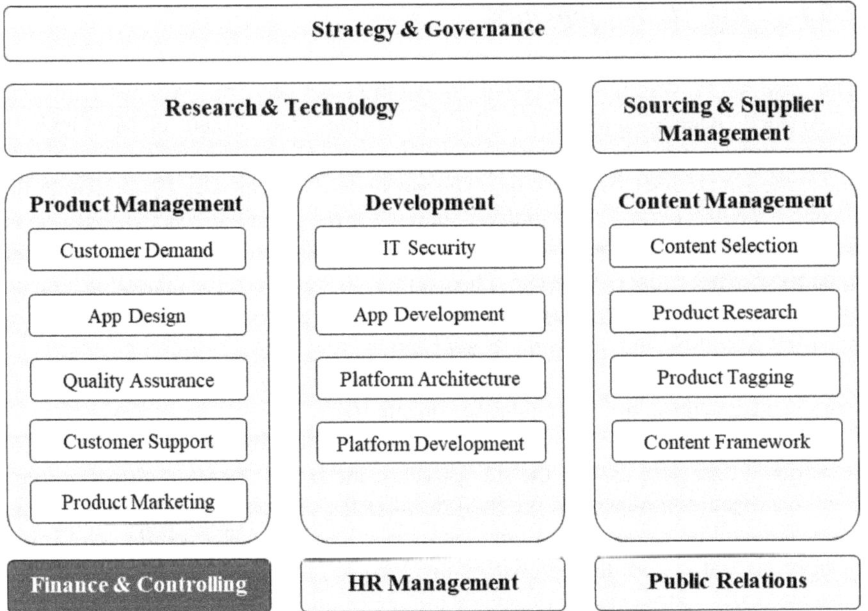

Fig. 2.6 MyScreen business capability map

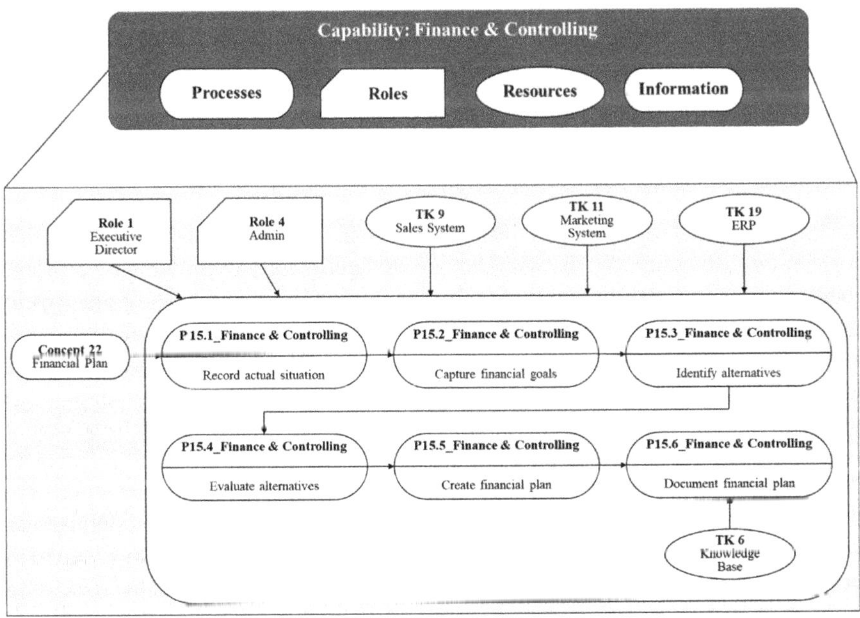

Fig. 2.7 Example for deriving enterprise architecture (EA) elements from a capability

2.6.1 Case Study Company

The case study was initiated by a cooperation between the Wismar University of Applied Sciences and the zoos in the German federal state of Mecklenburg-Vorpommern (represented by the *Landeszooverband Mecklenburg-Vorpommern e. V.*) and started in October 2018 [63, 64]. The association is formed by 19 companies that organize themselves as a community of interests in the so-called *Landeszooakademie*. There, zoos, wildlife- and nature adventure parks as well as tropical houses and aquariums work closely together. Their aim is to be places for education, science and environmental- as well as species- and animal protection within their animal breeding facilities and to promote appropriate measures. The companies record around 3 million visitors a year, with 60,000 pupils among them, who receive active nature and species conservation classes on site [64].

In cooperation with the Landeszooakademie, which has goals matching *DITP* scenarios 3 (see Sect. 2.4.1), the first phase of the *DITP* "*Analysis & Focus Intention*" has been completed. In order to develop the content for this phase, various analysis activities were carried out. Following the *DITP1*, the results were subsequently evaluated with the companies involved in order to create appropriate new measures or to adapt existing efforts that are relevant for *DITP2* (*Design Business*). The process is currently in this phase and first prototypes have been developed.

In the following, the analysis results of the 37 zoo gardening facilities are presented as an illustration for the focus areas of the *Analysis & Focus Intention* phase (*DITP1.1—What*, *DITP1.2—Who*, *DITP1.3—How* and *DITP1.4—Value*).

2.6.2 What—Vision, Goals and Strategies

An important goal for zoological institutions is that they harmonize and digitize their register and payment systems with the aim of significantly improving the customer journey. Typically, the zoo guests go through three phases during their visit, which differ in: arrival-, stay- and departure activities. In each of these phases, guests interact with different systems, which in the current situation do not necessarily have to be digital, but can still be available in analogue form, such as upon arrival manual turnstiles or the visual access control. Since sales of digital and mobile tickets are becoming increasingly important for zoological facilities, it is important that appropriate infrastructures and information systems (BAEoD—Sect. 2.4.2) exist that can process the corresponding data. A key measure for this is the establishment of integrated billing systems that digitally link machines, electronic turnstiles and the associated access control via various channels and make them evaluable. Due to the seamless transition from the arrival- to the stay phase, such an innovation will also have an impact on the guests within the facility. In this context, reference was made to cashless payments during the analysis (e.g. in restaurants, for vending machines and various additional services). Nevertheless, with the help of sensory measures, such systems offer the possibility of actively guiding customers through the zoo and thus avoiding wating times in the restaurant due to large crowds (*DITP1.4*). As the third and final phase of the customer journey, guests leave the zoological facilities. A digital connection there is possible by implementing services such as: payment machines in the parking lot, return machines for borrowed devices and automated exit controls. The information collected in the three phases is then combined in a corresponding cash register system. Within the framework of the value for the company (*DITP1.4*), this contains the advantage that the evaluation of this information enables the facilities to actively respond to the needs of customers by adapting offers accordingly.

2.6.3 Who—Target Group

Regarding the target group, it can be stated that in 2017 approx. 40 million people visited a Zoo in the *Verband der zoologischen Gärten e.V.* They generated a revenue of about 300 million euros [65]. Regarding the core target group, it was determined that about 2.87 million people visited a German zoo at least three times a year. In terms of segmentation, the largest customer group for zoos in 2013 with 61% are families with children, followed by adolescents aged 14–17 years with a share of 34% and young adults (18–24 years) with 29% [66].

For those target groups, considerations about relationships can be made using field research measures, e.g. population density, food ecology, population structure as well as spatial- and social behaviour are examined. It is conspicuous that most customers are so-called *digital natives*, as a generation that grew up in the digital age [67]. It is assumed that it is relatively easy for them to deal with technological innovations, as they are often in contact with digital innovations, both in their private and professional environment. Concerning the social behaviour, the use of social media, such as Twitter, Instagram, Facebook, Pinterest and YouTube, is an integral part of daily communication. The necessity of the digital approach as well as the adapted management of the zoological facilities to improve the customer journey is confirmed by the string expression of the *digital natives* within the guest structure for zoos.

2.6.4 How—Digital Potentials and Actions

The object of knowledge of this process step is the definition of digital potentials and the subsequent actions. Based on the documented results of *DITP1.1* and *DITP1.2*, research for digital trends is possible as the first activity of *DITP1.3*. Various data sources that deal with digital trends are suitable for this, such as *Gartner Hype Cycle* [52], *Technology Radar* [53], *TechTrends* [54] or *Producthunt* [55]. These trends can now be used, considering the status quo analysis, to determine best practices for zoological facilities. 38 different zoos (including the 19 zoological facilities of the

Table 2.1 Evaluation result of digital potential for zoo gardening facilities in M-V, based on [40, 41]

Digital potentials	Priority
Online forms and mobile donations	5
Uniform cash-register and access-system	4
Mobile zoo app	3
Cloud-based booking system	2

Landeszooverband Mecklenburg-Vorpommern e. V.) were examined for the existence of the following properties: *paperless e-tickets, online donation and sponsorship, interactive zooplan, tour suggestions + e-guide, virtual round tour, augmented reality services* and *social media marketing.* In addition, information about the animals, feeding times, prices, opening times, news, events, directions and highlights were also relevant.

The investigation shows that mobile websites that contain the previously mentioned information are widespread. In contrast, mobile applications (e.g. apps), which provide a large amount of the information mentioned and combine them with each other, are significantly more innovative and therefore also less common. This includes navigation options, interactive zoo plans, tour suggestions or point-of-interest navigation (to certain animal areas, feeding tours, restaurants etc.) as well as audio guidance applications. Innovations in the area of edutainment services could also be relevant by creatively combining entertainment and education. In this context, the zoological facilities are planning virtual maps in combination with information films about the animal in the facility, 360° tours and interactive educational games. The advantage of these services is that the respective target groups (*DITP1.3*) can also use them from home to prepare themselves accordingly for a visit to the zoo.

After the analysis results were determined, a participatory workshop [42] with several prioritization levels was organized. The aim was to evaluate the determined digital potential (Table 2.1) with the project partners in order to generate measures for the second phase of the *DITP* "*Design Business*". Finally, the assessment of the potential was presented at the *4th Symposium der Landeszooakademie Mecklenburg-Vorpommern* in 2018 with representatives of the zoological institutions (n = 19) and arranged according to priorities [63]. The result of this process is shown in Table 2.2. In order to ensure that the potentials were correctly recorded, an appropriate specification was made by questioning the participants on site.

2.6.5 Value—Gratification

In order to define and structure a selection regarding the digital potentials in the last step of the first phase of the *DITP*, it is important to delimit the benefits that the respective innovations generate for zoological facilities. For this reason, the potentials examined were subjected to a future analysis (*to-be*) [42]. It aims to derive problems

Table 2.2 Digital potentials and value for zoo gardening facilities based on [40, 41]

Categories	Digital potentials identified best practice	Value generated by	BAEoD (Fig. 2.4)
Sales	Paperless E-tickets	Digital parking cards, tickets, wallet, E-tickets	IS
	Online donation and sponsorship	Mobile donations, sponsorships, mobile payment integration	Infrastructure, IS
	Access and POS system	Cohort-analysis, spatial behavior, social behavior, population density, food ecology, population structure	Infrastructure, IS, data
Navigation	Interactive zoo plan	Information regarding POI as a Service	IS
	Tour suggestion + E-guide	Customizable routes, GPS-navigation, live-location guide, location based services (indoor/outdoor)	IS, data
Edutainment	Virtual round tour	360° services, E-learning-services	IS
	Augmented reality services	E-learning services	IS
Marketing	Social media marketing	Identify/activate/sensitize digital natives on social networks	

from the defined digital potentials that prevent an improvement in economic performance through innovation. The results of this analysis and the assignment of the *BAEoD* relevant for the innovation (Fig. 2.4) for the case study are summarized in Table 2.2.

Dealing with reviews or criticism, social media marketing and an integrative app are particularly relevant. On the one hand, internet users trust the inline ratings of their portals, on the other hand, thousands of users can be reached with these solutions at the same time. The basis for building all relationships and business obligations of all kinds is trust. This is especially true in the digital age in terms of branding and reputation. If a relationship of trust is established, the users take recommendations from their (social media) friends as an opportunity to buy certain products themselves or to visit a zoological facility. Such evaluations are important both for the selection of appropriate suppliers and for the differentiation and analysis of a specific market. Appropriate social media marketing is important for the zoological institutions for two reasons: firstly, there is a connection to the important target group of digital natives and secondly, the zoological institutions can react quickly to problems, inquiries and queries that arise strategically valuable knowledge about the target group.

Regarding the analyzes of the analysis phase, it was determined that, based on the participatory requirements, good solutions and best practices for exploiting the digitization potentials of the zoological facilities in Mecklenburg-Vorpommern already exist. Furthermore, as benefits for the zoos, they have determined their digital needs and respective requirements, by conducting the methodological examination. Taking those needs into account, the organizations can broaden their range of services and to secure new shares in the competitive market for leisure activities as part of diversification.

In the next step of the *DITP* ("*Design Business*") it is now important to professionally support the development of new or the further development of existing business models with the help of the *BAEoD* through the *DBA*. The intention is to help the businesspeople of the zoos to determine and coordinate the activities and architectures inside and outside of the facilities required for the adaptation or redesign of business practices. Regarding the success factors, it is essential for the second phase of the *DITP* that the involved zoos actively participate in the design process, as in the first phase. This contains that all members of the participatory workshop must be willing to make changes to the previous structures and/or business activities. If one or more member actively acts against the measures of the *DITP*, it is likely that the potential added value for the enterprise will only be realized partly or not at all.

2.7 Summary

This book chapter investigated digital transformation in enterprises from the perspective of required method support and competences. Digital transformation usually affects products and operations in an enterprise as well as the overall business model. The successful development and integration of digital business models requires a high degree of agility which is only reasonable from an economic perspective when the effects of the changes in an enterprise can be determined at any point in time. Basic prerequisites for digital transformation are innovative ideas, knowledge about the development of business models, capabilities necessary for the adaptation or restructuring and the architectures affected within the enterprise.

The case one of the start-up company myScreen in Sect. 2.5 illustrated that even for small companies it can be valuable to systematically connect business model development with capabilities required to implement them and enterprise architectures to structure the organizational and technical solutions. The integrative view of *DITP* supported the required process in myScreen.

The cooperation with the zoo gardening companies has shown that the spectrum of technological- or innovation-related changes is also very heterogeneous in existing companies, which in turn affects the approaches and methods required for this. In addition to the aforementioned case study one, which has gone through all phases of the *DITP* (creation of a business model, capability map, EA via enterprise modelling (EM) and management (EAM), this case has shown that analyzing the intention of a digital transformation initiative via a process, which goes hand in hand with subsequent business modelling and capability modelling approaches is useful to supports a structured and transparent planning for all stakeholders.

The common conclusion for both cases is that the digital innovation and transformation process caused a substantial amount of changes which are not easily implementable. To establish the necessary prerequisites, various methods and tools of different disciplines are necessary. These methods and tools were presented as elements of an integrated procedure of the *DBA* in this work.

In particular, the *DBA* profile could be taught to students of universities and higher education institutions, especially in business information systems studies, but also in addition to computer science and economic science. For this purpose, the procedure and its methodological contents in the area of electronic business, business model management, enterprise architecture management and business analysis can be used. But it could also serve for practitioners in enterprises for self-study or as manual and thus demonstrate first possibilities and approaches for the digital transformation.

References

1. Sandkuhl, K., Smirnov, A., Shilov, N., Wißotzki, M.: Targeted digital signage: technologies, approaches and experiences. In: Galinina, O., Andreev, S., Balandin, S., Koucheryavy, Y. (eds.) Lecture Notes in Computer Science, Internet of Things, Smart Spaces, and Next Generation Networks and Systems [S.l.], pp. 77–88. Springer (2018)
2. Kaidalova, J. Sandkuhl, K., Seigerroth, U.: How Digital Transformation Affects Enterprise Architecture Management—A Case Study, pp. 5–18, 3 June 2018 [Online]. Available: http://www.sciencesphere.org/ijispm/archive/ijispm-0603.pdf#page=9. Accessed on: 28 Feb 2020
3. Wißotzki, M., Timm, F., Stelzer, P.: Current state of governance roles in enterprise architecture management frameworks. In: Johansson, B., Møller, C., Chaudhuri, A., Sudzina, F. (eds.) Perspectives in Business Informatics Research: 16th International Conference, BIR 2017. Copenhagen, Denmark, 28–30 Aug 2017. Proceedings, vol. 295. Lecture Notes in Business Information Processing, pp. 3–15. Springer International Publishing, Cham (2017)
4. Wißotzki, M.: Capability Management Guide: Method Support for Enterprise Architectures Management. Springer Fachmedien Wiesbaden GmbH; Springer Vieweg, Wiesbaden (2017)
5. Plewka, T.: Die Entwicklung einer Fallstudie zur Integration von digitalen Geschäftsmodellen und Architekturen (2017)
6. Wißotzki, M., Sandkuhl, K.: The Digital Business Architect—Towards Method Support for Digital Innovation and Transformation. In: Poels, G., Gailly, F., Serral Asensio, E., Snoeck, M. (eds.) The Practice of Enterprise Modeling: 10th IFIP WG 8.1. Working Conference, PoEM 2017. Leuven, Belgium, 22–24 Nov 2017, Proceedings, vol. 305. Lecture Notes in Business Information Processing, pp. 352–362. Springer International Publishing, Cham, (2017)
7. Berman, S.J., Bell, R.: Digital Transformation: Creating New Business Models Where Digital Meets Physical, pp. 1–17. IBM Institute for Business Value (2011)
8. Hirsch-Kreinsen, H., ten Hompel, M.: Digitalisierung industrieller Arbeit: Entwicklungsperspektiven und Gestaltungsansätze. In: Vogel-Heuser, B., Bauernhansl, T., ten Hompel, M. (eds.) Handbuch Industrie 4.0 Bd.3, pp. 357–376. Springer, Berlin, Heidelberg (2017)
9. Heinemann, G., Gehrckens, H.M., Wolters, U.J.: Digitale Transformation oder digitale Disruption im Handel. Springer Fachmedien Wiesbaden, Wiesbaden (2016)
10. Rifkin, J.: The Third Industrial Revolution: How Lateral Power is Transforming Energy, the Economy, and the World. Tantor Media, Inc., Old Saybrook, CT

11. Warning, A., Weber, E.: Digitalisierung verändert die betriebliche Personalpolitik. Bundesagentur für Arbeit 12, 2017 [Online]. Available: http://doku.iab.de/kurzber/2017/kb1217.pdf. Accessed on: 28 Feb 2020
12. Depaoli, P., Za, S.: Designing e-Business for SMEs: drawing on pragmatism. In: Lazazzara, A., Nacamulli, R.C.D., Rossignoli, C., Za, S. (eds.) Lecture Notes in Information Systems and Organisation, Organizing for Digital Innovation, pp. 237–246. Springer International Publishing, Cham (2019)
13. Nambisan, S., Lyytinen, K., Majchrzak, A., Song, M.: Digital innovation management: reinventing innovation management research in a digital world. MISQ **41**(1), 223–238 (2017)
14. Tapscott, D., Ticoll, D., Lowy, A.: Digital Capital: Harnessing the Power of Business Webs. Harvard Business School Press, Boston (2000)
15. Perkmann, M., Spicer, A.: What are business models? Developing a theory of performative representations. In: Woodward, J., Phillips, N., Sewell, G., Griffiths, D. (eds.) Research in the Sociology of Organizations, vol. 29, 1st edn. Technology and Organization: Essays, pp. 265–275. Emerald Group Pub. Ltd, Bingley, UK (2010)
16. Zott, C., Amit, R.: Business model design: an activity system perspective. Long Range Plan. **43**(2–3), 216–226 (2010)
17. Gassmann, O., Frankenberger, K., Csik, M.: The Business Model Navigator: 55 Models That Will Revolutionise Your Business. Pearson, Harlow, England, London, New York, Boston, San Francisco (2014)
18. Wirtz, B.W.: Business Model Management: Design—Instrumente—Erfolgsfaktoren von Geschäftsmodellen, 4th edn. Springer Gabler, Wiesbaden (2018)
19. Antunes, G., Barateiro, J., Becker, C., Borbinha, J., Vieira, R.: Modeling contextual concerns in enterprise architecture. In: 15th IEEE International EDOC Conference Workshops: Proceedings: Helsinki, Finland, 29 Aug–2 Sept 2011, pp. 3–10. Helsinki, Finland (2011)
20. Chi, M., Zhao, J., Lu, Z., Liu, Z.: Analysis of E-business capabilities and performance: from e-SCM process view. In: 2010 3rd IEEE International Conference on Computer Science and Information Technology: (ICCSIT); 9–11 July 2010, Chengdu, China; Proceedings, pp. 18–22. Chengdu, China (2010)
21. Boonpattarakan, A.: Model of thai small and medium sized enterprises' organizational capabilities: review and verification. JMR **4**(3) (2012)
22. Brézillon, P., Cavalcanti, M.: Modeling and using context. Knowl. Eng. Rev. **13**(2), 185–194 (1998)
23. Stirna, J., Grabis, J., Henkel, M., Zdravkovic, J.: Capability driven development—an approach to support evolving organizations. In: Sandkuhl K., Seigerroth, U., Stirna, J. (eds.) The Practice of enterprise modeling: 5th IFIP WG 8.1 Working Conference, PoEM 2012, Rostock, Germany, 7–8 Nov 2012, Proceedings, vol. 134. Lecture Notes in Business Information Processing, pp. 117–131. Springer, Berlin, New York (2012)
24. Chen, J.-S., Tsou, H.-T.: Performance effects of IT capability, service process innovation, and the mediating role of customer service. J. Eng. Tech. Manage. **29**(1), 71–94 (2012)
25. Eriksson, T.: Processes, antecedents and outcomes of dynamic capabilities. Scand. J. Manag. **30**(1), 65–82 (2014)
26. Bērziša, S., et al.: Capability driven development: an approach to designing digital enterprises. Bus. Inf. Syst. Eng. **57**(1), 15–25 (2015)
27. Ahlemann, F. et al.: Strategic Enterprise Architecture Management: Challenges, Best Practices, and Future Developments. (Hg), Springer Science & Business Media (2012)
28. Wißotzki, M., Sandkuhl, K.: Elements and characteristics of enterprise architecture capabilities. In: Matulevičius, R., Dumas, M. (eds) Perspectives in Business Informatics Research: 14th International Conference, BIR 2015, Tartu, Estonia, 26–28 Aug 2015, Proceedings, vol. 229. Lecture Notes in Business Information Processing, pp. 82–96. Springer, Cham (2015)
29. Buckl, S., Dierl, T., Matthes, F., Schweda, C.M.: Building blocks for enterprise architecture management solutions. In: Harmsen, F. (ed.) Practice-Driven Research on Enterprise Transformation: Second Working Conference, PRET 2010, Delft, The Netherlands, 11 Nov 2010. Proceedings, vol. 69. Lecture Notes in Business Information Processing, pp. 17–46. Springer, Berlin, Heidelberg (2010)

30. Glissmann, S.M., Sanz, J.: An approach to building effective enterprise architectures. In: 2011 44th Hawaii International Conference on System Sciences: (HICSS); 4–7 Jan 2011, Koloa, Kauai, Hawaii, pp. 1–10. Kauai, HI (2011)
31. Sandkuhl, K., Simon, D., Wißotzki, M., Starke, C.: The nature and a process for development of enterprise architecture principles. In: Abramowicz, W. (ed.) Business Information Systems: 18th International Conference, BIS 2015, Poznań, Poland, June 24–26, 2015; Proceedings, vol. 208. Lecture Notes in Business Information Processing, pp. 260–272. Springer International Publishing, Cham (2015)
32. Johnson, P., Lagerström, R., Närman, P., Simonsson, M.: Enterprise architecture analysis with extended influence diagrams. Inf. Syst. Front. **9**(2–3), 163–180 (2007)
33. Johnson, P., Ekstedt, M., Silva, E., Plazaola, L. (eds.) Using Enterprise Architecture for CIO Decision-Making: On the Importance of Theory (2004)
34. Simonsson, M., Johnson, P., Ekstedt, M.: The effect of IT governance maturity on IT governance performance. Inf. Syst. Manag. **27**(1), 10–24 (2010)
35. The Open Group: TOGAF Version 9.1 (TOGAF), 10th edn. Van Haren Publishing, Zaltbommel, The Netherlands (2012)
36. Li, Y., Hou, M., Liu, H., Liu, Y.: Towards a theoretical framework of strategic decision, supporting capability and information sharing under the context of Internet of Things. Inf. Technol. Manag. **13**(4), 205–216 (2012)
37. Zimmermann, A., Schmidt, R., Jugel, D., Möhring, M.: Evolving enterprise architectures for digital transformations. In: Lecture Notes in Informatics, pp. 183–194
38. Patel, P., Cassou, D.: Enabling high-level application development for the Internet of Things. J. Syst. Softw. **103**, 62–84 (2015)
39. Chan, H.C.Y.: Internet of Things business models. JSSM **08**(04), 552–568 (2015)
40. Wißotzki, M., Wichmann, J.: Application of the digital innovation and transformation process in zoo gardening facilities. In: Joint Proceedings of the BIR 2019 Workshops and Doctoral Consortium Co-located with 18th International Conference on Perspectives in Business Informatics Research (BIR 2019). CEUR Workshop Proceedings, vol. 2443, pp. 83–97. Katowice, Poland, 23–25 Sept 2019
41. Wißotzki, M., Wichmann, J.: "Analyze & focus your intention" as the first step for applying the digital innovation and transformation process in zoos. Complex Syst. Inf. Model. Quart. (CSIMQ) **20**, 89–105 (2019). https://doi.org/10.7250/csimq.2019-20.05
42. Sandkuhl, K., Stirna, J., Persson, A., Wißotzki, M.: Enterprise Modeling: Tackling Business Challenges With the 4EM Method. Springer, Heidelberg (2014)
43. Martynov, V.V., Shavaleeva, D.N., Salimova, A.I.: Designing optimal enterprise architecture for digital industry: state and prospects. In: Proceedings, 2018 Global Smart Industry Conference (GloSIC): South Ural State University (National Research University), Chelyabinsk, Russian Federation, 13–15 Nov 2018, pp. 1–7. Chelyabinsk (2018)
44. Rusnjak, A.: Entrepreneurial business modeling. In: Rusnjak, A. (ed.) Entrepreneurial Business Modeling: Definitionen—Vorgehensmodell—Framework—Werkzeuge—Perspektiven, pp. 81–108. Imprint. Springer Gabler, Wiesbaden (2014)
45. Wirtz, B.W., Pistoia, A., Ullrich, S., Göttel, V.: Business models: origin, development and future research perspectives. Long Range Plan. **49**(1), 36–54 (2016)
46. Appelfeller, W., Feldmann, C.: Die digitale Transformation des Unternehmens: Systematischer Leitfaden mit zehn Elementen zur Strukturierung und Reifegradmessung. Springer Gabler, Berlin, Heidelberg (2018)
47. Gassmann, O., Frankenberger, K., Csik, M.: The St. Gallen business model navigator. Working Paper (2013)
48. Kofler, T.: Das digitale Unternehmen. Springer, Berlin, Heidelberg (2018)
49. Doz, Y.L., Kosonen, M.: Embedding strategic agility. Long Range Plan. **43**(2–3), 370–382 (2010)
50. Osterwalder, A., Pigneur, Y.: Business Model Generation: A Handbook for Visionaries, Game Changers, and Challengers. Wiley, Hoboken, NJ (2013)

51. Voigt, K.-I., Buliga, O., Michl, K.: Business Model Pioneers. Springer International Publishing, Cham (2017)
52. Gartner: Gartner Hype Cycle 2016 [Online]. Available: http://www.gartner.com. Accessed on: 19 Jun 2017
53. Thoughtworks: Technology Radar [Online]. Available: https://www.thoughtworks.com/de/radar. Accessed on: 19 Jun 2017
54. Deloitte: Deloitte Tech Trends 2017 [Online]. Available: https://www2.deloitte.com/global/en/pages/technology/articles/tech-trends.html. Accessed on: 19 Jun 2017
55. Producthunt: Producthunt Protfolio [Online]. Available: https://www.producthunt.com/. Accessed on: 19 Jun 2017
56. Schallmo, D.: Geschäftsmodelle erfolgreich entwickeln und implementieren: Mit Aufgaben und Kontrollfragen, 1st edn. Springer Gabler, Berlin, UA (2013)
57. Wißotzki, M.: Exploring the nature of capability research. In: El-Sheikh, E., Zimmermann, A., Jain, L.C. (eds.) Emerging Trends in the Evolution of Service-Oriented and Enterprise Architectures, vol. 111. Intelligent Systems Reference Library, pp. 179–200. Springer International Publishing, Cham (2016)
58. Aleatrati Khosroshahi, P., Hauder, M., Volkert, S., Matthes, F., Gernegroß, M.: Business capability maps: current practices and use cases for enterprise architecture management. In: Proceedings of the 51st Hawaii International Conference on System Sciences (2018)
59. Puspitasari, I.: Stakeholder's expected value of enterprise architecture: an enterprise architecture solution based on stakeholder perspective. In: 2016 IEEE/ACIS 14th International Conference on Software Engineering Research, Management and Applications (SERA): Proceedings: 8–10 June 2016, Towson University, Balitmore, USA, pp. 243–248. Towson, MD, USA (2016)
60. Schekkerman, J.: How to Survive in the Jungle of Enterprise Architecture Frameworks: Creating or Choosing an Enterprise Architecture Framework, 3rd edn. Trafford, Victoria, BC (2006)
61. Winter, K., Buckl, S., Matthes, F., Schweda, C.M.: Investigating the state-of-the-art in enterprise architecture management methods in literature and practice. In: MCIS, vol. 90 (2010)
62. Lundqvist, M., Sandkuhl, K., Levashova, T., Smirnov, A.: Context-driven information demand analysis in information logistics and decision support practices. In: Workshop on Contexts and Ontologies: Theory, Practice and Applications. Riva del Garda, Italy (2005)
63. Wißotzki, M., Schleifer, H.: Zoo digital—Eine bedarfsgerechte Kombination von Natur & Technologie: Potenzialanalyse neuer Erlebnisformate durch Digitalisierung. In: Viertes Symposium der Landes Zoo Akademie, pp. 10–19. Vilm (2018)
64. Landeszooverband Mecklenburg-Vorpommern e.V., Verband in Zahlen—Statistik [Online]. Available: http://www.landeszooverband-mv.de/verband_in_zahlen.php. Accessed on: 28 Feb 2020
65. Verband der Zoologischen Gärten e.V., Zoos erwirtschaften 300 Millionen Euro Umsatz. Accessed on: 28 Feb 2020
66. Statista.de: Besucherstruktur von Zoos und Tierparks in Deutschland in 2013 [Online]. Available: https://de.statista.com/statistik/daten/studie/261497/umfrage/umfrage-zur-besucherstruktur-von-zoos-und-tierparks-in-deutschland/. Accessed on: 28 Feb 2020
67. Dingli, A., Seychell, D. (eds.): The New Digital Natives: Cutting the Chord. Springer, Berlin (2015)

Chapter 3
Perpetual Evolution—Rethinking the Way Digital Transformations Are Managed

Oliver Bossert and Stefan Feldmann

Abstract To stay competitive in a world in which providing great customer experience has become paramount, companies in nearly every industry must continuously develop digital products and services as well as the business processes that support those products and services. To compete with digital-born companies, traditional firms need to adopt a much different approach to designing and managing enterprise architecture that will enable continual changes to and a modular design of business capabilities as well as the technologies behind them. In the following, we introduce a new "perpetual evolution" approach that allows for this by changing out elements of enterprise architecture quickly, adding new parts in no time, and incorporating the latest and greatest functionality. We also explore what is required to shift to this newer approach. Lastly, we demonstrate the model with examples and explain how companies can streamline their legacy business processes and mindsets.

Keywords Platform architecture · Enterprise architecture · Perpetual evolution · Digital transformation

3.1 Introduction

Why are some Internet retailers able to make crucial changes to their e-commerce websites in hours while it takes brick-and-mortar retailers three months or more to do the same? Why are only a few car manufacturers able to rapidly make online updates to their products in the field, such as to their infotainment systems or fuel and engine performance, even though this is crucial in a world of servitization? And how is the new wave of cloud-based enterprise software vendors able to make software

O. Bossert (✉)
McKinsey & Company, Inc., Taunustor 1, 60310 Frankfurt am Main, Germany
e-mail: oliver_bossert@mckinsey.com

S. Feldmann
McKinsey & Company, Inc., Sophienstraße 26, 80333 Munich, Germany
e-mail: stefan_feldmann@mckinsey.com

A. Zimmermann et al. (eds.), *Architecting the Digital Transformation*,
Intelligent Systems Reference Library 188,
https://doi.org/10.1007/978-3-030-49640-1_3

updates to their products in days or weeks, rather than the months it takes traditional enterprise software vendors to do so?

Given that they compete with "digital native" companies, established organizations must be able to operate quickly and agilely. But most of them cannot, largely because of their enterprise architecture or how they have designed their technology to support their business strategy. To compete with the best digital native companies, traditional organizations need to adopt a different approach to enterprise architecture. We refer to it as "perpetual evolution" [1] because it allows them to continually upgrade their digital business processes and the included technologies.

For organizations whose enterprise architectures lock them into legacy business processes and technologies for too long, competing against pure Internet companies with perpetual evolution architectures should raise red flags. Established companies in nearly every industry are racing to digitize their business models, product and service offers, and the business processes that support them. Some are spending enormous sums to ward off companies like Amazon, Netflix, Uber, and Spotify that started online and have rapidly gained share in sectors ranging from retail and entertainment to banking and transportation.

Because they are not required to connect their new digital systems to aging technologies, today's digital native companies can build "greenfield" digital business processes in online marketing, sales, distribution, and other areas. This is why they are thriving in this era of digitalization, product servitization, and dramatically reduced software release cycles—especially since these three trends thwart incumbents.

The findings drawn in this chapter mainly result from a study on enterprise architecture [2, 3] conducted in collaboration with Henley Business School since 2015. Respondents come from a number of countries and a variety of industries. The findings presented here are drawn from more than 150 responses collected in 2017. In this chapter, we explain what an enterprise architecture based on perpetual evolution is, how it differs to architecture approaches of the past, and why it has become necessary. Next, we introduce the main concepts being used in this chapter to ensure a common understanding (Sect. 3.2). We follow with an exploration of what perpetual evolution is (Sect. 3.3). Then we discuss what is required to operate with this new architecture (Sect. 3.4) and the implications for companies' future architecture teams (Sect. 3.5). Lastly, we conclude our findings in Sect. 3.6.

3.2 Explanation of Some of the Core Concepts Used Throughout This Chapter

What is the difference between microservices and a traditional SoA (service-oriented architecture)?

There is a technical and a more comprehensive answer to this question: Technically the SoA architecture is based on a heavyweight integration pattern—often an Enterprise Service Bus—that contains business logic. A microservice on the other hand typically connects to other services over a thin integration layer without any business logic in the layer itself but all logic being contained in the service itself. By looking beyond technology these two paradigms imply a very different business and IT operating model. With an ESB the business side is able to configure business logic between releases. This is important if the operating model is based on a waterfall process and there is plenty of time between releases. In an agile model, on the other hand, business and IT are collaborating closely and there is no need for configuration as the code can be modified directly.

What is the difference between a digital platform and a digital product or feature development?

We define a digital platform as the necessary toolset for a feature team to develop a new idea into a production-ready product or service for the customer. Quite often, we see feature teams that spend a substantial amount of time selecting the tools to develop, test, deploy and operate their products before they start to work on the actual features. A digital platform addresses this challenge and provides the feature teams with a toolbox to accelerate their workflow.

What are the six core elements of enterprise architecture?

For enterprise architecture, a framework consisting of six core elements has been developed by McKinsey around 15 years ago and is a reliable way to structure and communicate the complexity of architecture in large cooperation. The six elements of this framework are introduced in the following.

Business operations: Most companies think about the generation of value along processes. These processes could be customer-centric journeys or more internally focused business processes.

Business capabilities: Business processes should not be mapped directly to applications. There has to be a reference structure that balances the need for individual flexibility and reuse of the activities or services that create a process. This structure is called a strategic capability map.

Business applications: The services of the capability map are provided by software applications that cover a certain set of services or capabilities.

IT integration platform: Applications are connected by integration methods or integration patterns, such as a direct connection or an enterprise service bus.

Infrastructure services: All applications run on some kind of infrastructure service that is providing computational capacity, storage, etc.—ideally abstracted on a cloud platform.

Information and communication technology: At the lowest level, all infrastructure services rely on some hardware and physical structures.

3.3 From the Old to the New: Perpetual Evolution

A good way to understand the evolution of enterprise architecture is to consider how companies have traditionally treated the six core elements—business operations, business capabilities, business applications, the IT integration platform, the IT infrastructure services, and the underlying information and communications technologies (Fig. 3.1). How would these elements differ in a perpetual evolution model? We discuss this in detail in the following section.

3.3.1 *Business Operations: Designed with an Outward View*

Business processes and digital systems must be designed with an outward view—i.e., the customer experience (online and offline)—rather than an inward view of a company's operations. Priorities have changed: while the customer used to be an element in a product- or company-centered process, today the products and services are an element of a customer journey.

This does not mean that companies' inward-focused view is obsolete. Enterprises need to keep their core transactional systems, such as accounts payable/receivable, order management, procurement, or something else. And they must also make sure those business processes and technologies remain efficient.

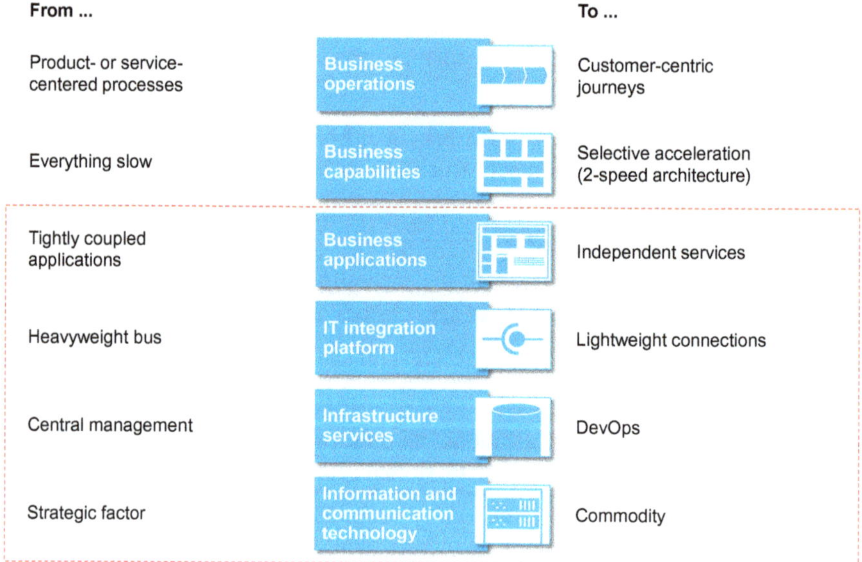

Fig. 3.1 Overview of the six core elements of enterprise architecture

However, increasingly, an architecture that enables companies to perpetually change their businesses cannot focus only on the customer purchasing journey. For many companies, architecture needs to reflect a larger journey. Take, for example, an automobile insurance company's architecture: more and more policies are purchased on websites that aggregate the entire auto purchase, including the car and insurance.

These larger "ecosystems" are end-to-end customer journeys and are seen in many B2C industries in which consumers must buy several products and services to accomplish a single task, such as buying or renting homes or taking a vacation.

Note that B2B companies are not immune to the trend, especially those that embed digital technologies into their products (e.g., construction equipment, aircraft engines, power turbines, drilling equipment) to sell predictive maintenance, performance improvement, and other services. This is the servitization trend mentioned earlier that requires a company's enterprise architecture to encompass the entire time in which customers use their products—not just the moment of purchase.

Even B2C companies are getting into the servitization game. Take the case of manufacturers of high-definition, smart televisions. Previously, TV manufacturers designed their IT systems to follow the product to the customer (in this case, to a retailer). Today's digital TVs have become platforms for manufacturers to provide a range of TV-related services to the home, such as identifying shows consumers might want based on their viewing habits or more targeted advertising. This means that TV makers' enterprise architecture must now include the end customer's TV viewing experience.

3.3.2 *Business Capabilities: The Foundation of Digital Competition*

Digital systems that provide the narrow functionality (such as one-stop checkout) necessary for a superior customer journey are the foundation of an architecture of pervasive evolution. In fact, the best digital native companies organize their teams by distinct business capabilities and use this grouping method to deliver milestones in their target architecture [2, 3], (Fig. 3.2). Spotify, the online music powerhouse, groups teams into squads, each of which controls a single business capability or a cluster of them.

To compete in a quickly changing digital world, established companies must first move to a model for digital capabilities. But to adopt an architecture of perpetual evolution, they must be able to continually improve the digital business capabilities that comprise their customer journeys and add new capabilities when necessary.

That, in turn, means grouping systems into two categories: front-end systems of digital business capabilities, and applications based on existing back-end transaction systems (see our previous articles on two-speed architecture [4, 5]). These digital business capabilities become the basis on which companies compete in an online world.

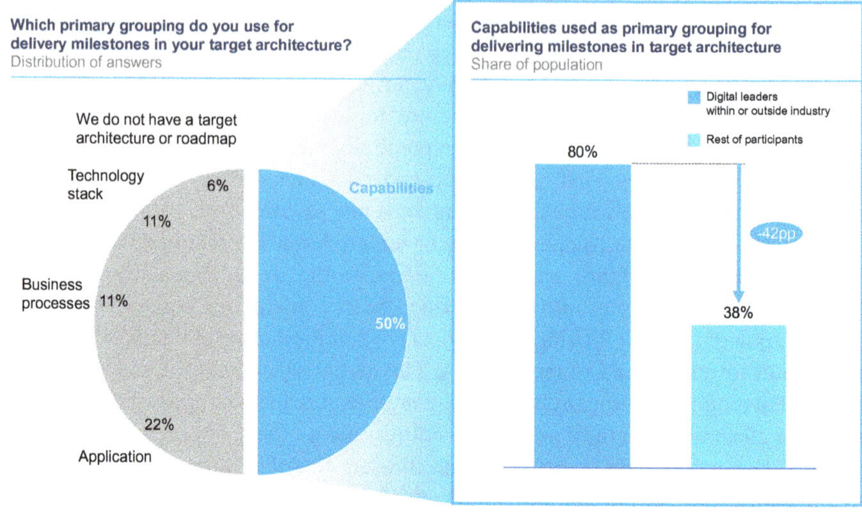

Fig. 3.2 Capabilities connect business and IT, and digital leaders use them more often [2, 3]

Consider a retail chain that sells a growing proportion of its products on its website. The company cannot take months to enhance its product recommendation engine when a digital native online retailer only needs days or weeks because they have an architecture that makes businesses capabilities systems agnostic. The kind of core transaction systems the retailer has does not matter; its new or enhanced product recommendation engine will work with any back-end system.

After moving to a capability-based model in the first two levels of business operations and business capabilities, enterprise architects need to ingrain a highly modular software skeleton in the four levels below them. We will discuss the business applications level first.

3.3.3 Business Applications: Shifting from Tightly Coupled Software to Independent Services

The goal in adopting a perpetual evolution architecture in this layer of the stack is to be able to upgrade core applications such as an ERP, CRM, and HRM module by module (or service by service) without having to upgrade to a whole new version of those applications (Fig. 3.3). This can be achieved in three steps:

- Step 1. Platform and services in one release container
- Step 2. SOA or microservices for the functional part
- Step 3. True modularity of the underlying platform (e.g., deployment, data management, analytics).

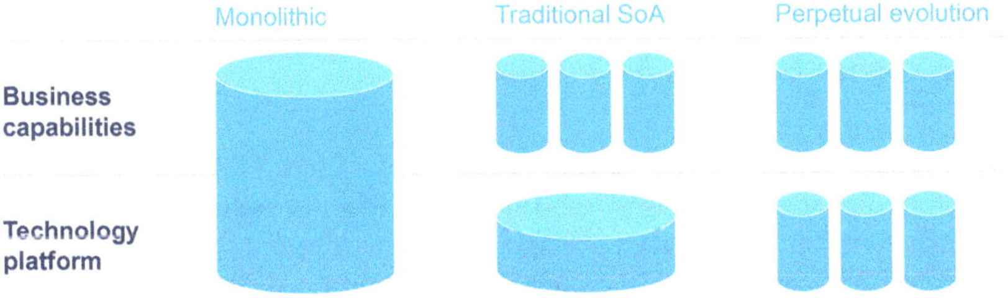

Fig. 3.3 Shifting applications to perpetual evolution architecture

3.3.4 IT Integration Platform: Lightweight Integration

Heavyweight buses such as an ESB are no longer the solution to every problem. Companies also need lightweight connections between their services. This is the only way to deal with the growing problem of latency—the time it takes for companies to deliver web pages to online customers who demand instant responses at every click. Without lightweight buses calling for an increasing amount of web services, IT integration platforms become major hang-ups in the customer experience and the main reason for online customer defection.

3.3.5 Infrastructure Services: DevOps Takes Down the Wall

The concept of DevOps—or grouping together IT professionals previously separated into the functions of software development and IT operations—has firmly taken hold in many companies. DevOps is central to a firm's ability to test new digital business capabilities and bring them to market rapidly. With DevOps, companies should be able to tweak business capabilities and put them online in hours or days, rather than the weeks or months it takes to develop an improvement, pass it to testing, and then drop it into a long production line.

However, to make the DevOps model work, companies need to put their DevOps team members into the teams charged with updating and building new digital business capabilities. In other words, there is one team for each business capability, and that team includes representatives from systems development and operations.

3.3.6 Information and Communications Technology: A Big Expense Becomes a Commodity

Technology services rented from public cloud vendors turn IT into an affordable resource for companies of many sizes. This means that the large, costly IT infrastructure that companies have taken years to build is no longer a competitive barrier. Start-up companies can jump into the game quickly, as firms like Netflix (with the help of Amazon Web Services) and many other digital native companies have done. Information and communications technology are now commodities, and prior investments are no longer large competitive advantages or barriers.

3.3.7 Summary

These six fundamental changes in the stack are why companies must move their enterprise architecture to one that supports perpetual evolution of their business and IT. That will enable them to loosen up their asset base of business and system capabilities—i.e., change out parts quickly, add new parts in no time, and replace other parts with the latest and greatest functionality. Certainly, such service-oriented architectures are complex and, therefore, must be carefully supported by architecture management [6, 7], especially for big data applications [8].

3.4 Establishing a Perpetual Evolution Architecture

So how do companies transition their architecture from the old to the new? How can they move from the design philosophy that guided their present IT asset base to an approach that governs the infrastructure they need now—one in which they can make perpetual changes to their business in a digital world?

The obstacles are many, but they all are surmountable. In helping companies across industries and regions adopt a perpetual evolution architecture, we have found five moves to be critical:

- Freeing up the team from unnecessary dependencies.
- Not overinvesting in one area by consistently eliminating dependencies.
- Regarding the platform as one of continuous change.
- Setting boundaries in technology—not on paper.
- Maintaining freedom within capability teams through strict separation from the platform.

We will discuss these five moves in the following section (see also Fig. 3.4).

Fig. 3.4 Overview of the five moves towards a perpetual evolution architecture

3.4.1 *Free up the Team from Having to Deal with Unnecessary Dependencies*

The concept of decoupling has a long history in technology. However, in the past, it could only be achieved with extreme difficulty and vast overhead. Today, all dependencies must be eliminated to allow for quick change of the pieces of digital products and processes and thus to keep the customer experience competitive. To do that, an architecture must eliminate dependences up and down the stack. Consider an auto manufacturer that has embedded digital technologies into its cars so customers can make online updates to the navigation, infotainment, and other systems. In order to sell new services à la carte, such as a certain navigation tool or infotainment feature, the automaker will have to design these systems in a way that capabilities can be offered separately.

Eliminating such dependencies is crucial to companies that want to design and sell new digital capabilities to small customer segments with different needs. Continuing with the car example, a manufacturer may decide to offer parents of a teenage driver a new service that monitors the speed and location of the teenager's car. If selling this new service requires a rewrite of the whole navigation system, it is likely to take the car company a long time to bring this one feature to market for a small (but profitable) market segment.

This flexible approach requires eliminating dependencies in coding, one of six aspects of software development (Fig. 3.5). In the carmaker's example, the design and coding of its navigation must eliminate dependences between features. The car

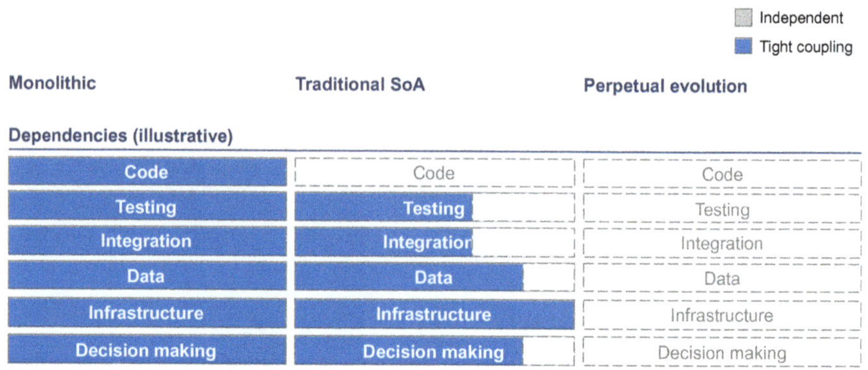

Fig. 3.5 Moving towards a perpetual evolution, dependencies have to be eliminated—not only regarding the technology but also regarding operational and decision making processes

company must be able to add new features (such as our hypothetical teenage-driver monitoring system) without having to change existing features.

3.4.2 *Do not Overinvest in One Area: Eliminate All Dependencies*

Coding is not the only place to worry about dependencies. Dependencies also crop up in testing, integration, data, infrastructure, and decision making. By the latter, we mean who must sign off on the testing of new business capabilities, either the team chartered to build and enhance them or senior management. If capabilities must be approved by senior management before being put on the market, it will take a long time for them to be available.

These kinds of dependencies, across all six aspects of a new piece of software, have slowed down companies for years. Such dependencies are a feature, in effect, of earlier approaches to enterprise architecture. They are what forced big technology changes to be made only periodically.

In monolithic architectures, all elements of the software stack as well as the code base of different modules were tightly coupled, requiring time-consuming dependency checks. Installing new software depended on the schedules of software testers and resources. Even when developers decoupled software functionality, they often coupled data, which created dependencies. Although they intended to decouple the integration layer from applications, teams too often hardwired pieces that created dependencies with the ESB or application programming interfaces. When software was ready to be produced, the handover from the development team to the infrastructure team often slowed things down. They were now working on the production team's schedule, competing with a long line of software releases. Perhaps most importantly, waiting for senior management approval for a new software system or functionality upgrade before it went into production could set things back by weeks. This is a decision making dependency.

To be sure, the last decade's move to service-oriented architectures (SOA) was a strong step forward and decoupled code from the other five elements. It helped companies design web services around specific business capabilities. Yet in most companies we are familiar with, the testing, integration, data, infrastructure, and decision making activities remained tightly coupled. That was why a new web service could take weeks or months before a company's top management approved it and IT production tested it and rolled it out.

In creating web services, also known as microservices, companies must make sure each service is independent of others as well as pieces in the IT stack. In fact, this should be companies' ultimate goal—not creating a focused service.

3.4.3 Set Boundaries in Technology (Not on Paper)

IT architects have often been stereotyped as "people drawing funny boxes in charts." Software developers, for their part, have been viewed as the people who write the code for the modules that those "funny boxes" represent. Unfortunately, this division of labor has all too often led to both groups operating in their own worlds rather than working closely together.

A company that wants to be digitally competitive will need enterprise architects more than ever. However, they can no longer maintain an arm's-length relationship with developers. Companies must work closely with developers to make sure the architectural rules of perpetual evolution—not just the code—are written into software.

That means architects need to be part of the teams focused on a business capability or group of related capabilities. They will find themselves working alongside product managers, developers, marketers, testers, production people, legal help, and others.

3.4.4 Make Sure Capability Teams Have Freedom (by Strict Separation from the Platform)

Teams working on specific business capabilities should not have to worry about whether their improvements affect their company's underlying technology platform.

Every company has a part of their IT architecture that they manage as a platform, and (increasingly) other parts that are organized by business capabilities (e.g., web or microservices). To shift to a perpetual evolution architecture, they must draw explicit boundaries between these two parts of the asset base. Then they must enforce those boundaries through strict oversight and other governance processes.

Because a company's digital business capabilities are what enable it to make rapid changes in digitally based products and business processes, we believe IT professionals must shift their focus along these lines as well. To start, that would mean defining the parts of the IT platform they support (e.g., customer management) in terms of business capabilities, rather than technologies. For example, defining an IT capability as a service integration will help a company identify the technologies in the organization with comparable functionality. But it will also help the company create more meaningful roles—e.g., Service Integration Architect rather than Integration Product XYZ Architect.

3.4.5 Regard the Platform as One that Will Continuously Change

By separating the boundaries between business capabilities and technology platforms, companies will be able to isolate the fast-moving parts of their infrastructure (business capabilities) from slower-moving ones (platforms). Nonetheless, they cannot ignore the need to continuously improve their platforms. Companies must make sure they can update pieces of their platforms continuously as well.

Crucial advances in platforms can provide key performance improvements. Capitalizing on them requires a platform whose parts are modular, decoupled, and organized for reuse. Only when the pieces of that platform are defined as business capabilities will companies be able to quickly implement improvements or new capabilities.

3.5 Implications for Architecture Teams

Clearly, speed is the new imperative of digital. It is essential for companies that want a competitive advantage to quickly bring new features to customers—however, architecture is traditionally considered an obstacle to, rather than a supporter of, speed. To allow for this perpetual evolution, a few things would need to change in today's architecture teams, specifically:

- Enterprise architecture. There is still a need for high-level architecture that connects business and IT. However, with all the digital changes, more flexibility is necessary—even on the enterprise level.
- Digital platform architecture. While each feature team should own their space, they all have to rely on a common technology standard. Architecture plays a critical role in facilitating the exchange between teams and establishing a common standard in this area.

- Solution architecture. Finally, architecture capabilities have to be transferred to the team so that they have the full decision power within the boundaries of the platform architecture.

The necessary changes in these three areas are discussed in the following.

3.5.1 Changes in Enterprise Architecture

Even in a very agile, digital world, business strategy still needs to be translated into a target landscape and technology needs. This is the key task of the enterprise architecture team.

There are, however, a few very important changes. For one, architecture decisions are becoming top-management decisions. With the technology itself becoming a differentiating factor, it actually becomes an integral part of the business model. Therefore, the decision for the right data management or integration technology is no longer an IT-only decision, but should involve the broader management team.

In addition, communicating and articulating these changes requires the architecture team to create transparency on the ROI of architecture decisions and how they contribute to the business model. Communication must be more centered on the ROI and monetary benefits and adopted to the board level—instead of traditional boxes and lines. In essence, our survey [2, 3] has shown that digital leaders inside and outside the industry ensure that enterprise architecture interacts mostly with the board (Fig. 3.6), focusing communication on strategic priorities.

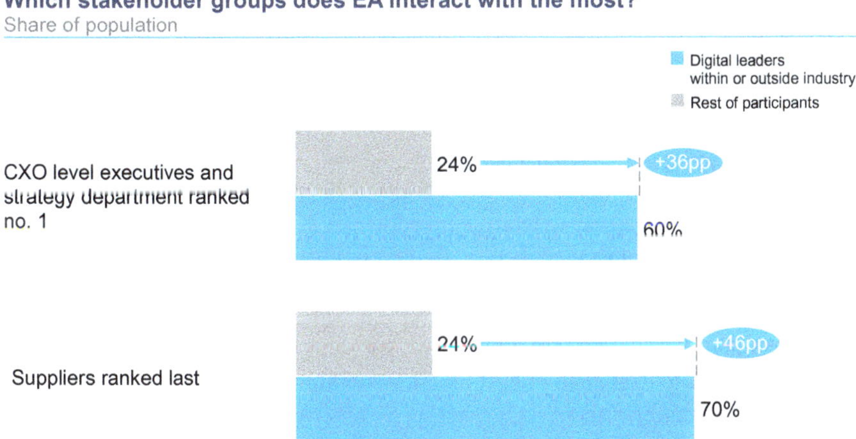

Fig. 3.6 Digital leaders prioritize proactive communication with the board and demote connection with suppliers [2, 3]

Enterprise architecture has to adapt to change more quickly. With the fast adoption of new channels, business models, and technologies, the creation of a ten-year target picture becomes obsolete.

Finally, enterprise architecture must become more integrated with feature teams, platform teams, and the broader organization. Architecture still has to maintain companywide mapping of software and infrastructure assets to capabilities and teams (linked to HR systems and the collaboration platform); however, senior hands-on architects that teams can book for help must be coached and a self-service drill-down portal that links IT architecture components to teams must be provided.

3.5.2 Changes for the Digital Platform Architecture

In order for individual feature teams to work effectively, companies must establish a digital platform. The architecture of this platform must be steered and maintained, but not necessarily by the architecture team.

Managing this platform is different from the traditional architecture approach in some ways. For instance, the architecture must have very strict digital platform design but allow maximum freedom on how to use the platform. Quite often, freedom is misunderstood as the freedom to modify and change the platform, try different cloud providers, and accumulate a myriad of tools and technologies. This typically leads to a very inefficient setup. One organization that used this method ended up deploying software in three different cloud environments without an efficient way to work across these newly created divisions. If a company decides to use cloud platform X, teams should not deploy in platform Y just for the fun of it.

Digital native companies, on the other hand, are typically very strict when it comes to managing the digital platform itself, but not in terms of limiting the development of features on the platform. This ensures flexibility when necessary and cost efficiency when an individual team cannot create a competitive advantage.

The rules of the platform should be carved in code and technology rather than lengthy rule sets (e.g., having each team running in its own cloud environment enforces service communication). We believe that maintaining the companywide mapping of software assets to teams (linked to HR systems and collaboration platforms) is essential for the success of a digital platform. Some companies misunderstand this concept and give their feature teams complete freedom to develop their own platform. For instance, one company found more than 50 different version control systems in their landscape, each with their own access rights. This clearly prevents a collaboration and exchange of best practices within the company.

The digital platform team has to act as a center of excellence for engineering practices and establish an external reputation as a true engineering company. It is no longer feasible to split responsibilities for the technology platform and the operating model and usage of the platform. This team will hence be responsible for:

- Organizing companywide external exposure (e.g., engineering blog, participation in conferences, open-source office with community managers)
- Organizing DevOps days
- Organizing "lunch and learning" events to share engineering successes and failures
- Providing a coaching pool that teams can book for help
- Participating and presenting at external conferences
- Engineering blogs and publishing at other online resources
- Managing an open-source office to promote and support internal software to be open-sourced and contributions to open-source projects that are being used internally.

In order to play a role beyond supporting the strategic direction setting, architects will have to be involved in the day-to-day practical implementation work of the platform much more than in the past.

3.5.3 *Changes for the Solution Architecture*

For many companies at the beginning of a digital transformation, the architecture team is involved in the solution architecture work. This is because companies often look at established digital native enterprises, realize the limited involvement of a central architecture team in the solution architecture process, and decide to adopt this as a best practice. This overlooks the maturation and length of time these companies went through to establish the current processes. In digital native companies, most teams have gone through several implementations similar to the current challenge in the past. Based on this experience they have a good understanding of what to do and what not to do. While these teams do not need the support of a central architecture function to avoid the implementation of point-to-point interfaces, the situation is very different for an incumbent that is implementing its first digital project.

In this situation, a central architecture coaching team can help the feature teams implement (minimal) and very practical process standards that teams need to adhere to and monitor usage of:

- Definition of being done (e.g., not only produce apps, but also update integration testing environments, produce documentation, provide test stubs, and carry out handover to the help desk and support teams)
- Integration of patterns that should be used to connect different applications
- How reference and other technical documentation is produced. We can imagine an architecture function producing automatic reports with top-ten teams that adhere to standards and the bottom ten (or some other gamification aspect)
- How postmortems are logged or used.

They can also help identify and drive opportunities for code reuse (and trigger setup of new DevOps teams to manage/own these or assign them to existing teams):

- Libraries that can be generalized
- Services that can be merged
- Services that can be defined based on existing code
- Libraries/products that can be open sourced
- Services that can be offered as an external API.

While this guidance is necessary at the beginning, the ultimate goal of the architecture team should be enablement of the decentral feature teams to become independent.

3.5.4 Summary

When feature teams and a digital platform are established, architecture plays a pivotal role in their success. To enable them, they have to change their approach. For one, they have to start a discussion on architecture decisions in the board room. In addition, they should either own or drive the development of an overarching digital platform. Finally, they have to work closely with feature teams in order to ensure successful implementation of the new concepts. A discussion on what it needs to make agile organizations hum can be found in [9]; the impact of enterprise architects on the success with digital transformations is discussed in [10].

3.6 Conclusion

To stay competitive in a world in which providing a great customer experience has become paramount, companies in nearly every industry must continually innovate digital products and services, as well as the business processes that support those products and services. They can gain greater agility if they abandon rigid enterprise architecture management practices of the past and adopt a new approach. This shift in methodology can help traditional companies keep pace with digital-born competitors.

In this chapter, we introduced the concept of this perpetual evolution—a new enterprise architecture that allows companies to change out elements of enterprise architecture quickly, add new parts in no time, and incorporate the latest and greatest functionality. We identified what is needed to change from old to new approaches and integrate perpetual evolution.

Naturally, we cannot provide a "step-by-step" procedure of how to evolve from a traditional to a perpetual evolution approach. Nevertheless, we will focus on deriving "recipes" for different industry sectors to accelerate companies' transition within their digital transformation. Therein, it is especially essential to incorporate what it needs to make agile organizations hum [9]. In addition, reference cases for our perpetual evolution model are currently being developed as a starting point for such "recipes".

References

1. Bossert, O., Laatz, J.: Perpetual evolution—the management approach required for digital transformation. In: McKinsey Quarterly, 17 June 2017
2. Manwani, S., Bossert, O.: The challenges and responses for enterprise architects in the digital age. J. Enterp. Archit. **12**(3), 6–9 (2016)
3. Manwani, S.: The role of enterprise architecture in digital business. J. Enterp. Archit. **2**, 26–27 (2015)
4. Bossert, O., Ip, C., Laatz, J.: A two-speed IT architecture for the digital enterprise. In: McKinsey Quarterly, 14 Dec 2014
5. Blumberg, S., Bossert, O., Laatz, J.: Deploying a two-speed architecture at scale. In: McKinsey Quarterly, 16 Sept 2016
6. Zimmermann, A., Sandkuhl, K., Pretz, M., Falkenthal, M., Jugel, D., Wißotzki, M.: Towards an integrated service-oriented reference enterprise architecture. In: International Workshop on Ecosystem Architectures (2013)
7. Zimmermann, A., Schmidt, R., Sandkuhl, K., Wißotzki, M., Jugel, D., Möhring, M.: Digital enterprise architecture—transformation for the Internet of Things. In: IEEE 19th International Enterprise Distributed Object Computing Workshop (2015)
8. Zimmermann, A., Pretz, M., Zimmermann, G., Firesmith, D.G., Petriv, I., El-Sheikh, E.: Towards service-oriented enterprise architectures for big data applications in the cloud. In: 17th IEEE International Enterprise Distributed Object Computing Conference Workshops (2013)
9. Bossert, O., Kretzberg, A., Laatz, J.: Unleashing the power of small, independent teams. In: McKinsey Quarterly, 18 July 2018
10. Bossert, O., Laatz, J.: How enterprise architects can help ensure success with digital transformations. In: McKinsey Quarterly, 16 Aug 2016

Chapter 4
Enabling Rapid Digital Transformation

John Gøtze and Alex Romanov

Abstract This chapter introduces several methodologies to rapid digital transformation, including Segregated Business Services Design, based on pre-emptive identification and definition of eventually possible digital interfaces throughout the enterprise's business ecosystem. Adopted as a pattern-based modelling style from the get-go, this does not make the EA development heavier, but it makes the enterprise intrinsically ready for rapid changes, such as client-facing service digitalization, digital outsourcing, digital consolidation of internal services, digital interaction with suppliers, and generally digital-oriented business model change.

Keywords Digital transformation · Segregated business services design · Agility · Business architecture · Enterprise architecture

4.1 Imperative: Digital Transformation (and Do It Fast!)

4.1.1 Pre-requisites for Rapid/Agile Digitalization

The real digital advantage comes from having an organization designed to adapt with dexterity to rapidly advancing digital technologies and other complexity drivers. Digital dexterity is defined by MIT as "the ability to rapidly self-organize to deliver new value from digital technologies" [1]. Organizations that manage to both explore and exploit "the Digital" become what the organizational design literature for over

---

J. Gøtze (✉)
Business IT Department, IT University of Copenhagen, Rued Langgaards Vej 7, 2300 Copenhagen, Denmark
e-mail: jogo@itu.dk

A. Romanov
Avrolabs, Oakville, ON, Canada
e-mail: alex.romanov@gmail.com

A. Zimmermann et al. (eds.), *Architecting the Digital Transformation*, Intelligent Systems Reference Library 188,
https://doi.org/10.1007/978-3-030-49640-1_4

40 years have called ambidextrous organizations, from Duncan [2] over March [3] to O'Reilly and Tushman [4–6]. Digital transformation [7], IT ambidexterity [8], Bimodal IT [9–11] and two-speed development [12, 13] are variations of this in the digital perspective.

Today, in order to survive in its rapidly changing ecosystem, an organization needs a digital business strategy [14], which may include business model, business architecture, IT, security, socio-cultural, and other aspects in some combinations. This strategy must be neither rigidly carved in stone (leading to slow, too late reactions), nor panicky-spontaneous (fast, but often patchy) one. Both extremes are recipes for disaster.

The whole point of rapid, but precisely thought-through, digital adaptability, or "Business Agility" [15], is to:

- have a solid business knowledgebase (Business Architecture),
- have it structured to enable rapid change (crucial for business agility),
- be able to systemically explore it for digital opportunities,
- pre-define parts of architecture to be eventually digitally changed,
- and perhaps most importantly be ready with identified and pre-defined opportunistic would-be architectures to be rapidly implemented (exploited) when the business situation is ripe for that change.

4.1.2 Enterprise Readiness for Rapid Digital[1] Change

In order to be digitally adaptive, an organization needs a capability to anticipate, at least roughly,

- where in its business ecosystem, introduction of digital interfaces may become necessary,
- what business opportunities such introduction would create,
- what business model tweaks that would require, along with related socio-cultural changes or training, and
- what security/privacy/compliance issues this may cause and how to deal with these issues.

The key idea here is ensuring pre-emptive readiness of architecture, interface specs, and prioritized development requirements for a self-sufficient module—a specific component of the whole overall business architecture knowledgebase—for rapid digital change implementation. Pre-emptively revealing opportunities for digital transformations long before any technical implementations are considered, is an important function of business architecture. It allows to play with ideas, do

[1]This concept pertains to any kind of change. to change rapidly and efficiently, the enterprise's business architecture must be ready in the part concerned before the green flag drops to start the transformation project, so that the agile teams can release their clutches and rush to business victory being fully equipped with detailed gapless requirements and specs to implement. Now, that's real agility.

cost-benefit evaluations, create "to-be" versions of business architecture, map them to legacy systems functionality, estimate missing needs via gap analysis, and come with detailed advice for enterprise strategic decision makers.

Business Digital Agility is not about sprints, or other "agile" buzzwords. In order to be able to digitally adapt in a fast/agile way, the enterprise must be provided with

- a ready-to-use modular blueprint (Business Architecture),
- an ability to pull that blueprint and rapidly identify the modules involved in a specific business problem, which needs digital resolution or improvement,
- a method and know-how to rapidly modify the architectural blueprint only in the part related to the transformation, and
- a method to be pre-emptively ready with assessments of potential security, privacy, and compliance issues in case if pre-defined interfaces become exposed, and pre-emptive recommendations to deal with these issues.

4.1.3 Why Classic Business Architecture Is not Enough

Classic business architecture[2] inclines to develop detailed, complex holistic models around capabilities and/or business processes, reflecting the end-to-end complexity of the enterprise. If ever completed (which, alas, does not always happen), such models do reflect the "current" understanding of strategies, which existed when the business architecture exercise started. Often by the time of completion they are already obsolete due to changing business environment. This "classic" approach traditionally takes a lot of time, cost, thinking manpower, and includes lots of arguing about methods of depiction (i.e. frameworks, modeling styles, repositories), but often causes the business architecture endeavors to lag behind urgent enterprise opportunistic reactions to change demands.

Moreover, classic business architecture models built around whole-enterprise capabilities, or end-to-end processes views, just are not sharpened for revealing or helping to invent possible digital transformation opportunities. The classic excuse is being "system-agnostic", but when looking at a complex end-to end model of an organization, one does not see any highlighted digital opportunities, except maybe the most obvious ones such as adding digitized customer interaction channels instead of existing "system agnostic" generic channels. That is not exactly disruptive digital innovation. It is this monolithic, hard to change, end-to-end complex business architecture, which has earned the reputation of a proverbial "ivory tower", and which makes organizations to drop business architecting altogether and divert to ad hoc agile IT development sprints.

Our view is the "business architected" versus "agile" controversy is the wrong framing of a problem. A modern competitive digital enterprise must be both well

[2]With this we mean business architecture approaches stemming from classic business-process management, such as the Business Architecture Guild's BIZBOK (https://www.businessarchitectureguild.org/) and OMG's BASIG (https://www.omg.org/bawg/).

architected, and agile/rapidly adaptable. Moreover, it must be architected for business agility and rapid adaptability. It not only has to "explore" and "exploit" its business architecture, but that knowledgebase must be by-design equipped with coherent proactive approaches for

- constant ongoing exploration of opportunities for rapid "digital" innovations in already established and exploited business model,
- rapid change of business architecture as reaction to changing market conditions,
- rapid inefficiency identification and improvement, and
- rapid output of sufficiently detailed prioritized requirements for agile IT sprints/implementations.

We do not try to diminish the value of classic business architecture, on the contrary, we recognize its huge value for an enterprise. However, in the fast-changing business environment of today, in many cases there is simply not enough time and resources to develop it in complete perfection. For today's organizations just starting (or considering starting) on the path of business architecture, it is important to understand that developing the agile and digital-ready business architecture requires new approaches—additional to classic business architecture understanding.

In this chapter we introduce several methodological approaches to developing business architecture, which allow to

- pre-emptively identify and define possible digital opportunities throughout the enterprise's business ecosystem, and
- create and maintain business architecture, which is easily re-designable, re-configurable, and allows to introduce the required changes fast, making the transformation architecture available-on-demand.

4.2 Identifying Digital Transformation Opportunities

4.2.1 Opportunity Areas for Digital in an Enterprise

To introduce any digital innovation, one must first define one or several interface(s) where the information flow or material resources are crossing some boundary. These interfaces may be digitized, and a new digital initiative involving changes to business architecture components (business model, business services, processes, etc.) may be built around them. In order to find digital opportunities, one must first understand around which existing, or implicit (yet to be created/defined) interfaces those opportunities can potentially emerge.

As seen on Fig. 4.1, an enterprise has three potential areas for introducing digital interfaces (we depict them as red boundary lines, across which some information or resources flow occurs, and for the sake of clarity call them "Type 1, 2, and 3"):

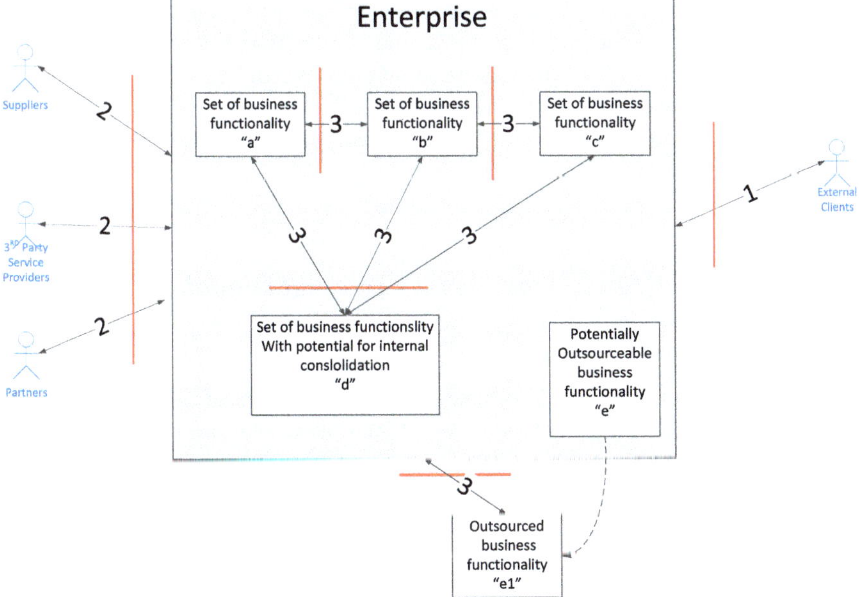

Fig. 4.1 Potential digital interfaces in an enterprise ecosystem

- *Type 1*: External—interfaces to external clients. These may exist, or be the "future state", and may currently operate via different channels—from over the counter—to snail mail, to truly digital.
- *Type 2*: External—already existing interfaces to 3rd party providers, suppliers, partners, etc.—the current ecosystem. These also may already exist or be envisaged in the future and may operate via different, including digital, channels.
- *Type 3*: Internal interfaces between the "*sets of business functionality*" inside of the enterprise. We use this term in order to stay independent of any specific architectural frameworks. Depending on how the specific business architecture is modelled, these sets may be defined as, e.g.,
  - cross-silo business services with their inputs and single business outputs working over service boundaries, (in which case the diagram on Fig. 4.1 would be replaced with the Service Integration and Management model - SIAM) [16],
  - business capabilities,
  - fragments of business processes, some of which could be reusable,
  - organization unit silos with their internal processes and "departmental" interface boundaries, or even
  - system interfaces between legacy systems.

Type 3 interfaces are sometimes obvious and explicit (as, e.g. interfaces between sets "a", "b" and "c" on Fig. 4.1), sometimes blurred and implicit; they may exist, but not be well defined, or maybe even not recognized as an interface—by the silo-minded enterprise (and as such—as a potential Digital opportunity). Such interfaces may be created as result of internal consolidation of redundant functionality ("d" on Fig. 4.1). Some pre-existing, or consolidated functionality set may be digitally outsourced, in which case the formerly internal interface would become external ("e" and "e1" on Fig. 4.1).

4.2.2 Digital "Quick Wins" Versus Internal Digital Efficiencies

Most of current publications [17–20] about "Digital Disruption" and "Digital" business transformations focus on types 1 and 2 interfaces. The same can be said about majority of "Digital" initiatives in organizations. There are two reasons for this: Firstly, that is what is on the surface, and usually represents the "low hanging fruit" of digital initiatives. Secondly, it usually follows the existing business model, and existing interactions, without their deep re-thinking for "Digital". Usually these interfaces were already established long ago, and only need some digitalizing (in fact, adding another channel), plus, maybe adding some additional digital marketing, maybe some content management, maybe some employees training, and of course, some IT digital channel implementations.

No doubt, focusing on the interfaces of type 1 and 2 and related digital opportunities is important for early digital initiatives, and they are talked about enough. But we'll leave them out of scope of this study and focus on efficiencies of internal digital transformation of an enterprise. The "type 3" interfaces need to be identified, sometimes invented. In order to make use of them, the business model of the enterprise often must be re-thought. The "type 3" interactions may belong to internal (support) services, which are invisible for clients. Addressing their efficiencies or transformations may be very fruitful, but they are frequently omitted being in the shadow of more obvious type 1 and 2.

In this chapter we focus on the business side of digital transformations happening inside an enterprise over the internal "type 3" interfaces—how to systemically discover this type of opportunities, how to utilize the discovered in the revised Digital business models of an enterprise, and how to do this early and fast,—to make the enterprise more business—agile.

We'll use the acronym IDIC for "Type 3" Internal Digital Interface Candidates hereinafter.

User experience (UX), client-facing service design, and socio-cultural aspects of Digital are out of our scope here.

4.3 Pre-emptive Identification and Definition of Internal Digital Interface Candidates (IDIC)

4.3.1 Real Life Scenarios Involving IDIC

Here are several examples of scenarios involving "Type 3" digital interfaces in business transformations from our practice (not an exhaustive list):

- *Example scenario 1*: Some business capability traditionally exists within the body of an enterprise, and the processes implementing it are deeply intertwined with hundreds of other internal processes. A decision is made to outsource this capability to a third-party provider—digitally. The interface needs to be established between this capability and the rest of the enterprise—by "cutting" a multitude of existing live processes and introducing information exchange modules into multiple points of previously monolithic body of business architecture. Assembled from these interface modules, the previously non-emphasized interface will become the one enabling digital transformation to interaction with an outsourcing provider[3]
- *Example Scenario 2*: Some business capability is redundantly repeated within multiple silos of an enterprise. A decision is made to consolidate it into an internal service, digitally providing its output to all previous internal (and possibly new external) silos/clients. A new, previously non-existing interface must be established in order to enable such digital transformation and enable establishing internal centralized business service[4]
- *Example Scenario 3*: A large enterprise may be in the process of changing its business model form an end-to-end monolith to flexible interaction of internal and/or external service providers. Some parts of the organization may be even not aware of this change happening in other parts of the same organization– changes in one line of business of the organization may be not very visible from other silos. In this situation the enterprise has a fractured business model, which may lead to severe inefficiencies: they may, for example, fund an expensive IT project updating internal functionality systems, which are redundant to other systems dedicated to outsourcing support, created by another department. It is extremely important to recognize such environment and to timely define appropriate unified digital interfaces for silos, which are engaging in internal digitalization as well as for those who don't (yet)[5]

---

[3]Based on a real-life Public-Private Partnership (PPP) project for a major permits-issuing ministry (public sector) digitally outsourcing all customer interaction of one of their lines of business to a concession-based network of private providers, while keeping the ownership of all data and decision-making business rules inside the ministry.

[4]Implemented an author in multiple real-life public sector initiatives leading to creation of internal centralized digital business services.

[5]Also based on a real-life situation and provided business architectural solution in a major public sector organization.

Below are several methods to pre-emptively identify potential "Type 3" internal digital interfaces working with business architecture—to have them defined by the time when the digitalization opportunity appears.

4.3.2 Anticipating IDIC by Analysing the Business Model

We assume the Reader is familiar with the concept of a Business Model (BM), and Osterwalder's Business Model Canvas [21]. Enterprise governance models typically do not define the BM as a mandated artifact. But the Business Model is the ultimate foundation of Enterprise Architecture and it is crucial for the clarity of a digital strategy. Neither Strategic business architecture, nor the rest of EA, nor any real business process improvements cannot be meaningful without clear business agreement on the BM first.

From the point of view of digital strategies, if the business model is undefined, poorly defined, obsolete, or, as in the Example Scenario 3 above, different silos of an organization have totally different implicit understanding of the BM, it won't help to define the general plans for Digitalization. Digital transformations must start from the business model revision/re-thinking, otherwise they will be only "scratching the surface"—without addressing true digitalization opportunities.

Analysis of the Business Model allows to identify Customer interactions suitable for being moved to 3rd Party, Reveals Digital Transformations opportunities, and prepares to serve "on demand" digital clients.

We find it recommendable to analyze the boundary between Key Resources and Key Activities on one side, and the Key Partners on another. Consider "Core" components—activities and resources, which would never be outsourced—for political, regulatory, information privacy, or security reasons, and "Non-Core" components, which may eventually find themselves on the "Key Partners" side of the boundary (in other words, may be eventually outsourced)—see Fig. 4.2.

The red line on this canvas illustrates the boundary, which may eventually evolve into a Digital interface. Understanding its role in Digital projects of the future in advance allows to:

- Foresee future possible digital opportunities and think about possible digital business model changes in advance.
- Understand data used on both sides of the boundary to define the interface, as well as privacy, confidentiality, and security requirements long before any actual "Digital" IT development is started, thus ensuring its quality, lower cost, and well-architected agility.
- Understand on which side of the interface business decisions are made, whether the interface needs to include requests/responses for business rules-based decisions, pre-plan decision patterns, and pre-architect efficient usage of rules engines.
- Evaluate the costs and IT requirements for (eventual) external outsourcing of a business capability, or establishing a similar internal interface, when redundant

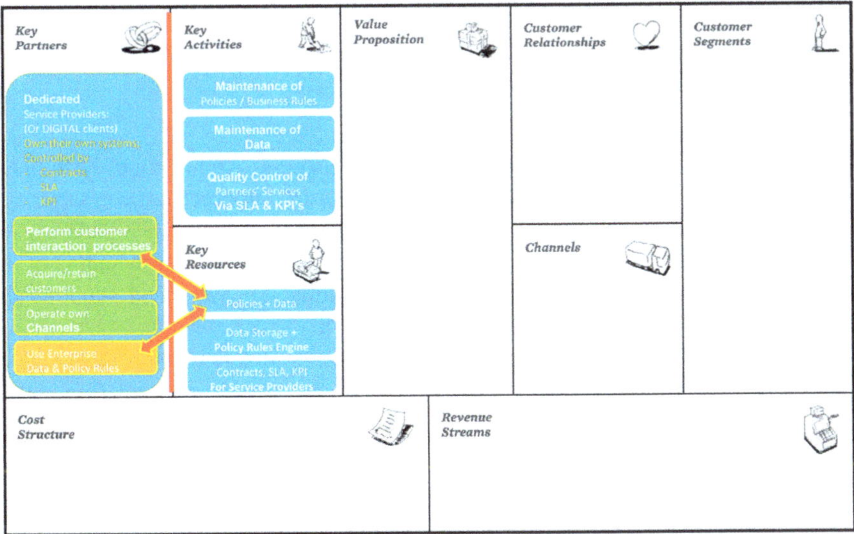

Fig. 4.2 An example of a potential digital interface (red line) identified right in a business model of an enterprise, which keeps maintenance of data and policy business decisions, while thinking to outsource the customer interaction to a SLA/KPI-driven 3rd party

silo'ed capabilities are consolidated into a single internal business service to serve the silos.

- Form the reusable enterprise-wide business process patterns library utilizing the core/non-core interfaces consistently across the enterprise, making it ready for on-demand outsourcing/on-demand digitalization of such interfaces. This means high business agility!

Architecting around core/non-core boundaries has these advantages:

- Once outsourced, the non-core business processes may be implemented in absolutely different ways, depending on service provider. This understanding may save time by skipping detailed BPM for this functionality. One may decide to consider the non-core functionality as a black box with an interface—and avoid wasting time on its detailed business architecture—the saving, which is not possible if one considers an end-to-end monolithic business architecture.
- Customer-oriented service design, as well as CX/UX design becomes the responsibility of a service provider, not "our" enterprise (except, maybe enforcing "our" customer service standards, if necessary. This again may save time on development of "our" detailed architecture.

4.3.3 *Anticipating IDIC by Establishing and Analysing the Business Service-Oriented Business Architecture (BSOBA)*

The concept of Business Service-Oriented Business Architecture (BSOBA), describing an enterprise via its client-facing and internal (support) business services,[6] is not new. It sits in the foundation of some business architecture standards, for example the GSRM (Government Services Reference Model) [22], which is in the foundation of some enterprise architecture standards, e.g. [23], and supports the concept of Service-Oriented Enterprise (SOE) [24].

Principles of BSOBA/SOE make it convenient for systemic identifying the IDIC:

- Business processes are defined only within each service.
- Business services may span multiple organizational units, however it's necessary to define which one is accountable for a service.
- Each service has a clear output, which either satisfies the need of an external client (or target group) or serves as input into another service.
- There is a very clear set of criteria for defining services, and distinguishing them from processes (criteria defined, for example in Sect. 3.5.5 of [23].
- SOE-style business architecture does not contradict more widely known approaches, such as Capability-based, or process-based business architecture. The concepts described here are applicable to different frameworks, and there is no intention here to argue which one is better. For the purposes of SOE, we define a capability as a "measured service output".
- The map of services, their interaction, and accountabilities are presented by a SIAM [16] (Fig. 4.3 is a simple example of a SIAM).
- A value chain may be formed by multiple services feeding their outputs into other services.
- It means only that the clear output (or value) handovers must be defined on the boundaries of services. This in turn is the pre-requisite to defining interfaces on the boundaries of services and preparing these boundaries for digitalization.

Building the whole enterprise business architecture the BSOBA/SOE way has many benefits: It helps to identify redundant functionality in multiple org units of a legacy enterprise, it reveals "unexpected" actors from silos, nobody thought about, it helps to realize the existence of "forgotten" service outputs, forgotten dependencies, overseen accountabilities, and helps to form high quality requirements for IT initiatives.

When done, and the in- and out-interfaces are properly defined with information or material resources moving across the service boundary, and trigger events, then

---

[6]It is important to emphasize that BSOBA and SOE business architectural model pertains purely to business services, this must not be confused with the "Service-Oriented Architecture" (SOA) which is the implementation architecture. At some circumstances the boundaries of Business SOE and SOA services may coincide, but far not always.

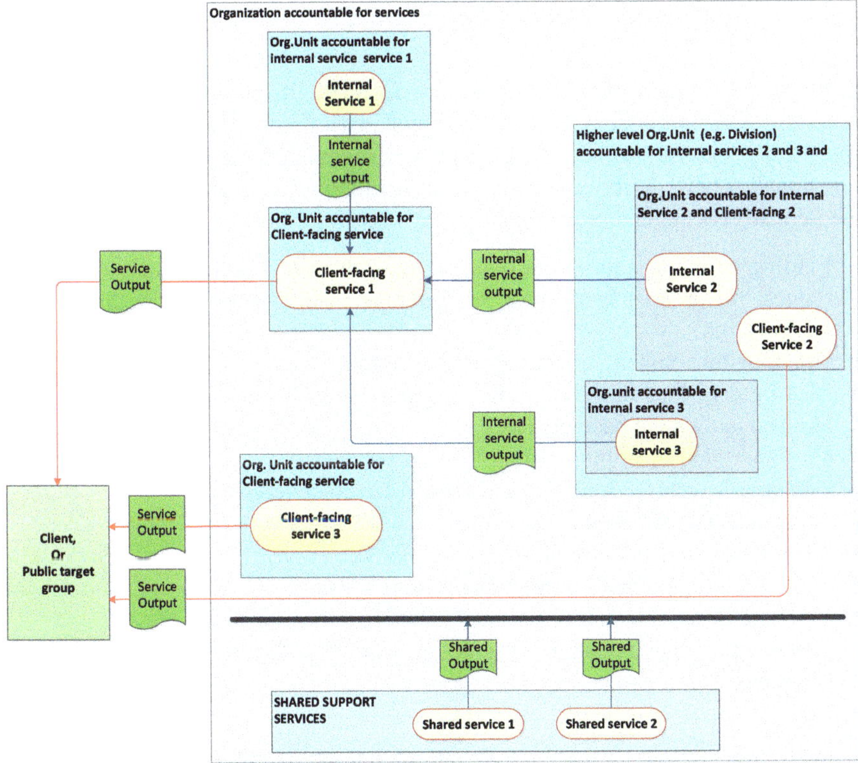

Fig. 4.3 Example of SIAM with internal and client-facing services. Each service boundary (red borders of yellow ovals representing business services) is a ready pre-defined digital interface for future digital transformation considerations

each service represents a ready architected "construction block", which can be digitally outsourced externally, or digitally separated internally and serve the rest of the enterprise.

Then in fact the SIAM provides the map of already defined possible digital interfaces, around which the BM changes and digital initiatives may be designed in a rapid and agile way, whenever the time is right to use service boundaries in the digital transformation initiative.

4.3.4 Anticipating IDIC by Segregated Service Design (SSD)

Segregated Service Design[7] (SSD) is a powerful method to pre-emptively identify and develop possible interfaces for building digital innovations inside of the enterprise. It specifically looks at business situations where digital outsourcing of some business functionality, or its digital consolidation into an internal digital service) may be (even theoretically) beneficial. Such functionality may, for example, include:

- Finding/retrieving appropriate data (e.g. customer profile) in internal storage upon request for processing (from a 3rd party, or from another service in the same organization)
- Updating that data as per the processing results
- Making decisions based on policy business rules for a requestor, which is not allowed to own the business rules
- Customer identification for other internal services
- Customer's service eligibility verification for other services
- Order fulfillment for other services
- Service output quality monitoring
- Service health and performance monitoring
- Customer's address change initiated from any other service
- Providing, managing, and routing notifications for other services
- Document management, for other services
- Etc.

Such services may be initially present redundantly in different silos of the organization. Consolidating them into one efficient service, which can digitally serve the rest of the organization, internally or externally, means introducing a digital efficiency.

Consider the first three bullets (referring to the prior Business Model example). Imagine a typical end-to-end process shown in the upper part of Fig. 4.4. Imagine that as result of digital outsourcing transformation it's being changed so that the actual customer interaction, data input and processing happen on the 3rd party site, while data ownership and policy business rules never leave "our" organization. In other words, the process now works as shown in the lower part of Fig. 4.3: The non-core activity does not just obtain some data, it explicitly inquires it across the interface from the core area, and the core "Inquire data" activity provides it back across the interface. The same happens with decision. Instead of executing the if-then-else by the non-core process, it requests the decision from the core area (there is some kind of business rules-based decision mechanism there), which makes the decision based on the request + business rules + embedded decision patterns and returns the decision without explaining its rationale to the initiating process.

[7] Note that typical "Service Design" exercises frequently focus on customer experience [25–28], i.e. deal with client-facing services, and digital business transformations of related external "type 1" and "type 2" interfaces. The SSD looks deep inside of the enterprise for multiple "type 3" opportunities for digital efficiency improvements, which typical service designers do not identify.

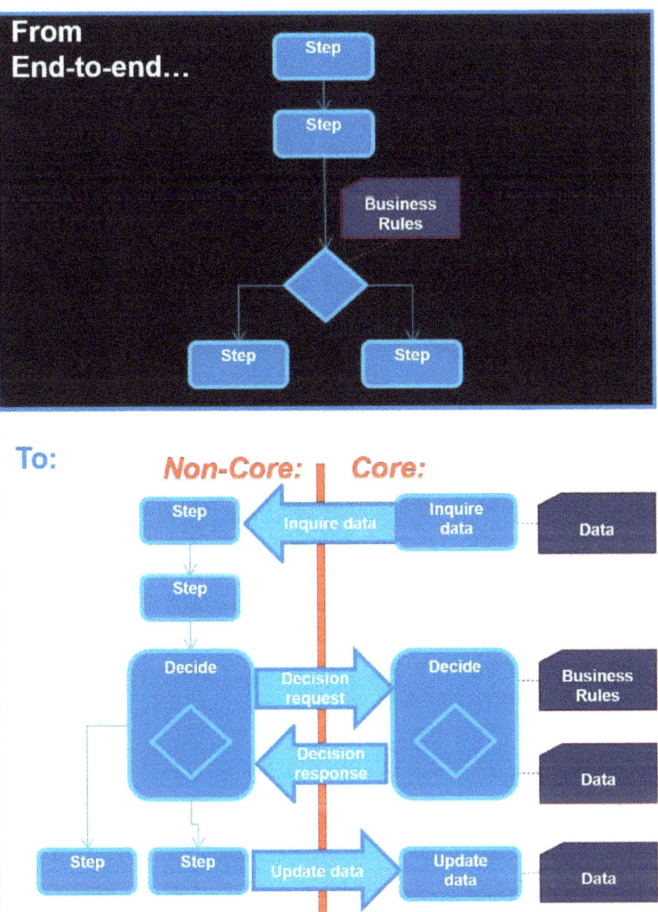

Fig. 4.4 From usual process model to pre-segregated core/non-core process model

If an enterprise has invested into defining its complex business architecture in traditional end-to-end manner, converting it into the "digitally outsourced" version would require significant rework and search of all points where the over-the-interface activity would occur upon outsourcing.

The idea is to consider at the time of modeling anything that can *potentially* work over a digital interface (even if it currently does not) as if an interface existed—pre-segregate it. In every business process the potential boundary can be outlined, moving from standard end-to-end process modeling to segregating potentially outsourceable/consolidable functionality and pre-defining the potential interfaces. Currently, the red boundary may not even explicitly exist. However, if the process is defined as described above, the definition of the digital interface is ready to be on-demand implemented as soon as Business decides to go ahead with the digital transformation.

Deliberate by-design segregating such services pre-emptively, and pre-defining the interfaces, as well as triggering events, the data flows and constraints, the business architecture is essentially prepared for thoughtful on-demand "Type 3" digitalization.

4.4 Business Architecture Ready for Business Agility

Besides pre-emptive understanding of potential interfaces, which can become the pre-defined "seeds" for rapid digital transformation initiatives, the enterprise business architecture must be generally ready for rapid changes. Typical end-to-end complex models usually require a lot of time-consuming re-modelling in case of desire to change. Below are some methods to structure business architecture in such a way that its changes can be done rapidly.

It's important to understand that if Business Architecture is developed from scratch, it takes about the same time and effort (or, if the reference model used is well developed—much less time and effort) to structure it using approaches described here, than in a "classic" end-to-end manner. However, when the enterprise needs urgent change, these methods enable far greater speed of adaptability.

If there is a legacy end-to-end architecture to deal with, it makes sense to invest time into gradual restructuring it into agility-ready one—to be prepared for rapid changes when the time is ripe.

4.4.1 The Principle of Modifiable Pattern-Based Modularity

This method calls for defining business architecture using

- As-simple-as-possible business architecture modules, based on reusable patterns, and
- Sets of hierarchical business rules modifying these simplistic modules to specific business situations, business lines, legislations, regulations, business models, etc.

We deliberately do not specify the nature of these modules. They can be capabilities, or groupings of business functions, or fragments of business processes. Specifics of what exactly approach on modules is taken, depends on the state of specific business architecture in a specific enterprise, in a specific industry domain, availability of an appropriate reference model, and the art of a business architect. Important is, that the modules are recognizable and self-sufficient enough to be used in separation of the rest of the architecture. Often the sets of such modules can be derived from an appropriate industry reference model. For example, for the public sector business, a convenient reference model is GSRM [22], for municipal business MRM [29], for insurance ACORD [30], or Panorama 360 [31], for banking BIAN [32], for many other industries APQC [33], there are Business Architecture Guild-published reference models [34], and many others. We do not have preferences regarding exact patterns of which reference model should be used—it depends on the situation, availability, cost of licenses.

For the purpose of anticipated Digital transformations, the most efficient approach is to define the modifiable modules in the same boundaries as the pieces of functionality surrounded with potential digital interfaces as described above.

Here's an example of how this approach works. Let's say we deal with a public service organization, which issues some kind of permission. It could be a driver license, or a building permit, or an annual vehicle usage permit, or a border crossing permit—any permit! In different countries the instances of these are called differently. But as per GSRM, all governments in the world do follow the same pattern for issuing permits. In any country, in any government, regardless of any politics, in order to issue a permit, one must execute more-or less these steps:

1. Authenticate a Client
2. Record Client's Request (maybe, selected from a product catalog)
3. Determine Client's Eligibility (by applying business rules)
4. Determine Fees (using business rules on fee tables)
5. Collect payment
6. Produce product (whatever permit it is) record
7. Assign Product to Client
8. Produce a temporary token (e.g. some kind of a printer printout), deliver to Client (using channel-related business rules)
9. Produce permanent token, (e.g. some kind of a card), deliver it to Client (using channel-related business rules)
10. Notify Client, Partners, 3rd parties (using business rules on whom to notify)
11. Monitor & Maintain the permission (Sanctions, Fines, Expiry, Replacement, withdrawals, etc.).

That's it. Only 11 reusable process *modules*, while all sorts of complications and details related to a specific line of business, country, legislation, regulations, etc., can be applied to the above simplistic process pattern as multi-level hierarchical business rules. The pattern steps shown above also produce/consume some data, and that data also may modify the pattern, based on its values, or validation. It's the rules and data, rules based on data, and rules based on external circumstances are the pattern "modifiers". The benefits of such architecture are:

- It is very simple and requires not much time to create.
- Business SMEs, typically insisting on the "uniqueness" of their line of business, usually recognize it instantly.
- It is easily modifiable, as the process almost always stays the same, while business rules around it can be changed fast.
- Because of this rapid changeability, it works very well for business agility, as well as for support of agile IT implementation projects.
- On the IT level, it easily translates into decision automation, rules engines. If implemented so, the IT changes of all sorts become also very agile—it all boils down to changing business rules.

If one looks at the enterprise business architecture in this modular manner, the idea to consolidate the similar items and group them outside of line-of-business silos comes naturally: either behind an internal digital interface (increasing the enterprise efficiency by consolidating certain functionality into a de facto internal service serving diverse lines of business by applying appropriate business rules for each LOB), or behind the external digital interface (increasing efficiency by outsourcing this repeatable functionality externally, to a 3rd party, or, say, to a self-serve service for mobile devices.

This approach allows to re-define business complexity in terms of "simplistic" modules, which are: highly reusable, easily customizable for multiple purposes, independently re-designable, and optimized in size—we want neither too many small modules (hard to manage diversity), nor too big/complex modules (hard to modify them).

Modeling business architectures based on simplistic process-or-capability pattern and business rules modifiers, PLUS pre-emptive definition of data exchange interfaces around such modules, when done routinely while creating the business knowledgebase, does not take too much extra time. The interfaces are defined and just sit there till their time comes. But when the time comes—they are ready for immediate action.

The benefit of creating well prepared opportunities for agile rapid digital consolidation or outsourcing—on demand is priceless. Now, that's business agility. Of course, the end-to-end business architecture models are still relevant and needed to provide the holistic view of the business, but now we present them only as conglomerates of modules assembled together and interacting between themselves via predefined interfaces. This way "the best of two worlds" is achieved: The Business has its classic strategic holistic view, and the architecture is modular and can be digitally re-arranged very rapidly on demand.

4.4.2 Governance of Pre-componentized Business Architecture

In order to be truly digitally agile, an organization needs to have not only the business architecture knowledgebase, but have it in such a form, which reveals possible digital interfaces, and is ready to support an IT-agile projects immediately upon the Business' decision. To achieve this, the background "Explore" continuous activity must be producing and managing the business architecture. Traditional governance [35–38] of the end-to end "monolithic" architecture is not convenient for this. Modular/segregated business architecture must be governed in the modular way too. Its governance must focus on:

- Architectural Modules/Modifiers/on top of a high-level architectural framework.
- Maintenance of modular business knowledge base, including continuous reflecting changes introduced by implementation projects.

- Potentially-exposed interfaces of modules, and potential related issues with security, privacy, and compliance.

The business architecture repository of an enterprise, whatever the shape it is in (an advanced modeling tool with the model database, or a stack of documents in Word/Excel/Visio) must be built of simple construction blocks—"highly modifiable modules" (e.g. capability patterns, or business-function patterns, or process patterns), "Modifiers" (business rules adjusting the modifiable patterns to specific business situations), pre-defined possible interfaces, and a light framework to tie together and align it all for specific business context.

In organizations requiring architecture review boards (ARB) and architecture approvals for project funds to be allocated, it makes sense to establish pre-emptive continuous pre-approvals for identified architecture modules and potential digital interfaces, so that when a transformation project comes along, the involved modules are already pre-approved and can be worked upon without further approvals and related reviews. This way, preparation of ARB approvals takes much less time of a project making the on-demand architectural project support much faster, and the enterprise respectively business-agile.

4.5 Organizing the Business Architecture Work for Digital Business Agility

4.5.1 Controversy of Digital Business Agility

There is an aspect of business architecture preventing "type 3" digital initiatives from being systemically identified and architected: Business architecture is a complex, costly, and time-consuming effort, which traditionally produced difficult to change end-to-end models. Digital transformation projects tend to always be urgent. Either it is a campaign initiated from within the enterprise, and it needs to show results fast in order to survive, or it is an urgent reaction to restore external competitiveness of the whole enterprise to survive. It is hard to blame those accountable for desire to be as fast as possible.

Frankly, nowadays attempting to build complex digital innovations in a classic waterfall way just does not work anymore (e.g. first architecture baseline, then architecture tweaks for the transformation project, then business requirements based on architecture, then delivery project planning and costing, then some approvals, then the actual development…). That is way too long and has very little chance to enjoy generous time allocation. Besides, executives are often sold on the promise of agile development—as the remedy against all known time-heavy factors. And if the enterprise had already spent years/millions on developing business architecture baseline, then "No time to re-architect, let's run!"—becomes the typical executives' mindset, and, in our opinion, the cause of digital initiatives to fail.

Of course, agile development is proven fast/successful on smaller projects, and approaches like SAFe [39] and Disciplined Agile [40] are embraced by even on larger ones, but as the Business Agility Manifesto (BAM) [15] states:

- Development of software faster is not sufficient for Business survival and growth
- Business change depends on a Business Knowledge Base
- The measure of agile business is how easily Business Knowledge can be re-configured
- Avoidable rework due to lack of Business Knowledge is waste
- Creation of value for silos has no value.

These BAM points are absolutely applicable to digital transformations. Sprints, SCRUM, user stories, or stand-up team meetings without a clear and knowledgebase-savvy digital strategy would not help the enterprise to rapidly react to changing market demands, "turn on a dime", launch new digital capabilities to dominate its market, mitigate commercial risks, or adapt to a new legislation/regulation. In other words, Agile IT methods are not enough to achieve the Enterprise Business Agility,[8] or in our case, Digital Business Agility.

Any decent strategy must be based on detailed knowledge of how exactly the enterprise and its ecosystem work, so in order to identify the digital opportunities, and properly weave them into the body of the enterprise, there must be the business knowledgebase, a.k.a. the business architecture.

Organizations are facing quite a typical controversial situation:

- They need to have a detailed business architectural knowledgebase to know about digital opportunities just in time for implementation
- Development of detailed end-to-end business architecture is very time- and effort-consuming
- Introduction of digital ideas, which are not based on detailed understanding of business architecture may only pick up some low hanging fruit and just scratch the surface of existing opportunities
- Most digital transformation projects are urgent, budget-constrained, and have no capacity to include serious business architecture exploration…
- … go to the 1st bullet point.

Many failed classic end-to-end enterprise and business architecture undertakings in enterprises have, sadly, proven [41–47] that just one-time (usually, very long time!) delivery of the set of strategic models and artifacts (though undoubtfully important)

[8]Business Agility Manifesto defines Business Agility as follows: "Business Agility" by its very name, would suggest implications far beyond the normal dimensions of an "Agile" discussion about software development or organization schemes. "Business Agility" envisions an ability to modify dynamically the concepts and structures of the business itself for maintaining relevance in the context of a dramatically changing, complex and uncertain operational environment. This ability necessarily must include dynamic reconfiguration of implementations, adolescent and mature, manual and automated, new and existing, to continue business operations as business changes are formalized [15].

would not grant the organization's rapid adaptability to urgent digital transformation needs.

Impatient dropping business architecture in favor of ad hoc quick solutions (risking endless iterations, misalignments, affecting seemingly unrelated processes, and budget overruns) is not an option. But waiting for detailed business architecture to be developed when the change is imminent and urgent is also not an option.

Resolving this controversy requires two sides of approach:

- Know how to structure the end-to end business architecture in such a way that it's architecturally ready for
  - rapid reactions to specifically digital business challenges—how to find and pre-define interfaces, which can eventually become the seeds of digital transformation initiatives (as discussed in Sect. 4.3).
  - rapid changes in general—by maximizing use of easily changeable, pattern-based, business rules-controlled modular constructs, and governance of business architecture in such manner (as discussed in Sect. 4.4).
- Know how to organize the work on business architecture to enable both:
  - long-term ongoing work on pre-defining the business architecture with pre-defined potential digital opportunities, possible interfaces, modules and their modifiers, and
  - short-term fast projects for rapid finalization of digital business changes based on selected pre-defined opportunities and rapid agile IT implementation of such changes.

4.5.2 *Two-Speed Business Architecture-as-a-Service*

We discussed the principles of Business Service-Oriented Business Architecture (BSOBA) and Service-Oriented Enterprise (SOE) in Sect. 4.3.3.

These principles are not specific to just Digital transformations. In fact, it's quite useful to apply them to organizing the very Business Architecture work in an organization[9]—as SOE services.

The SOE approach to business architecture for digital transformation is part of our broader approach to make business architecture not only strategic, but also highly visible and recognizable as continuously useful to diverse stakeholders and initiatives (including digital) in the enterprise's business ecosystem. In our view, stakeholders-consumers of business architecture do not have to wait months or years for its totally perfect completeness to enjoy its benefits. Its usefulness must be provided in real time, as opposed to the proverbial "strategic ivory tower" whose architectural outputs might

[9]Some good ideas about organizing enterprise architecture as-a-service can be found in [48–50].

become obsolete by the time of any implementation. Addressing only the organization's strategic, long term business architecture with all the complexity of its metamodels, frameworks, models, and alignments, is not enough nowadays. An efficient digital business architect must combine all the required classic models with efficient organizing, managing, and project-managing the business architecture development in such a way that the long-term nature of business architecture development is mitigated to allow very rapid, agile, and efficient support of digital initiatives, while the short-term nature of agile digital initiatives does not undermine development and maintenance of the long-term strategic business knowledgebase, and the strategic business architecture can be pre-emptively broken down into pre-defined and digital interfaces-equipped easily-modifiable modules (Sects. 4.3 and 4.4), which, while being developed within a "slow" strategic service, can be utilized for "rapid" agile digital development service. To satisfy these requirements, we suggest developing business architecture **as a set of two business services.** They provide their outputs to stakeholder clients in two different timeframes, serving respectively (Fig. 4.5).

The First, *Strategic*, Service

- Methodically develops the knowledgebase (strategic business architecture views of the enterprise), delivered in a long-term, relatively slow, detailed thought-through manner
- Manages business improvement ideas, including Digital
- Continuously integrates changes from projects to keep the knowledge base up-to-date.

Specific to digital transformations, this service also continuously:

- Analyzes business architecture for potential boundaries where digital interfaces could be created and later utilized for digital transformations.
- Uses the above analysis to constantly critically review current business models, business processes, capabilities, project portfolios, and generate innovative digital transformation business ideas.
- While outlining potential changes, provides continuous security and privacy assessments on potential digital interfaces.
- For regulated industries, it performs continuous compliance assessment.
- Manages and governs the business knowledgebase repository of pre-approved modifiable and digital interface-equipped modules.

The Second, *Tactical*, Service

- *Is* agile, on-demand, as-rapid-as-possible short-term quick transformations support.
- Is triggered by urgent business demand.
- Rapidly generates a transformation business case and case architecture on demand—It's a fast process, because all the architectural "billet" considerations are already done and formulated by the 1st service, which dramatically improves and speeds-up IT project scoping & cost estimation.

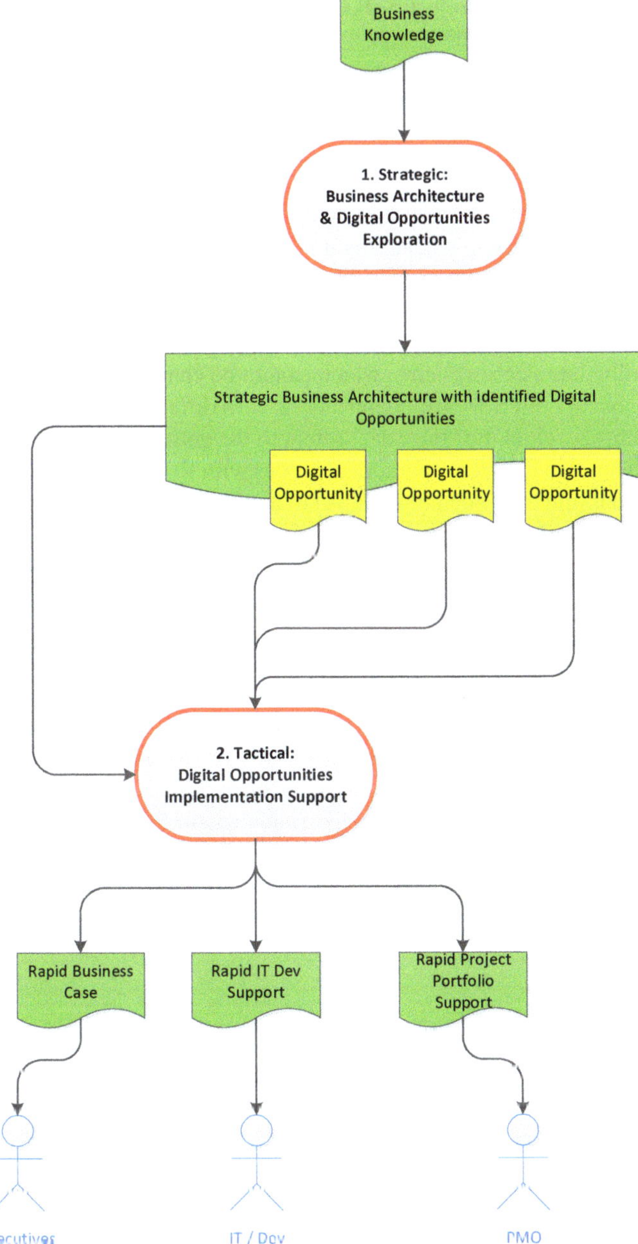

Fig. 4.5 Business architecture as two services of a service-oriented enterprise model

- Works directly with Project SDLC, SCRUM, Backlogs, etc. to optimize and prioritize agile IT delivery plans.
- Guides agile IT projects to use business architecture modules–based componentization as the basis for delivery prioritization.
- Ensures that the preliminary continuous exploration of potential security, privacy, and compliance issues are rapidly finalized and closed in the implementation phase.

The two-speed concept is generally not unique in modern EA or IT. For example [9–13, 51] discusses "two speed IT" versus "bimodal IT" organizations, essentially de-coupling a fast customer-facing, business digital innovation-oriented IT organization intended to react to rapidly changing customer needs from slower IT organization running large core systems, which cannot be changed or modified easily. The "2 speed" concept refers to two organizations with different speeds, governed as two separate entities, while the "bimodal" refers to the entity governed as one organization, which has two different operational modes with slow and fast speeds. Also, [52] talks of decoupling the Foundational EA stream in an organization, addressing capability gaps in foundational areas such as finance, project control, supply chain, from the Vanguard EA stream harnessing digital technologies—to improve efficiencies and increase technology adoption.

We focus on enabling two distinct business services within the same organization, addressing (1) Strategic/slow/systemic business architecture with pre-emptive digital business opportunities exploration and definition, and (2) Tactical/fast/burst business architecture supporting implementations of pre-emptively-identified digital opportunities. Ideally, both services should be governed by the same architecture principles and be driven by the same team of architects, who intrinsically know both the results and state of the Strategic Business Architecture and the discovered and pre-architected digital opportunities and related business views/requirements. As any well-defined business services, these can stay within the organization, or be outsourced to a 3rd party, say a consulting organization, if that is found efficient.

4.6 Conclusions

We've described a comprehensive conceptual model and a related set of approaches collectively enabling very coherent business-architecting for

- Building and maintaining the enterprise ecosystem business knowledgebase, and
- Supporting digital transformations in a consistently agile, but gapless manner.

This is a complex approach requiring long-term executive commitment, high quality business architecture work, serious discipline, good project management to ensure proper synchronization of slow and fast/burst processes and timely transfer of ideas and architectural requirements to IT development teams. On the IT side it also requires good agile project management to properly and timely prioritize and utilize the architectural outputs for rapid solutions delivery. But it did prove working well on multiple projects in very large and complex enterprises. It's not simple, but if the described concepts are managed to work like a clock, it's very efficient, and provides the enterprise with a smooth and steady flow of digital improvements and transformations, ensuring the organization's true business agility. On the contrary, the one-time "cavalry attacks" with all proper agile IT methodologies, SCRUM, stand-up meetings and a heap of user stories just cannot provide this methodical comprehensiveness, and unfortunately often make the initiatives to fail.

References

1. Soule, D.L., Puram, A., Westerman, G.F., Bonnet, D.: Becoming a digital organization. J. Digit. Dexter. Available at SSRN: https://ssrn.com/abstract=2697688 (2016)
2. Duncan, R.B.: The ambidextrous organization: designing dual structures for innovation. In: Kilmann, R.H., Pondy, L.R., Slevin, D. (eds.) The Management of Organization Design: Strategies and Implementation, pp. 167–188. North Holland, New York (1976)
3. March, J.G.: Exploration and exploitation in organizational learning. Org. Sci. **2**, 71–87 (1991)
4. O'Reilly, C.A., Tushman, M.L.: The ambidextrous organization. Hvd. Bus. Rev. 74–83 (2004)
5. O'Reilly, C.A., Tushman, M.L.: Organizational ambidexterity in action: How managers explore and exploit. Cal. Mgt. Rev. **53**, 1–18 (2011)
6. O'Reilly, C.A., Tushman, M.L.: Organizational Ambidexterity: Past, Present and Future. Academy of Management Perspectives. Rock Center for Corporate Governance at Stanford University Working Paper No. 142 (forthcoming). http://dx.doi.org/10.2139/ssrn.2285704 (2013)
7. Gerster, D.: Digital transformation and IT: current state of research. In: PACIS 2017 Proceedings, vol. 133. http://aisel.aisnet.org/pacis2017/133 (2017)
8. Lee, O.K., Sambamurthy, V., Lim, K.H., Wei, K.K.: How does IT ambidexterity impact organizational agility? Inf. Sys. Res. **26**, 398–417 (2015)
9. Horlach, B., Drews, P., Böhmann, T., Schirmer, I.: Increasing the agility of IT delivery—five types of bimodal IT organization. In: Proceedings of HICSS 2017, pp. 5420–5429. Waikoloa, Hawaii (2017)
10. Haffke, I., Kalgovas, B., Benlian, A.: The transformative role of bimodal IT in an era of digital business. In: Proceedings of the 50th Hawaii International Conference on System Sciences, pp. 5460–5469. Waikoloa, Hawaii (2017)
11. Drews, P., Schirmer, I., Horlach, B., Tekaat, C.: Bimodal enterprise architecture management: the emergence of a new EAM function for a BizDevOps-based fast IT. In: 2017 IEEE 21st International Enterprise Distributed Object Computing Workshop (EDOCW) (2017). https://doi.org/10.1109/edocw.2017.18
12. Bossert, O., Ip, C., Laartz, J.: A Two-Speed IT Architecture for the Digital Enterprise. McKinsey (2014)
13. Bossert, O., Laartz, J., Ramsøy, T.: Running Your Company at Two Speeds. McK Qtl (2014)
14. Bharadwaj, A., El Sawy, O.A., Pavlou, P.A., Venkatraman, N.V.: Digital business strategy: toward a next generation of insights. MISQ **37**(2), 471–482 (2013)
15. Burlton, R.T., Ross, R.G., Zachman, J.A.: Business Agility Manifesto. Business Agility Coalition. https://busagilitymanifesto.org (2018)
16. SIAM Body of Knowledge. Available: http://www.scopism.com/free-downloads/
17. Matt, C., Hess, T., Benlian, A.: Digital transformation strategies. Bus. Inf. Syst. Eng. **57**(5), 339–343 (2015)
18. Rauch, M., Wenzel, M., Wagner, H.-T.: The digital disruption of strategic paths: an experimental study. In: International Conference on Information Systems, Dublin (2016)

19. Skog, D.A., Wimelius, H., Sandberg, J.: Digital Disruption. Bus. Inf. Syst. Eng. **60**, 431 (2018)
20. Westerman, G., Bonnet, D.: Revamping your business through digital transformation. MIT Sloan Manag. Rev. **56**(3), 10–13 (2015)
21. Osterwalder, A., Pigneur, Y., Clark, T.: Business Model Generation: A Handbook for Visionaries, Game Changers, and Challengers. Wiley, Hoboken, NJ (2010)
22. GSRM: Government Strategic Reference Model. Governments of Canada. http://publications.gc.ca/collections/Collection/BT22-95-2004E.pdf
23. The Government of Ontario Information and Technology Standards (GO-ITS): https://www.ontario.ca/page/information-technology-standards#section-1 and 56.1 Defining Programs and Services in the Ontario Public Service Appendix A. https://www.ontario.ca/document/go-its-561-defining-programs-and-services-ontario-public-service-appendix
24. Poulin, M.: Ladder to SOE: How to Create Resourceful and Efficient Solutions for Market Changes within Business and Technology. Troubador Publishing (2009)
25. Brown, T.: Change by Design: How Design Thinking Transforms Organizations and Inspires Innovation. HarperCollins Publishers, New York, NY (2009)
26. Lockwood, T.: Design Thinking: Integrating Innovation, Customer Experience, and Brand Value. Simon and Schuster (2010)
27. Teixeira, J., Patrício, L., Nunes, N., Nóbrega, L., Fisk, R., Constantine, L.: Customer experience modeling: from customer experience to service design. J Serv. Manag. **23**(3), 362–376 (2012). https://doi.org/10.1108/09564231211248453
28. Zomerdijk, L.G., Voss, C.A.: Service design for experience-centric services. J. Serv. Res. **13**(1), 67–82 (2010). https://doi.org/10.1177/1094670509351960
29. Municipal Reference Model: https://ics2020.com/services-mrm/
30. ACORD Reference Model: https://www.acord.org/standards-architecture/reference-architecture
31. Panorama 360 Reference Model: https://www.insuranceframeworks.com
32. BIAN Reference Model: https://www.bian.org/
33. APQC Process Classification Framewords: https://www.apqc.org/process-performance-management/process-frameworks
34. Business Architecture Guild Reference Models: https://www.businessarchitectureguild.org/store/ListProducts.aspx?catid=677483
35. Wittenburg, A., Matthes, F., Fischer, F., Hallermeister, T.: Building an integrated IT governance platform at the BMW group. Int. J. Bus. Process. Integr. Manag. **2**(4), 327–337 (2007)
36. Cram, V.A.: Brohman MK and Gallupe RB (2015) Addressing the Control Challenges of the Enterprise Architecture Process. J. Inf. Syst. **29**(2), 161–182 (2015)
37. De Haes, S., Van Grembergen, W., Debreceny, R.S.: COBIT 5 and enterprise governance of information technology: building blocks and research opportunities. J. Inf. Syst. **27**(1), 307–324 (2013)
38. Wahab, I.H.A., Arief, A.: An integrative framework of COBIT and TOGAF for designing IT governance in local government. In: 2nd International Conference on Information Technology Computer and Electrical Engineering (ICITACEE), pp. 36–40 (2015)
39. Scaled Agile Framework: https://www.scaledagileframework.com/
40. Disciplined Agile Toolkit: https://disciplinedagiledelivery.com
41. Ahmadi, H., Farahani, B., Aliee, F.S., Motlagh, M.A.: Cross-layer enterprise architecture evaluation: an approach to improve the evaluation of enterprise architecture TO-BE plan. In: Proceedings of COINS Conference. COINS, Crete, Greece. https://doi.org/10.1145/3312614.3312659 (2019)
42. Hope, T., Chew, E., Sharma, R.: The failure of success factors: lessons from success and failure cases of enterprise architecture implementation. In: Proceedings of the 2017 ACM SIGMIS Conference on Computers and People Research. ACM (2017)
43. Jusuf, M.B., Kurnia, S.: Understanding the benefits and success factors of enterprise architecture. In: Proceedings of HICSS 2017. Waikoloa, Hawaii (2017)

44. Hauder, M., Roth, S., Matthes, F., Schulz, C.: An examination of organizational factors influencing enterprise architecture management challenges. In: van Hillegersberg, J., van Heck, E., Connolly, R. (eds.) Proceedings of the 21st European Conference on Information Systems, pp. 1–12. Association for Information Systems, Utrecht, The Netherlands (2013)
45. Kotusev, S., Singh, M., Storey, I.: Investigating the usage of enterprise architecture artifacts. In: Becker, J., vom Brocke, J., de Marco, M. (eds.) Proceedings of the 23rd European Conference on Information Systems, pp. 1–12. Association for Information Systems, Munster, Germany (2015)
46. Kotusev, S.: The history of enterprise architecture: an evidence-based review. J. Ent. Arch. **12**(1), 29–37 (2016)
47. Löhe, J., Legner, C.: Inf. Syst. E-Bus. Manag. **12**, 101 (2014). https://doi.org/10.1007/s10257-012-0211-y
48. Blevins, T.: Enterprise Architecture as a Service—Why? Open Group Blog Post. https://blog.opengroup.org/2018/07/03/enterprise-architecture-as-a-service-why/ (2018)
49. Blevins, T.: Enterprise Architecture as a Service—What? Open Group Blog Post. https://blog.opengroup.org/2018/07/10/enterprise-architecture-as-a-service-what/ (2018)
50. Blevins, T.: Enterprise Architecture as a Service—How?—Reach for the Stars. Open Group Blog Post. https://blog.opengroup.org/2018/07/17/enterprise-architecture-as-a-service-how-reach-for-the-stars/ (2018)
51. Horlach, B., Drews, P., Schirmer, I.: Bimodal IT: business-IT alignment in the age of digital transformation. In: Proceedings of MKWI 2016, pp. 967–978. Ilmenau (2016)
52. Bhaide, P., Rao, R.: Enterprise architecture as a framework for digital transformation in the EPC industry. J. Ent. Arch. **14**(1), 19–28 (2018)

Chapter 5
Digitally Transforming Live Communication—A Field Study on Services for Event Resource Management

Sven-Volker Rehm and Christian Coppeneur-Guelz

Abstract Live communication comprises all corporate activities for enabling personal contact with customers on occasions such as promotional events or trade fairs. The question of how the related event management can be digitally transformed has recently become a prominent issue. Firms face the challenge of analyzing how digital services can be deployed to intensify personal communications on-site, manage related marketing resources, or support strategic decision-making. We report from a field study on the use of digital services for live communication. We formulate performance criteria and indicators to study the effects of digital services on event management. Using a framework for Event Resource Management (EvRM) we provide insights on the services that support managing related assets. We discuss how digital services act as enablers for event management and how their use drives firm capabilities for continued digital transformations in event management.

Keywords Live communication · Event resource management · Event metrics · Digital event services

5.1 Introduction: From "Live" to "Digital" in Live Communication?

When was the last time you attended an event, say, a trade fair, or workshop, and had a conversation with company representatives or colleagues? Is this situation—in today's digitized world—still a viable format for transporting a firm's offerings and creating dialogue? We believe that, yes, it is. Yet, firms need to reflect on the

S.-V. Rehm (✉)
Ecole de Management Strasbourg, HuManiS UR 7308, Université de Strasbourg, 67000 Strasbourg, France
e-mail: sven.rehm@em-strasbourg.eu

C. Coppeneur-Guelz
WWM GmbH & Co. KG, Hans-Georg-Weiss-Str. 18, 52156 Monschau, Germany
e-mail: christian.coppeneur-guelz@wwm.de

A. Zimmermann et al. (eds.), *Architecting the Digital Transformation*, Intelligent Systems Reference Library 188,
https://doi.org/10.1007/978-3-030-49640-1_5

changing nature of communicating as part of their marketing function. This is why in this article we discuss in what ways live communication can be transformed digitally. Of course, it is not *communication* itself which is in question to be transformed digitally, but the question of how the mindful and intentional use of digital services can support the strengths of live communication, mitigate its weaknesses, and helps seizing opportunities and avoiding risks.

Live communication has emerged as a generic term for all kinds of corporate activities to enable personal and engaging contact with customers on occasions such as events, trade fairs, promotions, roadshows or exhibitions. These activities serve as communication instruments that focus on the personal encounter and the active, emotional and memorable experience of the target group with the brand. The idea is to create situations that lead to unique and lasting memories with respect to knowledge or attitude towards the brand. The staged, on-site, direct and personal interaction is the moment, i.e. place and time, to deliver an individualized service to the customer. We are focusing on how the entire value chain towards—and after—this 'moment' can be digitally supported.

Live Communication can be seen as part of a firm's communication policy, particularly as an element of event marketing. ***Event marketing*** aims at increasing the activation potential of (potential) customers through a set of experience-oriented communications. Engagement in live communication—and the planning and execution of related events—must thus be considered as integrated into the overall marketing concept and related processes of a company [2, 15].

The ***digitization of event marketing and event management*** has become a prominent issue in recent years [10]. Digital media and personal digital devices (both, for business and private use) are being used more widely and more intensely than ever before, changing the consumer decision journey—understood as interplay of consumers (or business customer representatives), brands, media and communications [4, 7]. There is also an extended opportunity to design functionalities and platform-based services, which requires rethinking approaches for designing integrated marketing communication (IMC) [8, 11]. The diverse set of available media need to be mixed and matched, involving display ads, social media, mobile, websites, e-mail and the like [1, 12]. Beyond communicating, with regard to managing, particularly, the optimum integration of marketing communications, processes, roles and assets requires attention [1, 6, 9, 18].

Firms face the challenge of analyzing how digital services can be designed and deployed to intensify and complement personal communications in on-site events, to manage related marketing resources, and to support strategic decision-making concerning event management, customer communication channels or target group management [15, 17].

To address this challenge, looking at the strengths, weaknesses, opportunities and threats (SWOT) linked to live communication today can be helpful; the following paragraphs summarize ours and our study participants' experiences, interpretations and expectations in this respect.

Strengths The particular strength of live communication is its inherent notion of personal, face-to-face, interaction. This becomes especially critical in times of digital media in which the power of this way of communicating is often overlooked. On the contrary, its significance for communicating convincing and winning arguments might even be increasing. Live communication for the time being might remain the only multi-sensory marketing activity, including our sensations of seeing, hearing, smelling, tasting and touching, and thus presumably will keep its complementary position to other marketing measures and communications.

Weaknesses Compared with purely digital communication measures, event-based communication remains scalable only in a one-to-many fashion; this is because of the physical nature of events and the limitations to resourcing this brings about [16]. This boundary condition also causes the relatively high effort and time consumption in planning and preparing 'the event' comprising tasks in resource and budgetary planning, logistics, team and schedule coordination etc. In addition, live communication raises relatively high costs (cost per lead), which ultimately need to be justified through the quality of interactions on-site.

Opportunities The event situation context allows focusing inter-personal communication in opposition to digital interactions. This creates an opportunity for designing emotional, experiential spaces and conveying narratives as part of storytelling approaches. As participants experience such contexts holistically, this allows for intensifying dialogues with a focus on the customer as message receiver. Currently, approaches to create 'hybrid events' try to enhance this offline interaction situation with digital services in order to increase the intensity of dialogue and to enhance its potential reach and content [5].

Threats A risk regarding consumers' readiness to participate to live communication events exists in the advances of technologies such as virtual and augmented reality or holographs. Such technologies promise cost reductions with an increased entertainment potential as compared to current digital media; yet they are not (yet) ready to simulate or replicate the multi-sensorial experience and thus the pivotal aspect of experience intensity of an on-site event situation. An additional boundary condition of events as of today is the potential risk associated with public, large or mass events; without anticipating a more general social discussion, this might prove a valid argument for some regional contexts to move to purely digital event formats.

The above SWOT discussion suggests that overall, live communication cannot defy the global trend of digital transformation. Yet, because of this condition, firms are increasingly required to confront the question what forms upcoming adaptations will presumably take. For each firm, this involves finding appropriate individual approaches for (a) keeping up a competitive edge in presenting themselves through the highly successful marketing means that live communication represents, (b) not losing ground against (maybe technologically, maybe otherwise) more advanced competitors, and (c) improving their own capabilities with respect to creating value from their marketing activities in an imperatively "online" world.

Our research thus started out with the questions *how digital services can support event management—and live communication in particular*, and *how this might lead to new opportunities for managing events and improving marketing capabilities*. In the remainder of this chapter, we report on results from a field study of digital services use in the live communication context. We first outline how digitally transforming event management can be understood with help of relevant performance criteria. In a second step, along these criteria we study in what ways digital services support major tasks of event management. We draw on our field study results to show that digital services serve as technological enablers as well as strategic drivers for digitally transforming live communication. In a third step, we discuss opportunities for engaging with digital services in the live communication context.

5.2 Field Study on Digitally Transforming Live Communication

5.2.1 Information on the Study

Basis for our report is a field study on the use of digital services in the live communication context [3]. The study was conducted in 2016 and involved 10 companies from several industrial sectors including health care, insurance services, telecommunications, auditing and mechanical engineering. In interviews and on-site visits, the firms answered questions about their strategies and processes, and we collected key performance figures and measured process figures. While the study does not claim to be quantitatively representative, the choice of companies reflects a diverse set of users of digital services that allows recognizing trends beyond specific settings—which is vital to identify potential novel value propositions of, or options for, digital services. All participating firms had operationally deployed digital event management services for at least twelve months, i.e., having passed through an initial ramp up phase in services' use. Our interviews and the process figures collected included the situations before and after service implementation in order to gain an impression of, and a base for comparing, the changes associated with the introduction of the services. As we learned from the interviews, during the introductory phase, many of the companies systematized their processes or adapted them significantly. In addition, various changes in work routines occurred.

Our analysis follows several steps. First, in order to approach the question how digitally transforming event management can be understood, we formulate a conception of *Event Resource Management* (*EvRM*) in Sect. 5.2.2. Subsequently, we define a set of six *performance criteria* that help us analyze the contribution of digital services to EvRM in more detail. Each of these criteria can be measured along several *key performance indicators* for which we find evidence in our study. We introduce these criteria and indicators in Sect. 5.2.3. The information technology support of event management activities can be conceived of as resembling the more general Enterprise

Resource Management (ERM). In Sect. 5.2.4 we therefore introduce a framework for *Event* Resource Management (EvRM). This framework comprises major management tasks related to event management and the marketing function in general. In our study, we use it as an analytic frame in order to discuss our study results in terms of qualitative contributions of digital services to event management. We present overall insights from our field study concerning the (quantified) performance criteria of EvRM in Sect. 5.2.5.

5.2.2 *Event Resource Management (EvRM)*

Event management comprises all business processes and activities for the strategic conception, coordination, planning and execution of events as part of a firm's overall marketing. It is usually connected to the use of *project management* approaches as well as *enterprise resource management* functions in order to safeguard implementation of strategic goals, efficient coordination of related resources and leveraging of learning effects from conducted events.

As an element of the marketing business function, event management is closely linked with the management of all related *assets and resources*, which is why we further address it as ***Event Resource Management*** (***EvRM***). If considered from an *enterprise* resource management perspective, various analogies can be drawn towards the software-support of comprised activities and information handling necessities. The scope of assets and resources includes people, processes, logistics, tangible resources such as marketing and public relations materials, trade fair equipment, data from and information about events and customers and the like, as well as intangible resources, involving the specificities of the corporate image to transfer or particularly, knowledge about the customer and others more.

5.2.3 *Performance Criteria for Event Resource Management*

In order to understand opportunities for digital support of event management and live communication, we draw on six performance criteria for event management. These factors comprise speed, scalability, information, agility, service quality and cost. Each performance criteria links with key performance indicators as illustrated in Table 5.1. We provide a rationale for each of the criteria below.

Speed Speed has proven to be of significance with respect to the effectiveness of live communication. While digital content can be provided with little time-delay online, this is more difficult for offline events to achieve. Reaching temporal proximity in addressing most recent events through content mediates consumers' intentions to participate to events, or to decide for stand visits on trade fairs. This involves also the capability for a comparatively quicker response to competitors' offerings.

Table 5.1 Performance criteria and key performance indicators of event management linked with digital service support

Performance criteria	Key performance indicators			
Speed	Planning time		Execution time	
Scalability	Number of events	Number of stakeholders		Degree of complexity
Information	Information availability	Cost transparency	Scope of data analysis	Scope of controlling
Agility	Temporal agility	Spatial agility		Autonomy
Service quality	Customizability		Parameterizability	
Cost	Non-recurring cost		Process cost	

One key performance indicator (KPI) to express speed is *Planning Time*, i.e., the duration from deciding on conducting an event, until starting the first operative actions. Being a second indicator, *Execution Time* describes the duration from starting the first actions on a particular event until start of the event itself.

Scalability While for purely digital services, scale can be safeguarded through increased (maybe last minute) accessibility to digital resources, in life communication and offline events this is comparatively difficult to achieve. In this context, scaling corresponds to the capabilities with respect to event planning (both short- and long-term). Specific KPIs that can be addressed through digital services comprise the Number of Events, Number of Stakeholders and Degree of Complexity. *Number of Events* simply refers to the total of accomplished events including live communication with respect to the resources used (i.e., independent from defined budget, but as conceived by the strategic marketing planning). *Number of Stakeholders* refers to the capability to plan and conduct event planning processes that include a high number of different stakeholders. The underlying assumption is that scalability grows with the ability and capacity to resolve more extensive coordination and involved decisions. The indicator describes the total number of roles involved into planning and execution of an event, including both active and passive (e.g., for approval tasks) roles. *Degree of Complexity* refers to the overall increasing complexity of event-related processes and logistics, event equipment, print products, hospitality, give-away articles etc., which are generally subject to internal and external disturbances and hold-ups. Here, particularities such as weight of materials or criticality of assembling equipment on-site can lead to glitches in processes and influence event success. Overlooking or misinterpreting critical measures might result in critical situations. The indicator thus comprises the perceived effort for successfully executing event-related processes.

Information For both, event planning and execution, information is an enabler, and safeguarding access to and availability of the right information empowers involved stakeholders to efficiently and effectively carry out their tasks. *Information Availability* refers to the capability to make decisions on basis of complete and correct information. Hence, collection, aggregation and distribution of data and information along and across the relevant processes. The indicator thus describes the completeness and timely distribution of information to involved stakeholders. *Cost transparency* is a second indicator; it involves explicit cost such as those documented in receipts and invoices as well as implicit cost, involving personnel cost for planning, calculation and execution of an event, but also cost for materials such as public relations (PR) giveaways that are handed out at diverse events. Hence, Cost Transparency describes to what extent the expenditure incurred for internal and external services is predictable and comprehensible. To some extent, this also involves the trust into the correctness and completeness of the collected data. The capability to analyze the collected data can be addressed through the indicator *Scope of Data Analysis*. This might refer to analytic functions but also to the ability to identify or formulate improvement recommendations with regard to processes. This indicator is particularly vital to achieve impact on the management level. The indicator *Scope of Controlling* focuses on the capability to align collected data with strategic and operative accounting measures. This relates to the question whether the aggregated data are meaningful to the overall interpretation, assessment and improvement of the event marketing and planning business processes.

Agility Agility has become a widely acknowledged performance criteria for organizational flexibility and performance. For event management, it can be understood as the propensity and capability to flexibly react to unforeseen or unanticipated exogenous or endogenous factors or variations. It conveys particularly two aspects. *Temporal Agility* means responsiveness of involved stakeholders. This indicator relates to the interrelated processes of event marketing, where internal and external stakeholders regularly need to coordinate, harmonize and synchronize their actions and decisions to varying extent. This involves the need to create spaces and opportunities for synchronous communication using appropriate tools. *Spatial Agility* means the location-independent availability of tools and information in event planning processes. While this has become a general imperative for any digital tool, in current event management software applications this has not yet been fully achieved. The indicator Spatial Agility thus describes to what extent event-specific processes and measures can be executed location-independently with help of adequate tools. The third indicator for agility refers to the *Autonomy* that event planners and managers own with respect to changing or adapting particular measures in (almost) real-time. This involves for instance the capability to control ad hoc for availability of PR materials, and whether they can be logistically scheduled or re-ordered for an event ahead. This indicator captures the perception of decision capabilities as well as the technical options available to make decisions about changes and implement them autonomously or in close coordination with related (and available) stakeholders.

Service Quality As it is the primary intent of live communication is to create memorable, emotional experiences and situations, the aspect of individualizing the narrative and direct personal interaction is of vital significance. *Customizability* is thus a strategic indicator that captures to what extent live communication measures can be specifically adapted to the expectations of a target group. It comprises three aspects, the effort to implement individual adaptations, the perception of individual offers by the addressee (as indicated after the personal interaction), and to what degree both of the previous aspects can be represented by digital tools. A sister indicator is *Parameterizability*, which refers to the adaptivity of operative processes for specific events, which might show as configurability of trade fair stands, selection of give-away articles from different sets or other operative tasks. This indicator thus captures the adequacy of the digital tool support, as well as coordination and representation of related tasks.

Cost Event management covers both, non-recurring cost as well as process cost. *Non-recurring cost* refer to one-time investments for a specific event format, e.g., for equipment that is used across the entire series of events. *Process cost* occur from the use of internal resources as e.g., the real time consumption of an event manager for event management. These cannot necessarily be linked to a specific event format.

When discussing the potential for digitally transforming live communication, we need to take into account that various interrelations exist between the mentioned performance criteria and KPIs and their implementation in a specific case. In this context, beyond the more general process-related measures regarding process efficiency, cost transparency and the like, finding an optimum route to digital service support will need to consider also the peculiarities of the firm in focus, its processes, stakeholders as well as the particular marketing strategies—i.e., the intended narratives and stories to be told on-site.

5.2.4 Digital Service Support to Event Resource Management

In order to analyze in what ways live communication and event management can be supported by digital services we use a framework for event resource management (EvRM) as outlined in Fig. 5.1 [3]. The framework orients itself at *enterprise* resource management frameworks and identifies relevant groups of services to support all aspects of event management processes and resources, and managing live

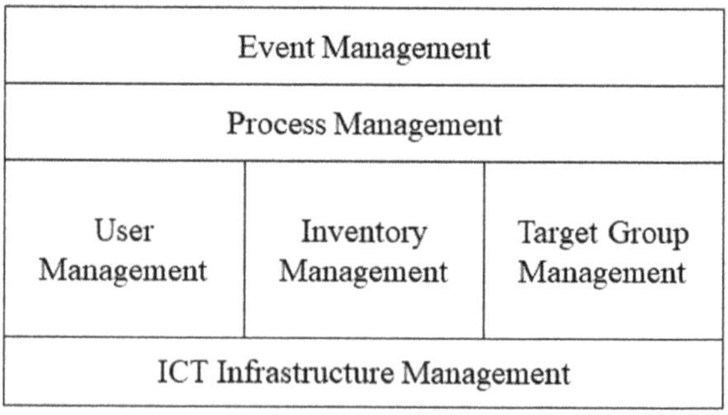

Fig. 5.1 Framework for event resource management (EvRM). Adapted from Coppeneur-Guelz and Rehm [3]

Event Management
Process Management
User Management
Inventory Management
Target Group Management
ICT Infrastructure Management

communication assets. The EvRM Framework works as a comprehensive reference to accommodate and discuss both, currently available and potential future digital services.

In the following sections, we discuss each group of services. Using insights from our field study, tables indicate for each of them in what ways digital services can contribute to the respective performance criteria.

ICT Infrastructure Management comprises planning, provision, implementation and maintenance of appropriate environments for digital services. Particularly, ensuring stakeholders' web-based and eventually cloud-based access to relevant services allows making all required services instantly available and promises efficiency gains. Table 5.2 outlines qualitative aspects relevant for digital support with respect to the defined performance criteria. (Note: The tables in this section only discuss *relevant* contributions to the performance criteria, thus not all of the performance criteria might appear in the tables.)

User Management covers administration of stakeholder accounts and roles in EvRM processes. Safeguarding accreditation of internal and external stakeholders and granting access to adequate functionality (parameterization) is vital to reap benefits in process support (see Table 5.3).

Inventory Management includes functionalities for planning resource production (respectively supply or purchase), resource consumption, availability planning,

Table 5.2 ICT infrastructure management contributions to EvRM

Criteria	Contribution
Speed	– Reduced waiting times in process coordination – Increased efficiency in multi-stakeholder coordination
Information	– Enhanced information availability through cloud-based access to data and software
Agility	– Increased temporal and spatial flexibility as well as autonomy through improved accessibility of services
Cost	– Avoiding hidden coordination cost caused by non-integrated services

Table 5.3 User management contributions to EvRM

Criteria	Contribution
Speed	– Increased speed of interactions through reduction of discontinuities in accessibility of services and systems – Decreased planning and execution times through more efficient coordination
Scalability	– Providing a central user administration builds a basis for ensuring service accessibility
Information	– Increased information and process accessibility
Agility	– Enhanced temporal and spatial flexibility through easier inclusion of users to workflows across services
Cost	– Decreased process cost through centralization of role administration

Table 5.4 Inventory management contributions to EvRM

Criteria	Contribution
Speed	– Reduced overhead searches and coordination effort through increased transparency (AP)
Scalability	– Extended analysis capability (IT) and improved planning reliability (AP) through better transparency, e.g., to increase number of events
Information	– Increased efficiency and effectiveness in calculating resource use and cost (IT, AP)
Agility	– Increased reliability of resource disposition creates options for flexible and autonomous decisions (IT)
Service quality	– Improved preliminary information about resource condition supports event planning ("look and feel" of resources) (IT)
Cost	– Improved decisions on investments (e.g., resource supply) or disinvestments (resource disposal); reducing inventory cost (IT)

managing the resource inventory, and resource maintenance. The functionality *Inventory Turnover* (IT) refers to managing the use of resources that represent investment goods (e.g., trade fair stands, banners, counters); here, particular benefits in assuring transparency and planning effectiveness can be reached. This is supported through *Availability Planning* (AP), which concerns the resource use across event series (see Table 5.4).

Target Group Management refers to functions supporting the handling of data and information about event visitors and interactions with customers (e.g., post-event). *Visitor Logging* (VL) enables the documentation of contacts (dialogues) with event visitors. The documentation might cover the type of interaction or incentives exercised on-site, such as competitions, product tests or games of chance; follow-up information; and quantified information such as time stamp, event ID, contact addresses/ID. *Visitor Tracking* (VT) conveys technologies for collecting data about the visitors' behavior, such as the number of unique users per period, length of stay, ratio of repeated visits etc. *Online Retargeting* (OR) refers to functions for using information from the previous services to reconnect with event visitors for instance through online communication activity on social networks (see Table 5.5).

Process Management constitutes a horizontal task in the EvRM Framework. Representative services include Visual Event Planning, Budget Management, and Supply Chain Management among further possible services. *Visual Event Planning* (VP) refers to the integration of data from Computer-Aided Design (CAD) application software to the event planning process. Together with data about the particular event space occupied, such data can be used to visualize the event installation or setup. If linked with Availability Planning functions (see section Inventory Management), the availability of required equipment can be evaluated and the stand reconfigured accordingly. *Budget Management* (BM) refers to definition, administration and monitoring of event management-related budgets, and commonly comprises definitions for types of budgets (events, series, responsible roles etc.), budget volumes and periods, and their correspondence to, or clearance for, specified processes in event

Table 5.5 Target group management contributions to EvRM

Criteria	Contribution
Scalability	– Improved process definitions of event planning and on-site processes because of extended data availability (VL) – Enhanced objective comparability of events (VL) and event formats (VT) on basis of data as prerequisite to scale
Information	– Increase in all KPIs, transparency, availability, and analysis capability (VL) – Decrease of effort for post-event or on-site information collection and manual analysis (VT) – Improving the focus and enhancing the scale of retargeting measures through better data availability (OR)
Agility	– Improved focus in retargeting through creating the option to dynamically and immediately create content focused on the addressee's interest (OR)
Service quality	– Improvement of services (short to mid-term) through better knowledge of visitor interests (VL) and event benchmarking (VT) – Improved precision in profiling visitors and estimating visitor numbers to increase target group focus (VT) – Improved sustainability and robustness in consumer focus in online retargeting channels through increased individualization and parameterization options (OR) – Improved specificity of company presentation as relevant to events, as spill-over effect from better online retargeting measures (VT)
Cost	– Reduction of cost for visitor profiling compared to other target group analysis services (e.g., market research) (VT) – Development of a unique data reference base for future use (e.g., in event planning) (VT)

management, which for instance allows to assign marketing materials to events. *Supply Chain Management* (SM) services involve relationships with (systems of) external service providers. These cover for instance the configuration and execution of insurance services for equipment or logistics services (see Table 5.6).

Event Management establishes the primary horizontal leadership task in the EvRM Framework. Particularly, it encompasses the performance measurement system that allows leadership and management roles to address and analyze all data relevant to event management. It comprises Event Controlling (EC), Quantitative Key Figures (QuanF), Qualitative Key Figures (QualF), and Integrative Key Figures (IF) as ***event metrics***. *Event Controlling* refers to the comprehensive documentation of all cost items related to event management processes, such as actual and implicit costs. *Quantitative Key Figures* comprise data about utilization and consumption of resources such as the volume of promotional gifts consumed per event. This information allows to initiate improvement processes with respect to better focused usage of marketing materials for specific target groups. *Qualitative Key Figures* play a role for estimating brand awareness or visibility of measures after an event, related data might be collected in form of walking or proximity patterns and can be used to qualitatively re-assess marketing measures. *Integrative Key Figures* integrate the previously mentioned quantitative and qualitative figures to create aggregated analyses;

Table 5.6 Process Management contributions to EvRM

Criteria	Contribution
Speed	– Reduced planning time through visualizations and reduced coordination times (e.g., between event planners and designers) (VP) – Reduced effort for approval processes (asynchronous communication) through functional integration (BM) – Reduction of transaction cost and effort with third-party service providers reducing planning and execution time (SM)
Scalability	– Increase of autonomy by reduction of coordination efforts and integration of visual and logistic planning (VP) – Decrease in the number of required stakeholders (BM) – Increase in event planning capacity because of process automation (SM)
Information	– Improved information availability through visualizations (VP) – Potential for automation of accounting procedures (BM) – Automation of cost item recording increases transparency for analysis (SM)
Agility	– Enhanced automation of information distribution, e.g., by providing sketches to logistics personnel for stand construction on-site (VP) – Reduction of planning time, as parallel evaluation of different event layouts becomes possible (VP) – Potential for instant reaction to budgetary requests (BM) – Reduction of lead times in interactions with service providers (SM)
Service quality	– Increased detail of layout and design allows to improve focus and reduce errors (VP) – Reduced risk of information loss (SM)
Cost	– Reduction of variable cost if VP service is integrated to EvRM system, as third party services are avoided (VP) – Process cost is reduced through reduced overhead effort (BM, SM)

this comprises indicators such as cost-per-passage or cost-per-visit. Such indicators are the basis for continuous improvement processes as they allow benchmarking events or measures against each other, balancing cost and target group focus; taking strategic decisions about when to arrange for particular event happenings—or which events to select at all (see Table 5.7).

5.2.5 *Impact of Digital Services on EvRM Performance Criteria*

In this section, we summarize the results of our field study with respect to the defined performance criteria of event management. The detailed quantified study results are documented and discussed in Coppeneur-Guelz and Rehm [3].

Overall, we have seen that there are only very few significant *direct* relationships between implemented digital services and the identified perceived and measured improvements in KPIs. Only in the cases of the criteria *speed* and *scalability*, we could identify several distinct services that have vitally contributed to the perceived

Table 5.7 Event management contributions to EvRM

Criteria	Contribution
Scalability	– Improved data reference base for scaling decisions because of increased information density (QuanF) and better transparency of measures (e.g., visitor patterns) (QualF, IF)
Information	– Improved transparency through aggregation of data for analysis; increased robustness of the data reference base (EC) – Increased information density (QuanF) – Extended information sourcing becomes possible through qualitative measures (QualF) – Generation of knowledge about event performance and success of past improvement measures (IF)
Agility	– Decreased decision risks because of improved reliability of data reference base (EC, IF) – Improved data reference base for ad hoc decisions because of increased information density (QuanF) and extended indicator reference base (QualF, IF)
Service quality	– Improved target group focus because of better analysis data reference base (QuanF) and enhanced base of indicators (QualF) – Improved decision making about strategic aspects such as event and event format selection because of aggregated information for benchmarking (IF)
Cost	– Improved understanding of cost factors and causes (EC) – Potential mid- to long-term cost decrease due to better analyses because of higher information density (QuanF) and because of better indicators for decisions/improvement processes (QualF)

change (see below). Further reported changes occur from *synergetic uses* that become possible through linking the services' novel functionalities. For instance, we see that functionality for aggregating and benchmarking diverse event metrics improves the decision basis for adapting event formats and event management processes.

Speed Particularly the services Availability Planning, Visual Event Planning, Supply Chain Management and Budget Management contribute positively to capabilities for reaching "speed". These services increase the overview of which resources are available and help to retrieve current inventory levels. They also allow for compiling visuals and for controlling budgets and costs in real-time. Furthermore, voting and approval loops are avoided and response times of process stakeholders decrease.

Scalability Particularly the services Supply Chain Management, Visitor Logging and Availability Planning have shown a direct relevance for scalability. These services support integration and coordination of stakeholders, which allows to plan and implement more and more complex events. In general, digital services support process standardization, which fosters efficiency. The companies leveraged improved process coordination through more meaningful task distribution; deciders benefitted from greater autonomy, decreased workload and more efficient decision-making.

Information Particularly the indicators Scope of Data Analysis and Scope of Controlling were positively influenced through enhanced availability of data and information. Many firms reported an increased awareness towards collecting and actually utilizing data for analytic purposes, particularly concerning the handling of resources, costs and the documentation of individual activities. Increased information access also motivated the companies to rethink their allocation of responsibilities, e.g., data access authorizations, allowing to retrieve information about resources or processes. Respondents also reported increased cost transparency, leading to more strategically approaching issues of event planning, keeping budgetary limits, and reducing costs.

Agility All three indicators of "agility", temporal and spatial agility as well as autonomy, overall showed a positive increase through digital services. We see the reception of an improved agility in *perceiving* synergetic effects across various services that show in extended functionality for process stakeholders and an increased standardization of processes. Particularly in making decisions, stakeholders perceive an amplified temporal-spatial independence.

Service Quality The opportunity to individualize and parameterize tasks *across* services has been reported by interviewees as perceived increased freedom for resource design, spontaneity in event planning, analyzing cost aspects or engaging in strategic considerations. All these aspects overall contribute to an increased capability for focusing on how to communicate the brand through the "right" service.

Cost While from our study we cannot draw insights on long-term cost effects, we see some immediate trends in our respondents' accounts. Throughout our set of companies in the study, a robust reduction of internal process costs with almost constant non-recurrent costs was reported. What for us seems to be of higher importance however, is an altered awareness about event *resources* in general; which shows in the significantly changed internal process structures as reported by the companies. Further cost-relevant effect have appeared in reduced time efforts for planning, coordination and implementing events. Some firms used this in order to create novel, target group-specific events or to engage more in strategic considerations.

Figure 5.2 demonstrates our study results; the inner dashed hexagon represents values without using digital services, the outer one those including the use of digital services. The underlying values were surveyed with help of a 5-point Likert scale that asked for the key performance indicators of event management; the farther outside the values lie, the higher the reported capabilities. While the figure here is intended only as a conceptual illustration, it expresses the perceived increase in capabilities for event management at the responding firms across all performance criteria.

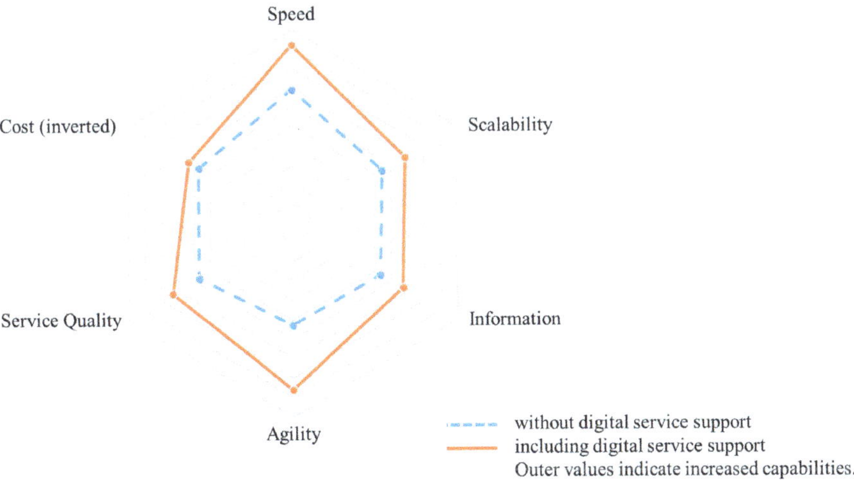

Fig. 5.2 Influence of digital services on EvRM performance criteria in our field study. Adapted from Coppeneur-Guelz and Rehm [3]

5.3 Digitally Transforming Event Management

5.3.1 Digital Services as Enablers and Drivers for New Capabilities

Our study results indicate that digital services act in two ways on a firm's capabilities in event management. On the one hand, digital services act as ***enablers***. In this capacity, which is particularly driven by technology, the services provide functionalities that allow to better manage processes and information in order to support efficiency and effectiveness within the defined scope of a marketing strategy. This includes the development and information technological implementation of a firm's KPIs with help of event metrics in order to monitor and assess strategy implementation and operational performance.

On the other hand, digital services act as ***drivers*** for (continuous) improvements concerning procedural, informational and analytical ***capabilities***. We have encountered this second effect of *using* digital services in our study respondents' increased awareness of the assets that undergird performativity in event management.

We therefore conjecture a *double loop virtuous learning cycle* when it comes to leveraging benefits from digitally transforming live communication: Those firms that opt into the use of digital services directly benefit in terms of ensuring event management performance, while at the same time building the capability to learn from, and reflect about, their services use in order to improve their overall approach to marketing communication. This second learning loop might prove vital to accommodate future technology-driven developments such as the currently discussed, 'intelligent events' or 'online pre-targeting', which presumably will only offer value propositions when linked with other services in a holistic manner.

Due to our study's interpretive nature, we cannot derive representative suggestions for "how-to" digitally transform in live communications—but we hope to set some signposts for future efforts to be taken in practice.

5.3.2 Conclusion

In spite of the overall increase of digital communication, live communication will keep its raison d'être. At the same time, event management, as precursor in managing, planning, organizing and handling of live communication resources, assets and measures must deploy new, digital technologies and services in order to ensure and expand its future viability.

In today's platform economy [13, 14], it is conceivable that further automation in data collection, e.g., visitor tracking, and flexibility in service provision by external stakeholders, e.g., through platforms and Application Programming Interfaces (APIs), will occur. Thus, firms now need to build the capacity to integrate and manage related technology and services, and to learn from their use in order to improve their adaptation capabilities. This way, event managers will presumably transform from assuming the role of an event coordinator to becoming knowledge managers responsible for creatively engaging with transporting a firm's *personality* through live communication.

References

1. Batra, R., Keller, K.L.: Integrating marketing communications: new findings, new lessons, and new ideas. J. Market. **80**(6), 122–145 (2016)
2. Bladen, C., Kennell, J., Abson, E., Wilde, N.: Events Management. An Introduction. Routledge, New York (2018)
3. Coppeneur-Guelz, C., Rehm, S.-V.: Event-Resource-Management mit digitalen Tools. Schnell—skalierbar—messbar: Wie die Digitalisierung die Live-Kommunikation verändert. Springer Fachmedien Wiesbaden, Wiesbaden (2018)
4. Court, D., Elzinga, D., Mulder, S., Vetvik, O.J.: The consumer decision journey. In: McKinsey Quarterly (2009)
5. Dams, C.M., Luppold, S.: Hybride Events. Zukunft und Herausforderung fuer Live-Kommunikation. Essentials, 1st edn. Springer Gabler, Wiesbaden (2016)
6. Devine, J., Lal, S., Zea, M.: The human factor in service design. In: McKinsey Quarterly (2012)
7. Edelman, D., Singer, M.: The new consumer decision journey. In: McKinsey Quarterly (2015)
8. Hoban, P.R., Bucklin, R.E.: Effects of internet display advertising in the purchase funnel: model-based insights from a randomized field experiment. J. Mark. Res. **52**(June), 375–393 (2015)
9. Keller, K.L.: Unlocking the power of integrated marketing communications: how integrated is your IMC program? J. Adv. **45**(3), 286–301 (2016)

10. Knoll, T. (ed.): Veranstaltungen 4.0. Konferenzen, Messen und Events im digitalen Wandel. Springer Gabler, Wiesbaden (2017)
11. Kumar, A., Bezawada, R., Rishika, R., Janakiraman, R., Kannan, P.K.: From social to sale: the effects of firm-generated content in social media on customer behavior. J. Market. **80**(January), 7–25 (2016)
12. Lamberton, C., Stephen, A.T.: A thematic exploration of digital, social media, and mobile marketing: research evolution from 2000 to 2015 and an agenda for future inquiry. J. Market. **80**(November), 146–172 (2016)
13. McAfee, A., Brynjolfsson, E.: Machine, Platform, Crowd. Harnessing Our Digital Future. W.W. Norton & Company, New York, London (2017)
14. Parker, G., van Alstyne, M., Choudary, S.P.: Platform Revolution. How Networked Markets are Transforming the Economy—And How to Make Them Work for You. W.W. Norton & Company, New York, London (2017)
15. Preston, C.: Event Marketing. How to Successfully Promote Events, Festivals, Conventions, and Expositions. Wiley Event Management Series, 2nd edn. Wiley, Hoboken, NJ (2012)
16. Raj, R., Walters, P., Rashid, T.: Events Management. Principles & Practice, 3rd edn. Sage Publications, London (2017)
17. Smith, K., Hanover, D.: Experiential Marketing. Secrets, Strategies, and Success Stories from the World's Greatest Brands. Wiley, Hoboken, New Jersey (2016)
18. Thaichon, P., Surachartkumtonkun, J., Quach, S., Weaven, S., Palmatier, R.W.: Hybrid sales structures in the age of e-commerce. J. Pers. Sell. Sales Manag. **38**(3), 277–302 (2018)

Chapter 6
Digital Transformation of Public Administration

Václav Řepa

Abstract Application of IT in public administration is usually understood as a way to higher performance of existing public administration authorities, digitization of existing documents, easier spreading of information, and better control over citizens than as a way to the transformation of the public administration business model. However, the digital transformation of public administration is nowadays more significant and acute than in any other area as it is also stated in famous OECD E-government Imperative: "E-government is more about the government than about the "e"". In the chapter, we analyze the essential factors of public administration business and propose the way to the application of the idea of digital transformation in public administration based on the idea of process-driven management as a representative of the essence of digital transformation.

Keywords Public administration · Process-driven management · Life events · Public administration services · Public administration processes · Object life cycles

6.1 Introduction

Public administration (PA) is in general one of the areas least positively influenced with the technology development. There are more reasons for that; low effectiveness of investments to the information technology (IT) in this field is not the most critical one. It is rather the consequence than the cause of this state. The most critical reason for so hardly achievable progress with the use of IT in PA is the objectively more complicated nature of this area in comparison with other areas of human activity. First, PA is hardly complex and abstract. Compared with the market-oriented enterprise, the complexity of PA organization mission is substantially higher. What is explicit and simple in the regular market company, it re-quires a great portion of abstraction in the PA organization. For instance, the role of client, which is explicit in market organization, is an abstract concept in PA organization. In PA, the concept of client

V. Řepa (✉)
Department of Informatics, ŠKODA AUTO University, Mladá Boleslav, Czech Republic
e-mail: vaclav.repa@savs.cz

A. Zimmermann et al. (eds.), *Architecting the Digital Transformation*,
Intelligent Systems Reference Library 188,
https://doi.org/10.1007/978-3-030-49640-1_6

covers besides traditional citizens many other social roles and groups including organizations of various forms and also informal societies even if their members do not realize their membership. Consequently, the number of different ser-vices relevant to different kinds of clients and related complexity of their mutual relationships are also dramatically bigger in PA. All above-mentioned aspects make the meaningful application of IT in the field of PA extremely hard.

The development of IT is a fundamental factor of the development of business. In early 1990s when this idea became unquestioned, the related idea of **process-driven management** was born [1]. Almost thirty years of following experience undoubtedly confirmed that to exploit the value of technology, the management of organization must be primarily oriented on business processes. **The value of the new technology can be exploited only by changing the business practices—business processes.** Therefore, the management of organization must be based on the explicit knowledge of its business processes and be prepared for the immediate implementation of their relevant changes allowed by the new technological phenomena. Such behavior requires system of management completely different from the traditional one. At first, the traditional primary orientation on organization structure in the organizational management must be left; the management must be primarily oriented on business processes using the organizational structure as their infrastructure.

We believe that the transformation of the business is the only way of exploiting the value of technology. **With meaningful application of IT in PA we mean the application that primarily leads to the transformation of the PA business itself.** Implementation of IT preserving the current way of performing the business means just wasting money and energy as no substantial contribution of the use of IT that would balance the costs can be achieved this way. This is especially important in the field of PA, as there is no other business field where the investments to IT could be so extensive and benefits so hardly quantifiable at the same time. It is obvious that even in PA, the main problem is in the management itself. The famous definition of E-government [2] also outlines this idea: "E-government is the use of ICT as a tool to achieve better government". McKendrick explains: "The impact of e-government at the broadest level is simply better government "e-government is more about government than about "e"" [3]. This idea lately became widespread as follows from other sources like [4] or [5].

From the previous paragraphs follows that digital transformation of public administration, hidden in the concept of e-government, primarily requires the application of the process-driven management in this field. In this chapter, we outline our idea of **application of the process-driven management concepts to the PA**. This idea is based on the **identification of life cycle events of PA objects**, which then support the **definition and analysis of key and support PA processes**. The chapter is organized in five sections. In the section immediately following this introduction, we briefly describe the main ideas of the process-driven management as a deep background of our approach to the application of IT in PA. In this section, we also introduce the basic ideas and principles of the underlying Methodology for Modeling and Analysis of Business Processes (MMABP). In the third section, we pay the attention to the specifics of the PA business field mainly from the point of view of the idea of its digital transformation. The crucial fourth section integrates the topics of both previous section and outlines the idea of the application of the process-driven management ideas in the field of public administration as a basic way of its digital transformation. In this section, we also outline the main features of MMABP approach to the e-government using the specialization of the MMABP meta-model for the field of PA. In the final section, we then conclude from the previous sections and discuss the most important aspects of the process-driven public administration together with their consequences in terms of the digital transformation of PA.

6.2 Process-Driven Management

Figure 6.1 describes the main factors of the process-driven management of an organization. It shows the organization (the white area) placed in its environment (the gray area)—a socioeconomic system. As the general importance of the organization always comes from its meaning for other actors in the system, its primary function must be always targeted at the goals from its environment, not inside the organization. Primary function represents the values, which the organization provides other system actors with. Primary function originates from the business processes of the organization, from what the organization does. For that purpose, the organization has the information and organization systems as basic infrastructures. Information and organization systems represent the secondary function supporting the primary function of the organization. Information technologies allowing the performance of both infrastructural systems then represent the tertiary function. Nevertheless, at the

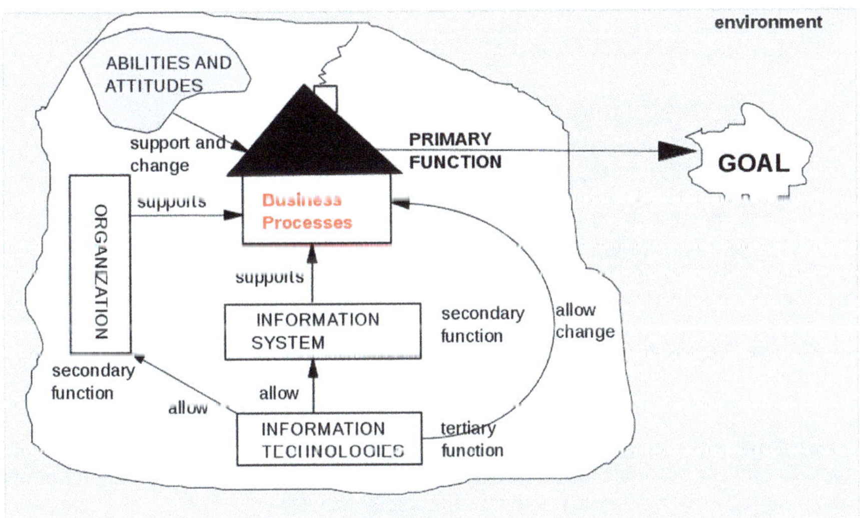

Fig. 6.1 Process-driven management of organization

same time the information technologies work as a trigger of changes in business processes. This double role of IT in the organizational development characterizes the essence and main purpose of the process-driven management; technology development allows us "doing things in a different way" [1] which should be understood as simplifying the processes towards their natural substance.

The crucial task in the process of building the process-driven organization is the proper decision about the structure of business processes which perfectly fulfills the primary function on one hand and which is as close as possible to the natural substance of processes (i.e. as simple as possible in the borders of the current technological possibilities) on the other hand. The first step of this task must be the clear understanding of the natural substance of the given business in terms of the system of conceptual business processes independent of the technological, organizational, and legislative and other real-world constraints.

The conceptual analysis of business processes is elaborated in detail in our background methodology MMABP.

Figure 6.2 shows the MMABP procedure for building a process-driven organization. The first two steps of the procedure represent the initial analysis of process substance of the given business: identification of the natural chains of activities, which are necessary to fulfill the defined business goals (step 1), and analysis of the primary function in order to identify so-called key processes i.e. those chains of activities which directly fulfill the primary function. In the third step, the uncovered key processes are analyzed in order to cleanse them from the support activities. This way the process content of the business system is clarified; key processes are focused primarily on the management of the chain of activities pointed to the business

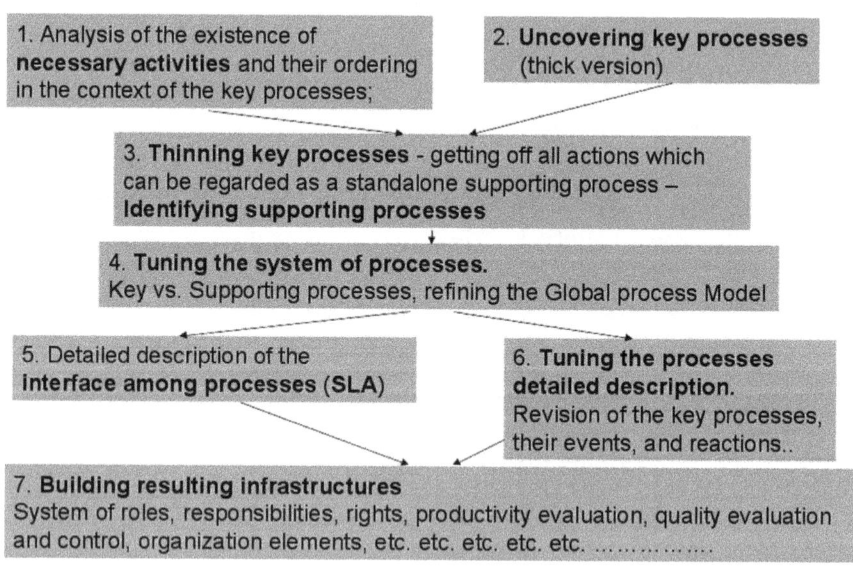

Fig. 6.2 MMABP procedure for building a process-driven organization

Table 6.1 Key versus support processes

	Key process	Support process
Customer's needs	Fulfilled directly	Fulfilled indirectly via key processes
Responsibility	Management—oriented Responsible for the context of the whole business case from the client's point of view	Production—oriented. Responsible for the quality of its service, not for the context in which it is used
Dynamics	Very dynamic, often changing, permanently developing, individually oriented	Rather static, stable, offering standardized and multiply usable services

goals and freed from the support activities that then form so-called support processes specialized in the internal logic of needed supporting services.

This division of processes according to their business role (key/support) represents the important value of the process-driven management as a sophisticated way of exploiting the effect of specialization. The specialization is a powerful tool for increasing the effectiveness and efficiency. In traditionally managed organizations, the specialization is exploited via the organizational hierarchy what breaks key processes to fragments according to the organization functions, specializations. Organizational structure then fixates each key business process as a sequence of tasks, which can be changed only with the change of the organizational structure. This way the organization can exploit the effect of specialization but at the cost of the complete loss of flexibility. Business goals can be achieved only the way predefined by the organizational structure. Process-driven management offers the way of exploiting the effect of specialization without the loss of the ability to immediately change any process if needed. In this approach, just the processes are specialized which makes specialization independent of the organizational structure. Table 6.1. shows the essential differences between two main kinds of a business process in the organization: key and support processes.

Key processes specialize in the management of the whole business case. Key process represents the primary function what means it is primarily responsible for the delivery of the given value for the customer. It is focused on the production of this value as directly as possible using the services of support processes. In other words, key process orchestrates services of support processes in order to produce the final value for customer. Its responsibility covers the whole business case—the whole way from the very first identification of the customer's needs up to the final delivery of the value. On the other hand, key process does not need to be focused on nor responsible for the professional services which the production of the value for customer consists of; they are delivered as services of support processes. Key process is also very sensitive to the possibilities that the technology development brings as the change in the key process directly influences the performance of the whole organization and consequently its competitiveness. In these terms, the key processes represent the essence of the organization's flexibility as well as individuality.

Support processes specialize in professional services. Support process is triggered via requests of other (supported) processes and it is responsible for the delivery of service(s), which it specializes in, including the way in which the service is produced and delivered. Support processes are not responsible for the context in which their services are used which means that they are not so closely tied with the overall performance of the organization as the key ones. Unlike the key processes which are intended to be as much flexible as possible support processes are intended to be rather stable, important for the quality of services instead of the flexibility. From the point of view of the organization, the support processes represent the standard part of its operation. Standardization of them is very important as it allows the organization to utilize the effect of outsourcing unlike in the case of key processes, which are important for their individuality, and their outsourcing does not make a sense in principle.

Distinguishing between the key and support processes is a basic way of the application of specialization in the process-driven organization. At first, it separates the management of the key processes (alias primary value chains according to Porter [6]) from support activities, which allows extra specialization of the key processes to the management of the value chain. At second, it frees the support activities from the responsibility for the context of their use what allows paying the proper attention (specialization) to the internal logic of providing support services and the maximal standardization of the support activities/processes/services. Such specialization does not influence the integrity of key processes, which is a critical problem of the traditional hierarchical organization.

Orientation on processes is necessary for keeping an active control over the business case in the current turbulent world. Active control means not only controlling the process but also instant improving the process based on the new technology possibilities. It requires two essential attributes of the business system: **resilience** on one hand and **stability** on the other hand.

Resilient business system in this context is such system, which can immediately react to any relevant change. There are two basic kinds of needed change: *change of client's preferences* and *new technology possibilities.* Ability to immediately accommodate the process to the specific client's needs requires the complete control over the key process on the level of client, which is not possible in the functional organization where the key processes are unconditionally determined by the organizational structure. To make the business simpler and more effective thanks to the new technology, the organization has to perceive its business as a system of mutually interconnected processes, which determine all other organizational features including the system of competences and responsibilities.

Stable business system is a system as much as possible resistant to unnecessary changes. Building such system thus requires at first clear understanding of the essence of possible changes in order to be able to decide about their objective necessity. To do so, one must primarily respect the essential principles and qualities of the objective reality as a basic source of objective and thus certainly necessary needs for changes. Informatics traditionally calls such objective reality the Real World.

While resilience of the business system requires orientation on business processes, its stability has to be based on the precise ontological (conceptual) model of the corresponding business domain.

6.3 The Nature of Public Administration Business

In the previous section, we discuss the idea of process-driven organization as a basic way of creating resilient and stable organization what is necessary for keeping an active control over the business case in the turbulent world. Application of this idea in PA requires at first clear understanding of specific features of PA organizations in comparison with the market organizations. In this section, we analyze the nature of the PA business, especially focusing on such features of PA that are important for the application of the process-driven style of management in PA organizations.

The main difference between market-oriented and public organizations is the *absence of the market in PA*. Public administration is based on the natural monopoly coming from the territorial division of the local authority from authorities in other territories. Local authority is the only ruler in the given territory and its power is limited only by legislation, not by concurrent authorities. This absence of the competition in PA causes fatal difficulties with the quality and costs of PA services. Unlike on the market, in PA is missing the natural mechanism determining the proper cost of the service. The price, as a complex category, including all hard and soft, general and specific, global and local, technological and cultural, and all other even unknown factors, is incomputable in principle. Only the market, by means of the competition, could naturally take into the account all factors and determine the proper cost of the PA service. In the monopolistic environment of PA, the costs of the services must be approximated in other ways. The same can be said also about the quality of services, which is a natural counterbalance to the price. The certain way of overcoming this critical insufficiency can be *taking the market into the game of PA as much as possible.* This may be still effective, as many current PA services and their components do not have to be necessarily implemented in a monopolistic way. Nevertheless, there always remain a certain number of PA services or their components, which naturally cannot be the subject of competition.

Overcoming the problem of the absence of the market in PA requires the precise analysis of the essence of PA business. General ontological/conceptual model of the field then can serve as a basis for the decisions about possible use of the market and also helps to uncover the essence of the needs for the PA services which is necessary for determining the necessary service components and consequently for the approximation of their proper cost. Regarding the abstract nature of the PA values, it is obvious that the analysis of the origin of the needed PA services requires the abstraction of the higher order than in the market environment.

Another important difference relates to the concept of *client*. Unlike on the market where the meaning of the customer is self-evident, the client of PA organization must be understood more abstractly. In our approach, PA client is an *abstract entity, which may be a physical person (citizen) or any legal entity (public or private) with the mission, purpose-orientation or goals, operating in the region. It may be a purpose-built civic association, a group of people with the same goals or problems even if they do not know each other,* etc. *It may be even the whole society (community) in general (for instance in the prevention of crime or environmental issues).* This feature of PA is especially important regarding the necessary orientation on client as one of the basic principles of process-driven management. Identification of the client and its interests is the very first step in designing the system of processes. In PA, such identification is extremely complex and abstract task. Another direct consequence of so complex nature of the client is the dramatically higher number of key processes (i.e. processes directly connected with clients) in PA in comparison to the market organizations.

Fuzzy nature of the PA client discussed in the previous paragraph is closely connected with the next problem of the PA business, which is the *identification of primary function.* The primary function of organization expresses the meaning of the organization from its clients' point of view; it represents the value, which the organization provides its clients with. Regarding the nature of PA client, the primary function of PA organization represents many different values with highly general meaning at the same time. It requires on one hand the precise *conceptual analysis of the PA business* in order to identify the general, universal values, which the PA authorities should care of. On the other hand, the identification of primary function in PA requires also respect to the individual values for clients as everybody is in the specific "*life situation*". For individualization of the primary function, we propose in the following section the approach based on modeling of so-called *life cycles* of PA objects (clients).

One of the essential principles of the process-driven organization is so-called "*strategic anchorage*". Strategic anchorage means the permanent alignment with the strategic values of the organization to ensure the needed stability of the process system on one hand and to allow its changing according to the relevant objective changes, which have to be followed like the development of technology for instance, on the other hand. In PA, it means the permanent alignment with the overall strategy of society. Strategy of society is usually implemented as a set of rules, mainly in the legislative form. Uncertainty of the stability of leaders caused by mandatory periodical elections in the democratic model of the society leads to the essential need for ensuring the permanent validity of general democratic principles. In specific contexts, this may contrast with the natural need for the evolutionary change. Regarding this natural feature of the democracy, one can say that democratic governance is an art of achieving the harmony between the principally unchangeable basic principles of democracy on one hand and the maximum space for natural, evolutionary changes on the other hand. Process-driven management undoubtedly helps to achieve such harmony by creating the maximal space for possible changes in processes preserving at the same time the stability of the whole system by means of its anchorage to the strategic values. Nevertheless, building of such flexible management system requires the dramatic changes in the system of competencies as we also discuss in the previous section. The traditional hierarchical organization structure with determinately defined competencies for functional positions must be replaced with the variable system of competencies derived from the defined processes. Such managerial idea is very far from the current style of the management in PA with the key role of the strong hierarchical organization. This part of the digital transformation of the managerial system of PA remains the biggest challenge and it is an important subject of the discussion in the final section of this chapter.

After discussing the features of the process-driven management in previous and the specifics of the PA business field in this section, we are going to focus in the following section on the question how to apply the process-driven management in public administration organizations.

6.4 Application of the Process-Driven Management Ideas in Public Administration

The background methodology MMABP perceives the business (real world) system as equilibrium of intention and causality and consequently recognizes two basic dimensions of the business system model: the *ontological* one and the *process* one. Each dimension has specific character and rules, and requires specific modeling approach incompatible with the second one. Even in the field of PA as a special kind of business system, the analyst must pay the proper attention to both dimensions. The basic features of MMABP are formally defined in [7] and explained in [8]. In this chapter, we focus on the use of this methodology for the modeling of PA system in order to analyze it for the purpose of its digital transformation.

6.4.1 MMABP Approach to the Modeling of PA Business System

Figure 6.3 characterizes the MMABP approach to the modeling of PA business system. *PA object* (such as "citizen") always exists in some *external environment* that is a source of basic rules, necessities, possibilities, obligations, and other limitations, which the object must respect. These objective factors are always commonly valid for all public administration objects. At the same time, the object influence also its *internal factors* coming from the context of its life: its current life situation, its history as well as its possible future. This context of internal factors we call the "life cycle" of the object. In its life, the object experiences various *life situations*. Each life situation is given by the external, objective situation together with the internal state of the object's life.

The objective, environmental factors are driven by objective logic of the external world. There are three basic sources of crucial life events (see Fig. 6.3): **Physical environment** represents the natural and other physical values of the given territory.

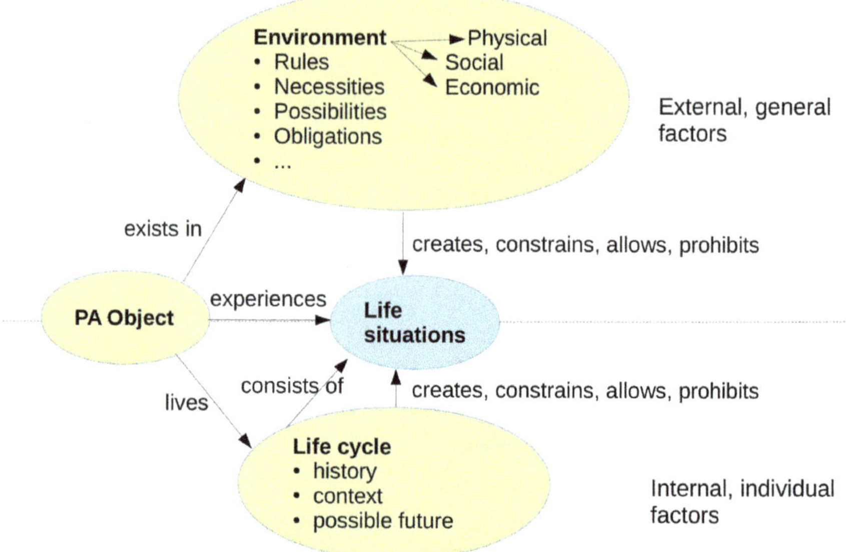

Fig. 6.3 Life situations as basic reasons for PA processes [9]

It consists of the natural and physical attributes of the territory such as raw materials, nature and its values for recreation and tourism, and infrastructure and other values of physical environment important for people as well as for business. **Social environment** consists of people and their personal, cultural, and social values. It includes not only the individual personal characteristics of people but also global attributes of the society itself usually called the "culture' of the society. **Business environment** includes organizational and cultural values necessary for realization of opportunities. This area includes all business entities: enterprises, entrepreneurs and their quality, employees and their qualification, as well as legislation and quality of public administration services supporting the business.

Similarly, even the internal life cycle factors are driven by the general logic of the life cycle, which reflects the real-world causality, and thus it is objective in the same way as the general logic of the external world. Therefore, the life cycles must be described on the same level of preciseness and are as important as the logic of the external world.

6.4.2 *Life Events as Triggers of PA Processes*

Meta-model at Fig. 6.4 is derived from the general MMABP meta-model as a specialization of the general meta-model to the field of public administration. Concepts from

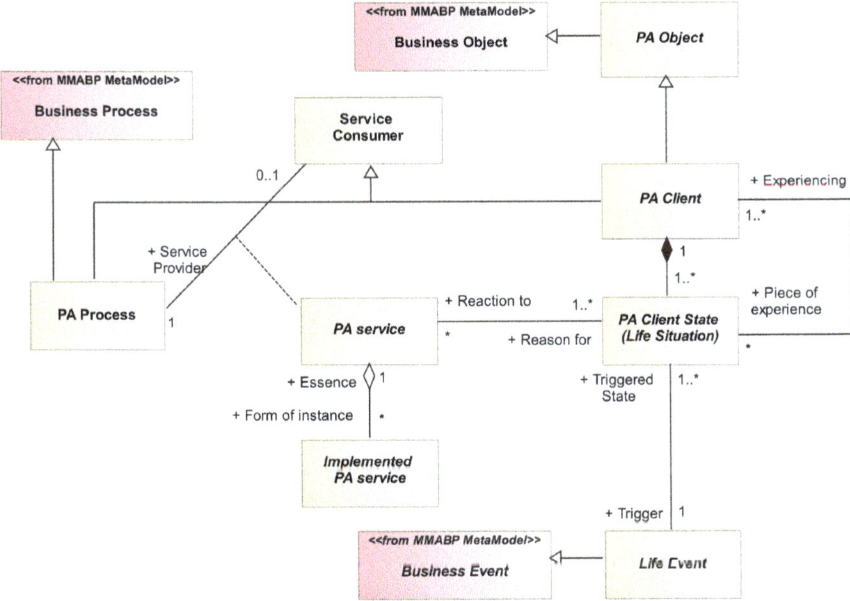

Fig. 6.4 PA business specialization of the MMABP meta-model (a fragment)

the original MMABP meta-model are in red color and are labeled with the stereotype <<from MMABP MetaModel>>.

Other concepts are specific for the PA field. Either each specific PA concept is a specialization of the general MMABP concept or it is derived from another PA concept as its specialization, part or relationship. In other words, this meta-model is an extended sub-model of the general MMABP meta-model. This fragment of the complete PA-specialized MMABP meta-model is focused on the life cycle of the PA client as a source of essential reasons for the relevant PA services. *PA service* is understood here as a relationship between *PA Process* as a service provider and *Service Consumer*. Service consumer is either another *PA Process* or *PA Client*. The concept *PA service* represents here a general content of the service, which may be implemented in various ways. Each kind of implementation of the service is called here *Implemented PA service*. The right side of the model describes the concepts related to the life cycle of the *PA Object*. *PA Client* is a special kind of *PA Object*, which is important in this context because its life cycle is understood in the methodology as a basic source of the reasons for PA services. In these terms, *PA Client* is a collection of *PA Client States* alternatively called also *Life Situations*. Each state is triggered with some *Life Event* (PA related special kind of the general *Business Event*) and is a reason for some *PA services*, which are understood in this context as reactions to the life situations. Based on this idea the PA-specialized MMABP uses the **analysis of the life events** derived from PA objects life cycles as a **main tool for the identification of objectively needed PA services**.

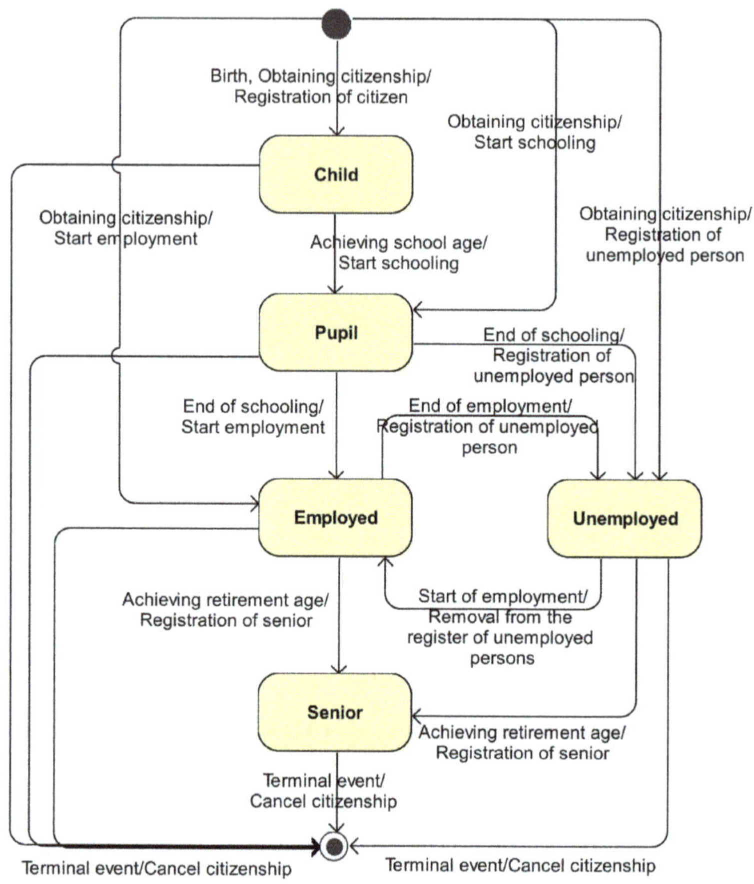

Fig. 6.5 Life cycle model of *Citizen*

The concept *PA Client* as a collection of life situations represents the *life cycle* of the client. At this point, MMABP completes the standard conceptual model with the algorithmic description of the life cycle in the form of the UML State Chart (State Machine Diagram) [10] what allows enriching the standard conceptual model of the Real World with the dimension of causality. State Chart describes causal relationships of states as their algorithmic structure.

The example at Fig. 6.5 shows the life cycle of one sub-type of PA client—the *Citizen* (see also the classification of concepts in the ontology model at Fig. 6.6). The model describes the whole "life" of a citizen from the very beginning to all possible ends.[1] The life cycle is structured as a set of possible states (life phases, life situations) and possible transitions among them. Each transition is described as

[1] It should be noted that the meaning of the concept *Citizen* is not equal to the meaning of the concept *Physical Person*. So, the "end of life" of the citizen does not mean necessarily the physical death but any possible end of citizenship.

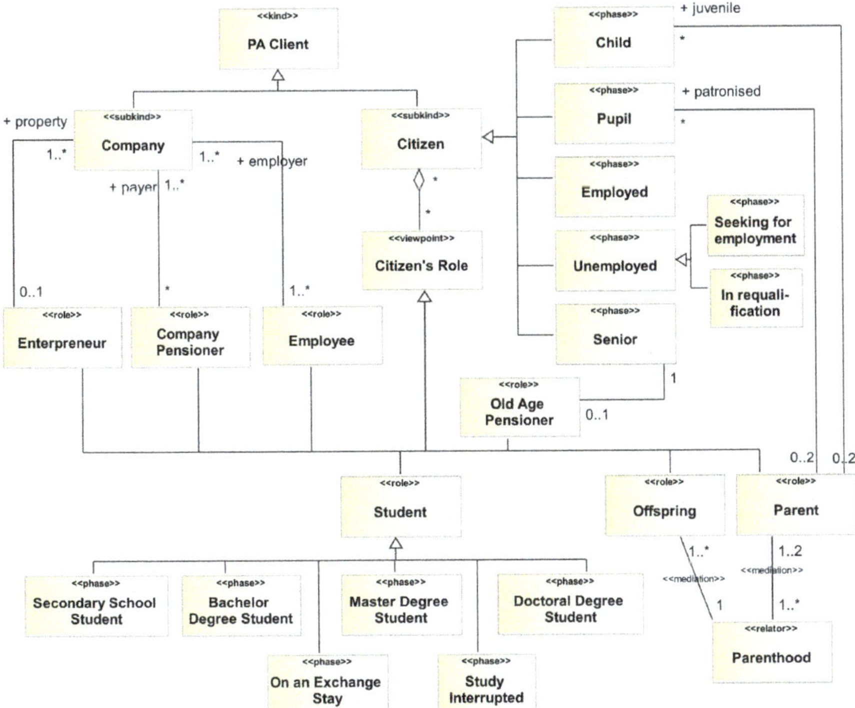

Fig. 6.6 PA ontology model (a fragment)

a couple cause/consequence (event/method in the terminology of UML). This way, the model describes the causality related to the given object and it is a complement to the static description of the general modality of given domain in the conceptual model. It can be seen at Fig. 6.6 how the life cycle of the *Citizen* looks from this static, modality-oriented point of view. Life states are expressed there as special sub-types <<phase>> of the *Citizen*. Such generalization, where sub-types represent rather the phases in the life of one object than independent objects, is in literature called "dynamic generalization" [11, 12], and causes also so-called "problem of identity" in UML [10] which the language still does not successfully cope with.

Figure 6.6 shows the fragment of the PA ontology model, which defines the basic real-world object classes, their classification and their mutual relationships in terms of the modal logic of the Real World related to PA. The first version of the model has been created in the analysis project managed by the Czech Ministry of Interior focused on the revision of the official view of life events (life situations) in order to make it consistent with the Czech e-Government Architecture. For more details about the project and used methodology, see [9]. The model methodologically follows the OntoUML standard [11] extending its basic set with the <<viewpoint>> stereotype needed for the correct extension of the conceptual ontology model with the State

Chart models of life cycles of crucial object classes. Detailed explanation of the purpose and the way of use of this extension can be found in [13].

Besides the "subkind" class *Citizen* also the "role" class *Student* is expressed as a generic concept covering sub-classes of the type <<phase>> which points to the fact that life cycle of this class expressing the real-world causality related to it should be described as well. This life cycle (see Fig. 6.7) describes the causality related to the role of student from the point of view of PA; identified life phases are important for the social and health care as well as tax and other reasons relevant for PA. Besides the standard degrees of education for that reasons there are also important the states *Study Interrupted* and *On an Exchange Stay*. Modeled transitions between states then determine possible causal relationships of relevant facts: events and needed related actions of PA.

Non-existing transition between two states means that such causal relationship cannot exist. For instance, there is no direct way from the secondary school to the higher school master study; the only way goes through the bachelor study. Similarly, the citizen cannot transit from the life phase *Child* directly to the phase *Employed*

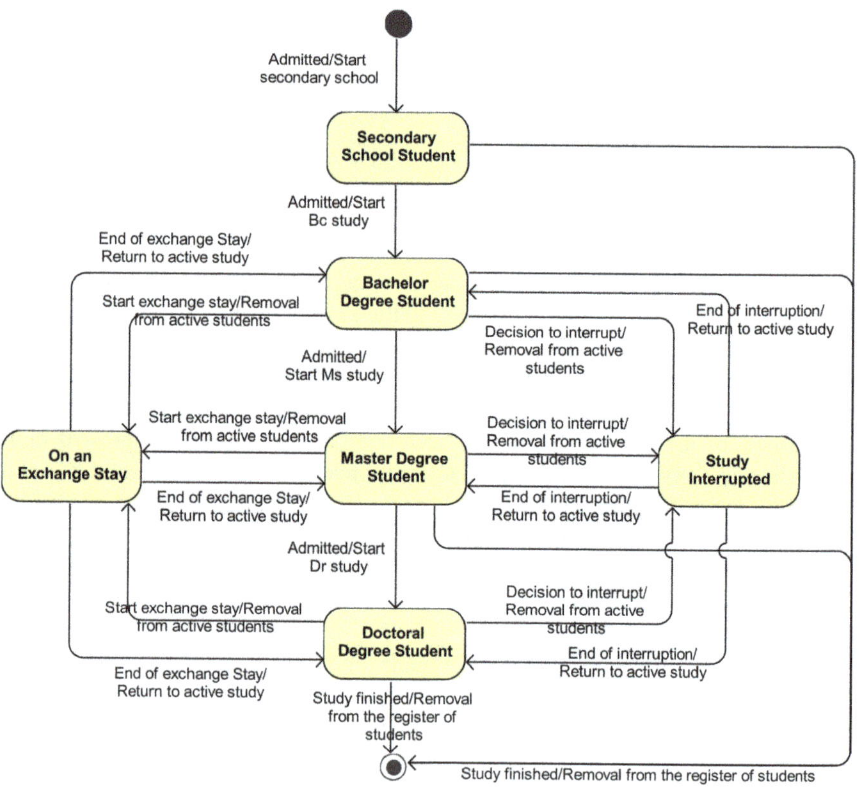

Fig. 6.7 Life cycle model of *Student*

skipping the phase of compulsory school attendance. Presented life cycles are mutually independent. Nevertheless, as student is a role of the citizen, both life cycles can be valid simultaneously for the same citizen. Similarly, the life cycle of any other role of the citizen can be described if needed and all are valid at the same time. Each particular instance of *Citizen* thus lives a number of parallel lives, each from the specific <<*viewpoint*>> represented by the given <<*role*>>. This is allowed by using the <<*viewpoint*>> pattern introduced and explained in [13]. *Citizen* is an aggregate of several *Citizen's Roles*, each role can represent the specific life cycle. Each of parallel life cycles is guaranteed valid (independently of the other ones), and all are valid at the same time as it is warranted by the aggregation of all viewpoints.

The examples show how the life cycle models represent the compendious definition of an essential causal context of real-world events and corresponding PA actions/services. Moreover, the life cycle provides us with the information about the context of each cause—consequence couple. The model shows what can or must be the next step, and which step could or had to be the previous. Even moreover, as the life cycle models are organized according to the context of objects (one life cycle can be described just for one object, it is impossible to describe more life cycles in a single model) the system of PA services derived from such description of the causality is always perfectly consistent with the ontology of the domain.

Meta-model at Fig. 6.4 shows how the MMABP approach uses this compendious information from described life cycles of crucial PA objects for the **derivation of objectively needed PA services**. *PA Process as a reaction to the PA Client State* (*life situation*) *produces the PA Service*. The *Life Event*, which causes the Life Situation, is the starting event for the corresponding *PA Process* (to avoid an information redundancy this fact is not directly modeled in the meta-model because it follows from the above-described set of relationships). These facts allow the **creation of general PA processes** producing the needed services identified as needed reactions to life situations. Nevertheless, the concept *PA Service* represents also the relationship between *PA Process* and *Service Consumer*. The identity of the service is thus related not only to the producing process but also to the consumer what means that services must be "customized" for consumers. This fact underlines the essential need for modeling the life cycles of PA clients as a way of the **individualization of general services in PA processes** according to the context of the life of the given client as we discuss in the previous sub-section.

Table 6.2. contains examples of PA key processes, their source areas (environments) and their initial events identified in life cycles of corresponding crucial PA objects. Mentioned processes as well as objects and their life situations are very general. Particular instance of life event always belongs to the particular object and its life situation. This example is also an illustration of the diversity of PA objects, roles and their life situations which obviously requires the precise information not just about the general context of the given life event but also about the momentary contents of the life cycle of the PA object.

Table 6.2 Examples of PA key processes

Source area	Key process	Initial life event (object)
Social environment	Social incident management Execution of the social development action	Social incident (Citizen) Need for social development change (social system role)
Physical environment	Execution of the Ground plan change Environmental incident management	Need for change in the ground plan (ground plan) Environmental incident (environmental role)
Business environment	Help with implementation of business idea	Business idea (entrepreneur)

6.5 Conclusions

In this chapter, we propose the application of the process-driven management in PA as a certain way to the essential transformation of the PA business model. Our approach is based on more that fifteen years of experience with the application of process-driven management in PA in Czech Republic, realized in seven long-term projects. The most valuable experience comes from the project [14] where the MMABP approach for PA took its current shape.

The literature describes many attempts to apply information modeling practices in public administration. They are mostly oriented on data modeling, usually for the integration of various PA applications, or the ontology modeling for achieving the interoperability of PA systems [15, 16]. One of the most comprehensive overviews of such approaches is in [15]. Almost all these approaches remain in the field of traditional conceptual/ontology static model, the attempts to enrich it with modeling the dynamics of objects with the use of life cycle modeling, like in [17], are very rare. The application of process modeling in PA is also widely popular. Nevertheless, instead of following the process reengineering ideas, such models usually only describe the traditional conception of PA processes: particular agendas structured according to the hierarchical organization of the given PA authority even if combined with the ontology model [18]. We have not found any reference to the use of the combination of the ontology model with process model for the reengineering of PA processes based on the model of the Real World causality in the form of life cycles of PA objects, which we propose in this chapter.

In the following text, we answer the questions how the proposed approach can change PA and help to solve the problems of PA, outlined in the third section and what are its most important consequences.

Process-driven management in PA leads to the model of flexible PA with maximum use of the power of specialization by means of outsourcing. The natural evolutionary consequence of such model is the permanent trend to downsizing and slimming via focusing to just essential part of the administration task—basic decisions. This trend also significantly contributes to the needed penetration of PA with commercial services anywhere it is possible and suitable which liberalizes the traditional monopolistic character of PA.

Above-mentioned *flexibility of PA strongly contrasts with the traditional, hierarchical model of PA*, which typically leads to the permanent weight gaining of PA in all parameters like number of employees, extent and complexity of bureaucracy, additional investments, etc. This trend is the great current problem of PA permanently even worsening by the turbulently developing IT if wildly applied in PA as it often can be seen in current approaches to the e-Government. Application of IT in PA, unlike in any other business field, critically requires to be provided hand in hand with the essential changes of the business model otherwise it reliably gets out of the control. Business transformation of PA, as a consequence of the application of IT, seems to be not just a chance but also a life necessity.[2]

Precise analysis of life events of the crucial PA ontology objects can lead to the creation of the *general-purpose definition of the needed PA services* at the objectively proper level of granularity. Such definition, if commonly accessible, can serve as a general standard for all potential providers of PA services and their IS/IT support. Such approach principally opens the market with E-government services what in turn causes the natural acceleration of the quality of services together with the natural lowering of their price. Moreover, this is another contribution to the needed public-private partnership.

The proper attendance must be paid also to the *possible risks of the solely technical management of PA*. Such factors are usually hidden in the beginning phases of the evolution spiral, waiting for later phases to become critical. Nevertheless, the earlier we take such factors into the account the better we can handle them. Particularly, the popular thoughtless technocratic "automation" of current PA "agendas" can conserve the current style of management in supporting IT and finally block its needed change to the process-driven model especially if supported by legislation.

Digital transformation is extremely hard, multi-dimensional and risky task. The process of the transformation must be managed very carefully with respect to the consequences of the change in all relevant dimensions. In the context of process-driven management, this problem led to the idea of the *organization maturity model*, which helps to manage the transformation process based on the knowledge of the relevant aspects and consequences of the change at the given level of organizational maturity. The first maturity model for the digital transformation by Nolan [19] led to several currently used models. The best elaborate model, based on the harmony of the process and organizational maturity, has been made by Hammer [20]. This model is also a basis of the methodology for the organizational development through the technology progress management [21] whose use can significantly help to handle the above-mentioned risks.

[2]Particularly in PA, the great problem is change of legislation, necessary for the application of changes in PA processes, competences and other related aspects. Digital transformation of PA is impossible without the legislative changes.

Although the most critical factor of the digital transformation in PA remains the active political support, without enabling by architecture and process management the political will only would not be of much use.

References

1. Hammer, M., Champy, J.: Reengineering the Corporation: A Manifesto for Business Revolution, 1st Harper Business Essentials pbk. edn. Harper-Business Essentials, New York (2003)
2. eGovernment: Information Society Commission, Dublin, Ireland. http://www.isc.ie
3. McKendrick: Making e-government more than a glorified service-delivery platform, SmartPlanet, 2009. http://www.smartplanet.com (2011)
4. OECD The e-Government Imperative: OECD e-Government studies. OECD (2003). https://doi.org/10.1787/9789264101197-en
5. The Obama Administrations Commitment to Open Government: Status Report (2012)
6. Porter, M.E.: Competitive Advantage: Creating and Sustaining Superior Performance, 1st Free Press edn. Free Press, New York (1998)
7. MMABP: Business System Modelling Specification. http://opensoul.panrepa.org/metamodel.html
8. Řepa, V., Svatoš, O.: Adaptive and resilient business architecture for digital age. In: Zimmermann, A., Schmidt, R., Jain, L.C. (eds.) Architecting the Digital Transformation. Springer International Publishing (2020)
9. Repa, V.: Life events: a crucial point of e-Government. In: Matulevičius, R., Dumas, M. (eds.) Perspectives in Business Informatics Research, pp. 33–47. Springer International Publishing, Cham (2015)
10. UML Superstructure Specification: v2.4, Object Management Group. https://www.omg.org/spec/UML/2.4 (2011)
11. Guizzardi, G.: Ontological Foundations for Structural Conceptual Models. Centre for Telematics and Information Technology; Telematica Instituut, Enschede (2005)
12. Heller, B., Herre, H.: Ontological categories in GOL. Axiomathes **14**, 57–76 (2004). https://doi.org/10.1023/B:AXIO.0000006788.44025.49
13. Repa, V.: Modelling life cycles of generic object classes. In: Linger, H., Fisher, J., Barnden, A., Barry, C., Lang, M., Schneider, C. (eds.) Building Sustainable Information Systems, pp. 443–454. Springer US, Boston (2013)
14. Optimization of Life Events: Project CZ.1.04/4.1.00/D9.00002, Czech Ministry of Interior, https://www.mvcr.cz/clanek/optimalizace-zivotnich-situaci-ve-vztahu-k-registru-prav-a-povinnosti.aspx?q=Y2hudW09MQ%3d%3d
15. Ryhänen, K., Päivärinta, T., Tyrväinen, P.: Generic data models for semantic e-Government interoperability: literature review. Innov. Publ. Sect., 106–119 (2014). https://doi.org/10.3233/978-1-61499-429-9-106
16. Flak L.S., Solli-Saether H.: The shape of interoperability: reviewing and characterizing a central area within e-government research. In: 45th Hawaii International Conference on System Sciences, pp. 2643–2652 (2012)
17. Vasista, T.G.K.: Thoughtful approaches to implementation of electronic rulemaking. Int. J. Manag. Pub. Sect. Inf. Commun. Technol. **7**(2), 43–53 (2016). https://doi.org/10.5121/ijmpict.2016.7205
18. Savvas, I., Bassiliades, N.: A process-oriented ontology-based knowledge management system for facilitating operational procedures in public administration. Exp. Syst. Appl. **36**(3), 4467–4478 (2009). https://doi.org/10.1016/j.eswa.2008.05.022
19. Nolan, R.L.: Managing the computer resource: a stage hypothesis. Commun. ACM **16**, 399–405 (1973)

20. Hammer, M.: The process audit. Harv. Bus. Rev. **85**, 111–119, 122–123, 142 (2007)
21. Řepa, V., Šatanová, A., Lis, M., Korenková, V.: Organizational development through process-based management: a case study. Int. J. Qual. Res. **10**(4) (2016). ISSN 1800-6450

Part II
Digital Business

Chapter 7
Requirements for IT-Support of Personal Services in the Digital Age

Michael Fellmann, Fabienne Lambusch, Birger Lantow, and Gregor Simon

Abstract Digitalization of work increasingly receives attention due to the wide spread of easy to use mobile computing devices. However, it remains unclear how the potential of ubiquitous information systems can be tapped in domains with less formal structured processes such as in personal services. Up to now, concepts such as *Adaptive Case Management* have been proposed and it has been investigated how such systems could be used in various domains. However, the specific requirements that originate from the domain of personal services are far less understood. We therefore empirically elicit requirements for improved IT-support in the domain of personal services by combining workshops, interviews, and shadowing techniques in a triangulation approach. We hope that our consolidated requirement catalogue will foster the design of information systems that provide an improved IT-support for people working in personal service contexts.

Keywords Requirements engineering · Personal services · Digitalization

7.1 Introduction and Motivation

Digitalization is one of the dominant topics of our time. Information and communication technologies have become an integral part of everyday life and lead to changes in many areas of work. Two fundamental aspects of Digitalization are the transformation of analogously stored information into digital data and the execution of more

M. Fellmann · F. Lambusch · B. Lantow (✉) · G. Simon
Information Systems Engineering, University of Rostock, Rostock, Germany
e-mail: birger.lantow@uni-rostock.de

M. Fellmann
e-mail: michael.fellmann@uni-rostock.de

F. Lambusch
e-mail: fabienne.lambusch@uni-rostock.de

G. Simon
e-mail: win.office@uni-rostock.de

A. Zimmermann et al. (eds.), *Architecting the Digital Transformation*, Intelligent Systems Reference Library 188,
https://doi.org/10.1007/978-3-030-49640-1_7

and more processes by machines or intelligent IT-systems [25]. As a result, there is the potential that companies fundamentally change the way they produce and deliver their products and services to their customers [6, 25]. At the same time, the adaption, alignment and generation of new business strategies based on digital developments is essential for the competitiveness of companies [5]. However, businesses are not yet making full use of the digital possibilities to support their workflows [31].

The central task for the future will be to develop concepts for digital change that support not only industries with highly structured processes, but also economic sectors with knowledge-intensive and weakly structured processes [17] and especially for work with a high degree of uncertainty like in personal services [3]. Due to the special characteristics and the increasing importance of personal services, innovative ideas and sustainable advancements are needed for the digital change in this sector. A major challenge is to determine how services that are primarily characterized by the interactive work of individuals [1] can be meaningfully supported by IT-systems. Therefore, it is very important to investigate what the specific requirements are that originate from the domain of personal services.

With our work we want to investigate such requirements in a systematic way in order to provide guidelines for the design of valuable IT-systems. To this end, triangulation of research methods was used to present requirements for an IT-based support for personal services in this chapter that were raised from practice. We therefore empirically elicit requirements by combining workshops, interviews, and shadowing techniques.

In the next section, background information and related work are presented. In Sect. 7.3 the procedure of investigation is described. To this end, the methodological instruments used to collect and analyze data are presented. Section 7.4 then describes the results from the requirements elicitation according to the different methods used, while Sect. 7.5 presents an integrated list of requirements that unites and structures the various requirements of the different survey techniques. Subsequently, we discuss the findings and provide an outlook in Sect. 7.6.

7.2 Background

In order to establish a common understanding, we first describe some background regarding services and explain what is meant when referring to personal services. Subsequently, related work is presented that considers requirements for IT-support in knowledge-intensive processes.

7.2.1 Personal Services

The service sector is gaining more and more importance. Although the economic crisis became visible in the service sector of the European Union through a short-term decline in the number of employees, the development of employment in services has shown a mostly steady increase even over the last two decades [9]. This trend illustrates the importance of addressing the specific challenges and resulting requirements of the growing service sector more comprehensively.

Services are especially distinguished from tangible goods. According to a literature review by Zeithaml et al. [32], the four consistently cited characteristics of services are *intangibility*, *inseparability* of production and consumption, *heterogeneity* (potential for high variability in the provision), and *perishability* (inability to be inventoried). The authors of this article furthermore call intangibility the fundamental difference from goods, as services are rather performances that cannot be sensed in the same way as goods. Although services are different to goods and characterized by intangibility, they can still be directed at goods or include tangible actions. A classification approach for services proposed by Lovelock [21] distinguishes between two dimensions of services. The first dimension represents the direct recipient of the service, which describes if the service is directed at people or things. The second dimension represents the nature of the service act that distinguishes between tangible and intangible actions. For services relating to people, the different proportions of tangible and intangible elements refer to whether the service rather affects the body (e.g. health care) or the mind (e.g. psychotherapy) of a person. Regardless of the proportion of tangible or intangible actions, both classes of services described above have in common that the direct recipients of the services are persons instead of things. Already Halmos [14] denoted those services as personal which are concerned with the change of the body or the personality of a client. Bieber and Geiger point out that a great challenge of personal services lies in the fact that the recipient of the service is at the same time in the role of a co-producer [3]. They state that personal services are characterized particularly by the close, indissoluble connection between persons who have to interact with each other. This interaction as a key element in personal services reduces the predictability and increases the uncertainty in the corresponding work processes. Since service activities that seem similar (e.g. working in a call center) can still differ in their degree of knowledge intensity depending on the specific field of activity, services can also be classified as knowledge-intensive or simple [3]. We refer here in particular to knowledge-intensive personal services.

Searching for a standard process model, such a reference model seemingly does not exist in the domain of personal services, while in contrast such models exist in other service domains such as the Technical Customer Service (e.g. [7]). However, Fließ et al. [11] describe a straightforward service model that we refer to instead. At its core, the model consists of the three phases *pre-service*, *service*, and *post-service*. Pre-service describes the preparation of a service, while service describes the actual provision of services in interaction with the customer, and post-service refers to the follow-up. We extend this model with two additional phases, which are described in the following. From the point of view of information processing, the registration of a client as the starting point for the relation between customer and service provider is also important; this is where the first information about the client becomes available. Therefore, the phase of *client intake* is added. Moreover, personal services are often

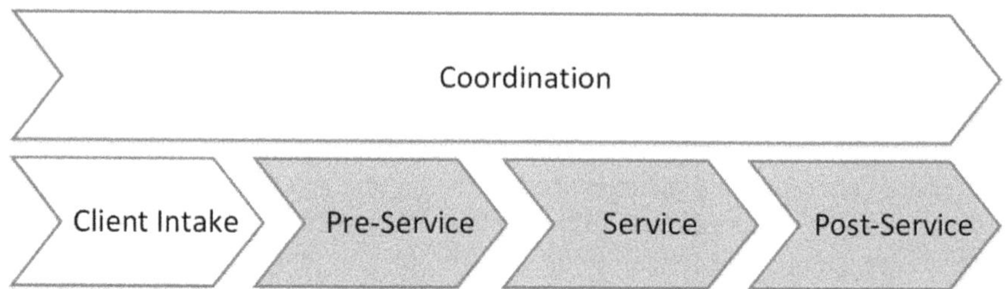

Fig. 7.1 Extended service phases

integrated into a longer-term process, which then also has longer-term objectives. One example would be the restoration of mobility in the context of physiotherapy. This means that phases of direct, short-term service provision may be intertwined with overarching coordination tasks to achieve objectives. Therefore, we add the phase of *coordination*. The extended version of the service phases is shown in Fig. 7.1.

The grey elements in the figure show the core phases proposed by Fließ et al. [11], which are passed through during each service operation. The two added phases, in contrast, are not necessarily executed during each service operation. In order to illustrate the phases, we use an example from a special field of personal services, namely social services, in the following. In this example, support is to be provided for a child who refuses to go to school. First, the service provider could be told about the problem and the family situation (e.g. by telephone). This can be assigned to the phase of client intake. In the pre-service phase, the service provider prepares for the consultation. During the service itself, a conversation takes place with the child and the parents. Subsequently, the documentation of the contents, duration and billing information is carried out in the Post-service phase. The service provider will call in the teacher for the next appointment. This activity falls into the coordination phase.

7.2.2 Requirements for Knowledge-Intensive Business Processes

Since we refer to knowledge-intensive personal services, requirements for knowledge-intensive business processes are relevant for our investigations.

While routine tasks often can already be automated by Workflow Management Systems or at least supported by IT, it is a major challenge to sensibly support knowledge-intensive work with the help of IT [25, 26]. Mundbrodt and Reichert derive in [26] 25 requirements of IT support for knowledge-intensive business processes. Though being very detailed, the authors just present a selection of requirements found during their investigations. Furthermore, a lot of the requirements target rather overarching process management than support during process execution which is in the focus of this work. Generally, the requirements by Mundbrodt and Reichardt are met by the *Adaptive Case Management* (*ACM*) paradigm which is currently the most prominent approach to the management and IT support of knowledge-intensive business processes.

Hauder et al. in [15, 16] as well as Michel and Matthes in [24] describe about 100 partly overlapping requirements for ACM which basically subsume the identified requirements for knowledge-intensive business processes. Thus, *ACM* solutions provide functionalities that might fit the requirements of the domain of personal services.

For comparability, a more general description of requirements is necessary. Kurz et al., for example, describe in [18] the core principles of *ACM* which can be transformed into requirements. First of all, there is adaption, meaning flexibility for deviations during process execution and continual improvement of process models. Furthermore, *ACM* application means supporting the collaboration of all actors by collaboration functionalities and central access to process data. At last, the control of case execution should be based on process goals using a data centric approach as input for decision making and recommendation with regard to progress in process execution. All defined requirements for ACM can be mapped to one of the principles.

Aside the process paradigm, there is also related work focusing on Knowledge Management in general and its requirements. Maier and Hadrich for example describe in [22] knowledge management system functionalities. These could also fit to the requirements of IT support for personal services.

Although *Adaptive Case Management* (cf. [19]) has been developed to support weakly structured, knowledge-intensive work, it is unclear to what extent the approach can be leveraged for personal services as it is not specifically geared towards personal services. The same is true for knowledge management systems.

In contrast to sectors with hardly any customer involvement, personal services stand out through the dependence on the interaction and cooperation between service provider and recipient. The challenge here is to take adequate consideration of the interaction as a key element of the service, as otherwise the introduction of new technology can lead to unfavorable results [3]. It is therefore necessary to determine the specific requirements that originate from the domain of personal services. Hence, we conducted a systematic analysis of requirements for the digitalization in personal services. The following sections describe the analysis procedure and its results. Section 7.5 will consolidate the results and compare it to requirements formulated for ACM, Workflow Management, and Knowledge Management Systems in order to assess the suitability of such solutions for personal services. As an aid to orientation, we structure our consolidated requirement catalogue along the typical phases of the personal service process as formulated in Sect. 7.2.1. Thus, it can be identified at that phase what kind of IT support is possible.

7.3 Methodological Considerations

In the following, the research process underlying this work is explained in more detail. Since we aim to specifically identify requirements for a potential IT-support for personal services, we focus on *functional* requirements. Functional requirements address the features and functions of the software, e.g. the number of supported file formats or reports provided. However, from the perspective of IT-based work support other aspects like information demands and software quality requirements connected to the specific characteristics of the personal services have been encountered during data collection too. Requirements regarding aspects such as quality or security are referred to as *non-functional* requirements. For the reason of completeness, these have been recorded too encountered. In order to identify requirements for IT-support, we use three different survey instruments: (1) workshop, (2) expert interviews and (3) shadowing. The use of different methods, also referred to as *method triangulation*, and the combination of different data sources, also referred to as *data triangulation*, has the decisive advantage that shortcomings of individual methods or singular data sources can be compensated [10, 28]. This approach is also described in literature as "systematic triangulation of perspectives", which is characterized by a combination of different perspectives with the result of methodological access and data source diversity [10].

In the workshops we conducted, different stakeholders from enterprises in the social sector come together to share and discuss ideas. The companies predominantly offer consulting to families, for example if children show an unwanted or even malicious social behavior or refuse to go to school or if there are severe conflicts between family members (such as between parents and grand-parents) or problems related to drug (mis-)use etc. In the course of this, different participants contributed their expertise and their individual expectations of IT-support. To document the results and suggestions developed in the workshop, a mind map was used, from which requirements were derived. The requirements deriving from this form are characterized by a high degree of creativity and innovation [30]. At the same time, this procedure ensures that the majority of relevant stakeholders in IT-support are involved in data collection. As added value, comprehensible requirements have to be derived for all interest groups [30]. The workshops and their results are described in more detail in Sect. 7.4.1.

We also conducted expert interviews as a further survey technique. In their implementation, the moderator used an interview guide to support a structured and at the same time flexible conversation process and to offer comparability within the conducted interviews. The interview guideline was inspired from the structure of the process optimization guidelines according to Best and Weth [2]. Experts were explicitly selected as interviewees because of their privileged access to relevant information, which is in line with [29]. The experts we interviewed are people working in the personal service domain for a single company engaged in services in the context of social services (the main service is family assistance). Due to their professional profile, they represent prospective users of IT-support. The further process of analysis was based on the recorded and transcribed statements of the interviewees following [8]. For the analysis of the scripted interviews we followed the guidelines for qualitative content analysis of Mayring [22, 23] and finally derived requirements. The result of the interview-based requirement derivation is described in more detail in Sect. 7.4.2.

We finally complemented two survey techniques by the use of shadowing. Shadowing is characterized by observing experts in their professional life. It provides insight into individual procedures and processes within the organization. The information obtained were documented in observation protocols. From these documents, relevant requirements for possible IT-support were analyzed. Parallel to the observation, this procedure offered freedom for the observed expert to provide further insight into his work independent of the current activity. This included, for example, the expression of ideas for new approaches or optimizing processes [30]. Shadowing and the results we gained using this technique are described in Sect. 7.4.3.

In the remainder of this paper, we present the derivation process and the results of each applied research method. We furthermore categorize the requirements and finally develop one comprehensive, consolidated catalogue of requirements (cf. Sect. 7.5) following the method of Nickerson, et al. [27].

7.4 Derivation of Requirements

In this section, requirement derivation is presented for each of the different methods introduced in the previous section. Requirements are elicited and then aggregated in a table at the end of each subsection.

7.4.1 Requirements from the Workshops

The workshops were carried out as part of a collaboration between the University of Rostock and various interested social enterprises from the Rostock area. For a time period of two years, they were organized and hosted by the University of Rostock. The workshops took place on an irregular basis every 2–3 months throughout this period. The overarching workshop topic was the IT-support of personal services. The number of participants varied, from smaller rounds of about 10 people (minimum) to larger forums with up to 30 participants. Among the representatives of the companies were both managers and employees of the organizations. Typically, representatives of 1–3 distinct companies participated in the workshops. While the experts of social work contributed their knowledge and experiences from their professional life, the experts of the University of Rostock contributed their knowledge concerning the possibilities of IT. The overall goal of the workshops was to identify how improved IT-support could be possible for practitioners working in the personal service industry. In order to document the results obtained, a mind map was developed and continuously improved

Table 7.1 Requirements from the workshop (WRE)

No.	Requirement	Explanation
WRE1	Flexible extensibility	The extensibility of the system includes the ability to add additional or complementary features without much technical effort, like plug-and-play
WRE2	Situational support of task execution	Depending on the particular situation and the associated individual circumstances, the sequence of processes can be flexibly adapted to the situation
WRE3	Client-specific task control	The system considers the client (i.e. the person that receives a personal service) and his or her individual influence on the process flows
WRE4	User-specific scope of functions	For the different users of the system, also different roles can be assigned that may be associated with different rights, e.g. access information and system functions
WRE5	User-specific task control	The user has the opportunity to adapt the process design within the system to his or her individual approach. The process is not fixed, it can be adapted by the user
WRE6	Recommendation function	Within the individual case processes, the system provides the user with recommendations for the further process steps. The recommended actions of the system are based on past, already completed cases
WRE7	Sensor function	Sensor technology such as GPS makes it possible to record additional information. This can in turn be used to provide additional functions, e.g. the provision of context-relevant information based on location

throughout the various meetings. As usual in mind mapping, the documented findings and considerations were reproduced in a networked structure with many small branches, which continuously increase the level of detail [30]. The mind map was then used to derive the requirements, whereby the top-level branches of the mind map were translated to high-level requirements of IT-support. The derived requirements can be found in Table 7.1.

7.4.2 *Requirements from the Interviews*

The second survey method represents the expert interviews. The basic aim of the interviews was to develop an understanding of the work performed by the interview partners. For the selection of the questions it was crucial that a good insight into the variety of the activities was made possible while also acquiring an overview of all activities. Thus, the interview guidance aimed at a balance between the required depth of detail for understanding process instances (e.g. help and coaching for a concrete family) and their respective specifics, and the necessary overview in order to be able to derive general requirements, detached from the specific consideration. Guided expert interviews were conducted for practical requirement elicitation. In these interviews, the experts described their view on the processes of personal social services, in line with Gläser and Laudel [12].

According to Liebhold and Trinczek [20], expert interviews are described with the expression of "closed openness", which illustrates the structured and simultaneously open style of this methodology. While the openness allows the interviewee to be able to tell freely and convey the desired information, the structure ensures that the objective of the interview does not go unnoticed and the interviews remain comparable.

For the selection of experts, their activities in personal services were the central criterion. Their specialist knowledge and their practical experience from everyday working life, including their knowledge of internal processes, should ensure meaningful answers [1].

All in all, 6 in-depth interviews have been conducted with experts with a broad spectrum of professional experience and know-how. To derive the requirements from the interviews, these were first transcribed. This was followed by a qualitative content analysis of the transcripts according to Mayring [23]. Table 7.2 illustrates the derived requirements sorted in categories.

7.4.3 Requirements from Shadowing

Shadowing or observation is the third method that has been applied to gather requirements. More precisely, field observation was used within this work. This is essentially based on the notion that the observer documents the activities and work processes of the observed [30]. The observer can ask questions in order to better understand unclear or unclassifiable observations. This complementary elicitation method has been used to develop a more in-depth understanding of the work done. The persons observed are essentially those who have already been interviewed, supplemented by the group of people present in the particular situation and interacting with the observed person. In sum, 6 shadowing sessions with a total of 13.5 h of observation have been made. During or after the observation, handwritten protocols were prepared on the basis of which the later requirements were derived. The aspects considered essential to the observer were simultaneously noted in the respective situations or completed as a memory protocol in a timely manner at the end. The requirements derived from the documentation are shown in Table 7.3.

Table 7.2 Requirements from the expert interviews (IRE)

No.	Requirement	Explanation
IRE1	**Information capture**	**Collection of information within the system and documentation**
IRE1.1	Consideration of various sources of information	The system enables the user to capture different types of documents, such as handwritten records and digital files
IRE1.2	Simplified documentation of processes	The system enables the user to easily transfer the required information to the system
IRE1.3	Integration of electronic tools for information acquisition	The system enables the capturing of information via electronic tools and link the information to the specific case
IRE2	**Information management**	**Storage of all case-related information**
IRE2.1	Storage of case-relevant information	The system must ensure that central access to the complete and structured file is ensured. The storage of the information takes place in a central location
IRE3	**Information use**	**Provision of information from the documentation**
IRE3.1	Provision of information	The system proactively provides the information previously recorded in the system in a clear manner. The transmission takes place, for example, as an e-mail for self-study before an appointment with the client
IRE3.2	Visualization of information	The system allows the retrieval of information, e.g. pre-defined goals of a coaching service, past events leading to the appointment or additional information on screens or mobile devices
IRE3.3	Central access to information	The system enables mobile access to the collected information
IRE3.4	Specify different access rights	The system offers the possibility to provide case-relevant persons with different access and editing rights
IRE4	**Information exchange**	**Exchange of information between case-relevant persons**
IRE4.1	Definition of persons who are responsible for the case	The system should allow to specify case-relevant persons in distinct roles such as case leader, case supporter and case supervisor
IRE4.2	Target-group-oriented information exchange	The system allows the information to be flagged in terms of its relevance to stakeholders and informs them automatically

(continued)

Table 7.2 (continued)

No.	Requirement	Explanation
IRE4.3	Coordination of acting persons	The system enables the organization of case-related activities with the help of system-internal communication tools
IRE5	**Help process**	**All requirements that directly and indirectly are related to the help process**
IRE5.1	Presentation of free capacities	The system needs to provide an overview of the free capacity of employees
IRE5.2	Flexible case handling	The system allows for a flexible sequence of processes in execution. The individual process steps can be freely combined and individually adapted to the specific characteristics of the case and its actors
IRE5.3	Capture of feedback	The system allows to capture feedback during documentation. Feedback includes the justified assessment of the social worker in respect to the goals achieved
IRE5.4	Appointment control	The system provides a feature for making appointments based on three dimensions (time, priority, location). Based on these parameters, a system-generated appointment proposal is made
IRE5.5	Time orientation in the help process	The system must be able to convey to the user the status of the help process in regard to time. This information should serve as an orientation for planning future activities
IRE5.6	Task control	The system allows to assign tasks to different people involved in the case
IRE5.7	Task status	The system must be able to provide an overview on the tasks currently being processed with their respective processing status. The user should be able to deduce what intervention is necessary

7.5 Requirements Consolidation and Generalization

To develop a holistic system model, the various requirements of the three survey methods were finally brought together. For this purpose, a catalogue of requirements was developed which combines and structures the results of the various survey techniques. This is described in Sect. 7.5.1. Based on this consolidated requirements catalogue, requirements are further analyzed regarding their specificity to social care as a sub-domain of personal services and the possibility of mapping them to general system approaches in the domain of knowledge-intensive business processes. Additionally, a general classification of the requirements is provided. These aspects are

Table 7.3 Requirements from shadowing (SRE)

No.	Requirement	Explanation
SRE1	**Coordination of different actors**	**Coordination of various actors within the help process**
SRE1.1	Coordination of actions	The system should enable users to reconcile their actions by providing communication tools (e.g., e-mail)
SRE1.2	Task assignment of the actors in the help process	The system should enable the integration of various participants in the process by task assignment
SRE2	**Exchange of information and documents among the actors**	**Empowering stakeholders to share information and documentation**
SRE2.1	Distribute cases by capacity	The system should allow to distribute new case requests only to employees who have free capacity
SRE2.2	Distribution of information by target group	The system should make it possible to pass on information only to those involved persons for whom this information is relevant
SRE3	**Management of information and documents of clients**	**System-supported management of collected information and documents about the client**
SRE3.1	Provide a centralized file for each client	The system should store and provide the information based on a central storage unit (file server, cloud storage etc.)
SRE3.2	Access from anywhere to the file	The system should provide access to information from anywhere, including mobile access. It thus should ensure a location-independent retrieval of all information and documents
SRE3.3	Assign different roles	The system should enable the user and the various case participants to take on different roles (case leader, case handler, case monitor)
SRE3.4	Assign access rights	The system should make it possible to assign different access and function rights with different roles
SRE4	**Create invoice**	**Preparation of the invoice to document the service provided to the client**
SRE4.1	Independence of internet connection	The system should allow the user to generate documentation independent of the Internet and provide evidence of provision, e.g. in the form of working hours offline
SRE4.2	Automatic synchronization with appointment diary	The system should automatically retrieve and synchronize billing information with the appointments entered in the calendar
SRE5	**Mapping of family systems**	**Illustrate inner-familial structures**
SRE5.1	Mapping the effects within a family system	The system should enable the representation of family systems and their interaction in the form of a genogram

(continued)

Table 7.3 (continued)

No.	Requirement	Explanation
SRE5.2	Use of different levels with varying levels of detail	The system must be able to distinguish different visualization levels in the representation of the family system. These include an overview layer for illustrating relationships and structures as well as a level of detail for the visualization of specific information about the persons within the family system

discussed in Sect. 7.5.2 and are the base for generalization and comparison of our findings.

7.5.1 *Consolidated Requirements Catalogue*

In order to classify the requirements, we use an extended version of the service phases proposed by Fließ et al. [11] as described in Sect. 7.2.1. According to this, our category system for the requirements contains the labels *client intake, pre-service, service, post-service*, and *coordination.* In addition to these phase-specific requirements for IT-support in personal social services, there are also general requirements for corresponding IT-systems resulting from the specific characteristics of the service processes under consideration. These can be subsumed under *general requirements.* They are mainly the result of the evaluation of the workshops. Table 7.4 shows the requirement catalogue where the origins of the consolidated requirements are listed in the rightmost column. The above introduced categories serve to structure the catalogue.

7.5.2 *Generalized Requirements*

So far, no distinction between different types of requirements has been made throughout the discussion. However, requirements are handled differently in systems development depending on their type. There are *Functional Requirements* that describe the behavioral aspects of a system [13]. They can be described in terms of possible inputs and desired outputs. Thus, a mapping to activities in service processes is possible because activities can be described in the same way. There are also requirements regarding *Data, Performance, Constraints*, and *Software Quality* in general [13]. Being diverse and sometimes not clearly distinguishable, these requirement types are commonly subsumed under the term *Non-functional Requirements.*

Table 7.4 Aggregated requirements

No.	Requirement	Source item
Client intake (CI) *Case is created in the IT-system and relevant information recorded*		
RE1.1	Consideration of various sources of information	IRE1.1
RE1.2	Presentation of free capacities	IRE5.1; SRE2.1
RE1.3	Storage of case-relevant information	IRE2.1
RE1.4	Defining persons who are responsible for the case	IRE4.1; SRE3.4
RE1.5	Specify different access rights	IRE3.4; SRE3.4
RE1.6	Modelling family systems	SRE5.1; SRE5.2
Pre-service (Pre) *Immediate preparation of an appointment with the client*		
RE2.1	Provision of information	IRE3.1
RE2.2	Central access to information	IRE3.3; SRE3.1; SRE3.3
RE2.3	Information exchange	IRE4.2; SRE2
RE2.4	Coordination of acting persons	IRE4.3; SRE1.1
RE2.5	Visualization of family systems	SRE5
RE2.6	Integration of electronic tools for information acquisition	IRE1.3
Service (S) *Direct interaction with the client*		
RE3.1	Visualization of information	IRE3.2
RE3.2	Central access to information	IRE3.3; SRE3.3
Post-service (Post) *Activities directly after the appointment*		
RE4.1	Simplified documentation of processes	IRE1.2
RE4.2	Integration of electronic tools for information acquisition	IRE1.3
RE4.3	Create invoice	SRE4
RE4.4	Target group-oriented information exchange	IRE4.2, SRE2.2
RE4.5	Capture of feedback	IRE5.3
RE4.6	Appointment control	IRE5.4
Coordination (Co) *Long-term process of service delivery*		
RE5.1	Coordination of acting persons	IRE4.3; SRE1.2
RE5.2	Target group-oriented information exchange	IRE4.2; SRE1.1
RE5.3	Time awareness in the care process	IRE5.5
RE5.4	Task control	IRE5.6; S3 RE1.2
RE5.5	Task status	IRE5.7
General requirements (Ge) *Implications from the general characteristics of personal service processes*		
RE6.1	Flexible expandability	WRE1

(continued)

Table 7.4 (continued)

No.	Requirement	Source item
RE6.2	Situational support of task execution	WRE2
RE6.3	Client-specific task control	WRE3; IRE5.2
RE6.4	Role-specific task control	WRE4; IRE5.2
RE6.5	User-specific task control	WRE5; IRE5.2
RE6.6	Recommender functionality	WRE6
RE6.7	Sensor functionality	WRE7

We start with the discussion of found *Functional Requirements* and their relation to functionalities of general system types in the domain of knowledge-intensive processes. Besides them there are also functionalities specific to family care due to the selection of the study participants (cf. Sect. 7.3). The latter are not applicable for personal services in general.

If we look into knowledge management system approaches, there is a broad consensus on the general functionalities that these systems should provide. Taking the architecture by Maier and Hadrich [22], knowledge management system service types and thus functionality areas on the application layer are defined. Table 7.5 is a contingency table that maps these functionality areas to the phases of service provision based on the aggregated requirements developed in Sect. 7.5.1. The first column contains the functionality areas whereas the following columns count the number of requirements that can be matched to the functionality areas.

The table shows that there are no requirements formulated regarding *Personalization* and *E-learning*. In the case of *Personalization*, a possible explanation is the little experience of the study participants with enterprise information systems. So far they only used an information system for the documentation and billing of performed work that shows a good Personalization. In the case of *E-learning* the explanation lies in the focus on the performance of service processes. Learning tasks

Table 7.5 Distribution of knowledge management related requirements

Functional Area	*CI*	*Pre*	*S*	*Post*	*Co*	*Ge*
Access	1	1	1	0	0	0
Personalization	0	0	0	0	0	0
Knowledge Discovery	1	1	1	0	0	0
Knowledge Publication	1	1	0	2	1	0
Collaboration	0	1	0	1	1	0
E-Learning	0	0	0	0	0	0
Knowledge Management in General (sum of above)	**3**	**4**	**2**	**3**	**1**	**0**

are not directly connected to these processes. Furthermore, it can be seen, that getting *Access* to knowledge and the *Discovery* of appropriate knowledge play a role in the first service phases, including the actual service provision. Knowledge *Publication*, meaning the capture of new knowledge, needs functional support in the phases *Client Intake*, *Pre-service*, and *Post-service*. The reasons for not including such functionalities in the actual service provision are the fear of disturbing the service situation by interruptions and of distraction through knowledge capturing activities and the wish for a personal reflection on the service situation in the *Post-service* phase as a means for a better quality of the captured knowledge. This might seem to be specific to family care scenarios, but disturbance by IT use can also be an issue in other settings. At last, *Collaboration* is needed in *Pre-service*, *Post-service*, and *Coordination* because multiple employees are involved in the service provision at the same time or at different service encounters. The involvement of multiple actors is a common characteristic of knowledge intensive work processes and not necessarily specific to social care.

Overall, knowledge management systems seem to address the requirements of personal services. Application potential lies in the *Discovery* of and *Access* to relevant knowledge in the early service phases including the actual service provision as well as in the *Capture* of knowledge before and after service provision. *Collaboration* functionalities are important if multiple actors are involved in the service provision.

However, knowledge management systems do not cover all functional requirements formulated. Besides family care specific functionalities there are also functionalities needed for process/workflow management. The coordination and control of task provision is an important issue considering knowledge intensive processes [19, 26]. There are classical workflow management functionalities, namely definition, execution and control of workflows. Furthermore, approaches to workflow management have been developed that address the special requirements of knowledge-intensive work processes. *ACM* is the most prominent of these approaches (cf. Sect. 7.2.1). Table 7.6 is a contingency table that maps the found functional requirements to general workflow management functionality and to *ACM* which adds run-time adaptability as a main functional area.

Since *ACM* subsumes classical workflow management functionalities, the table row containing the *ACM* mapping can serve for an overall evaluation. Workflow management functionality is required in all phases except for the actual service provision. Naturally, a strong correlation lies in the coordination of work in and between process instances. An important finding is the high relevance of *ACM* functionality in the general system requirements. Throughout the discussions with the

Table 7.6 Distribution of workflow management related requirements

	CI	*Pre*	*S*	*Post*	*Co*	*Ge*
Classical WfM Functionality	2	1	0	1	4	0
ACM Functionality	**2**	**1**	**0**	**2**	**4**	**6**

Table 7.7 Distribution of functional requirements

	CI	*Pre*	*S*	*Post*	*Co*	*Ge*
Knowledge Management Funct.	3	4	2	3	1	0
ACM Functionality	2	1	0	2	4	6
Family Care Specific Functionality	1	1	0	1	0	0
Overall IT Functionality (sum)	**6**	**6**	**2**	**6**	**5**	**6**

Table 7.8 Distribution of non-functional requirements

	CI	*Pre*	*S*	*Post*	*Co*	*Ge*
Data/Information Model	4	2	0	3	1	0
System Quality	1	1	0	0	0	2

study participants, the need for a flexible task coordination mechanism has been expressed.

Overall, most of the functional requirements are covered by knowledge management systems and *ACM* capable workflow management systems. The rest of the requirements are family care specific, namely the documentation and visualization of family systems (RE1.6, RE2.5) as well as the creation of invoices according to specific requirements of social offices (RE4.3). Concluding the discussion with regard to functional requirements, it can be stated that the required IT functionality in the domain of personal services can be provided by knowledge management systems and workflow management systems like *ACM* capable systems. Special functionalities that are required for the processing of family care specific data have to be considered separately because they are not applicable to personal services in general. However, in our study they made up only a small part of the overall functional requirements.

Table 7.7 shows an overview of the findings with regard to functional requirements. Potential for IT-support is seen mainly in the phases before and after service provision. Working on a certain process instance or client respectively generally requires knowledge management functionalities. The focus of an IT solution that supports personal services on company level should be on workflow management that implements *ACM* functionalities. An integration of both would be achievable with solutions that incorporate the data centricity principle of *ACM*, combining process instance knowledge (case data) with automated process execution support. However, finding method support for this approach is challenging [19].

Considering *Non-functional requirements* according to Glinz [13], there are two types of requirements present in the collected data—*System Quality* requirements and requirements regarding the *Data* that need to be handled by the IT system. The latter are closely related to functional requirements since functions can be described by a specification of input and output data. Table 7.8 shows the distribution of the found non-functional requirements along the phases of service provision.

Looking at the distribution of Information Model requirements, it is straightforward that the definition of information objects is related to knowledge intensive tasks and thus the phases of the service provision that require support by Knowledge Management Systems (see before: *CI*, *Pre*, *Post*). *System Quality* requirements can be tracked down to Interoperability (RE1.1, RE2.6) which is important in the *Client Intake* and *Pre-service* phases in order to allow the inclusion of different information sources for Knowledge *Discovery*. Furthermore, "Flexible Expandability" as a general system requirement can be decomposed to Adaptability and Interoperability. This is a very broad and coarse grained definition of requirements. However, in a qualitative view it can be concluded, that Interoperability is seen as an important feature of systems for the support of personal services. The reason lies in the great number of sources for personal data that are available. Adaptability is a central feature required for performing knowledge intensive work processes and thus also important for personal services.

7.6 Conclusion and Outlook

Digitalization is among the most discussed topics nowadays. However, it remains unclear how improved IT-support should be designed in domains with less structured processes such as in personal services. So far there is little IT-support implemented in the domain of personal services. In order to achieve the goal of a better IT integration in the course of the digitalization, we cannot rely on past experiences of domain specific IT-support. However, with our work we want to first collect requirements in a systematic way in order to provide guidelines for the design of appropriate IT systems. Regarding the identified requirements, a first observation is that only two *Functional Requirements* directly relate to the service provision, while all the other requirements are relevant for the other phases of the service processes. From this, it can be concluded that increased IT-support of services is not relevant for and should not interfere with the actual service work since it might disrupt the intimate relationship between the service worker and the client (cf. Sect. 5.2). Second, it can be concluded that increased IT-support is nonetheless highly relevant for *Client Intake* (6 *Functional Requirements*), *Pre Service* (6 *Functional requirements*), *Post Service* (6 *Functional Requirements*) and *Coordination* (5 *Functional Requirements*).

Regarding the process of requirement elicitation, it can be noticed that the different applied methods do not contribute to an equal amount to the identification of requirements along the service process. For example, the majority of requirements in the category *Post Service* can be traced back to interviews (5 *Functional Requirements*) and only a few requirements originate from shadowing (2 *Functional Requirements*) with an overlap of a single requirement. Similarly, requirements in the category *General Requirements* have been identified most often by workshops (6 *Functional Requirements*) and less often with interviews (3 *Functional Requirements*) with an overlap of three requirements. It thus can be learned that applying three different analysis methods is fruitful for requirement elicitation, since it gives a more holistic set of requirements.

Overall, 31 *Functional Requirements* and 14 *Non-functional Requirements* have been elicited. From further analyzing these requirements, two insights can be made. First, the large majority of *Functional Requirements* (15 out of 31) belong to the *Adaptive Case Management*-subcategory. The second largest group of *Functional Requirements* (13) can be attributed to *Knowledge Management Systems.*

It therefore seems adequate that increased IT-support should be based on systems developed in the *Adaptive Case Management* (*ACM*) and *Knowledge Management* context. However, finding sophisticated *ACM* solutions and the integration of *Knowledge Management* and *ACM* are still challenges [19].

Only 3 features are assigned to the *Domain Functionality*-subcategory. This supports the possibility of generalizing the results for the digitalization of personal services from the family care scenario.

Looking at the *Non-functional Requirements*, a main conclusion is that a focus should be on knowledge modeling because knowledge or *Data* in general resembles important in- and outputs of personal service processes. Furthermore, Interoperability and adaptability will play a major role with regard to *System Quality*.

A major limitation of our work is the small amount of workshops, interviews and shadowing sessions conducted with in sum approx. 30 experts of three companies in the personal social services domain in Rostock. Also, there was an overlap between participants in the three studies (workshop, interview, shadowing). Our results should therefore be interpreted with caution. However, using a triangulation-based research model partly serves as a remedy since combining signals received using different methods makes the conclusions drawn more robust.

Another limitation of our study is the focus on the service provider. Service consumers as important actors within the service process should also be considered in future investigations. Especially the fact that little potential for IT-support in the actual service provision is seen by the experts should be reflected from consumer side. There is a demand for anonymized IT-based personal services like coaching. This aspect has not been covered so far in our investigations. Still, some experts indicated that personal encounters are essential for the achievement of a good service outcome.

Our next research step is to apply and validate the requirements catalogue by designing an improved IT-support tailored to practitioner's needs. For this purpose, we already have defined coarse-grained use cases each of them spanning multiple requirements. We also have designed a prototypical user interface of a system that implements the requirements. In our future research, we will describe the method of design and evaluate the proposed system based on feedback acquired by interviews. This feedback again will serve to evaluate our requirements catalogue as well as to conduct an ex-ante evaluation of the improved IT-support we envision.

Acknowledgements This contribution was possible thanks to the domain expertise of our partner companies engaged in family care—GeBEG mbH, Wohnen ohne Barrrieren GmbH, and Barrierefreies Rostock gGmbH from Rostock, Germany.

References

1. Bauer, R.: Personenbezogene Soziale Dienstleistungen: Begriff, Qualität und Zukunft. VS Verlag für Sozialwissenschaften, Wiesbaden (2001)
2. Best, E., Weth, M.: Geschäftsprozesse optimieren: Der Praxisleitfaden für erfolgreiche Reorganisation. Springer Fachmedien Verlag, Wiesbaden (2009)
3. Bieber, D., Geiger, M.: Personenbezogene Dienstleistungen im Kontext komplexer Dienstleistungssystemen- eine erste Annäherung. In: Bieber, D., Geiger, M. (eds.) Personenbezogene Dienstleistungen im Kontext komplexer Wertschöpfung: Anwendungsfeld "Seltene Krankheiten", pp. 9–50. Springer, Wiesbaden (2014)
4. Bogner, A., Littig, B., Menz, W.: Interviews mit Experten: Eine praxisorientierte Einführung. Springer VS Verlag, Wiesbaden (2014)
5. Bundesministerium für Wirtschaft und Energie: Den digitalen Wandel gestalten. Retrieved 03 Apr 2019 from https://www.bmwi.de/Redaktion/DE/Dossier/digitalisierung.html
6. Cardona, M., Kretschmer, T., Strobel, T.: ICT and productivity conclusions from the empirical literature. Inf. Econ. Policy **3**, 109–125 (2013)
7. Däuble, G., Özcan, D., Niemöller, C., Fellmann, M., Nüttgens, M., Thomas, O.: Information needs of the mobile technical customer service—a case study in the field of machinery and plant engineering. In: Proceedings of the 48th Annual Hawaii International Conference on System Sciences (HICSS), pp. 1018–1027. Manoa (2015)
8. Dresing, T., Pehl, T.: Praxisbuch Interview, Transkription & Analyse: Anleitungen und Regelsysteme für qualitativ Forschende. Eigenverlag, Marburg (2013)
9. Eurostat: Labour input indices overview—statistics explained, data sets: labour input in industry—quarterly data & labour input in services—quarterly data. Retrieved 23 July 2019 from https://ec.europa.eu/eurostat/statistics-explained/index.php/Labour_input_indices_overview and http://appsso.eurostat.ec.europa.eu/nui/show.do?dataset=sts_selb_q (2018)
10. Flick, U.: Triangulation: Eine Einführung. Qualitative Sozialforschung. VS Verlag für Sozialwissenschaften. Springer Fachmedien GmbH, Wiesbaden (2011)
11. Fließ, S., Dyck, S., Schmelter, M., Volkers, M.J.: Kundenaktivitäten in Dienstleistungsprozessen—die Sicht der Konsumenten. In: Fließ, S., Haase, M., Jacob, F., Ehret, M. (eds.) Kundenintegration und Leistungslehre, pp. 181–204. Springer Fachmedien Verlag, Wiesbaden (2015)
12. Gläser, J., Laudel, G.: Experteninterviews und Qualitative Inhaltsanalyse als Instrumentere konstruierender Untersuchungen. VS Verlag für Sozialwissenschaften, Wiesbaden (2010)
13. Glinz, M.: On non-functional requirements. In: 15th IEEE International Requirements Engineering Conference (RE 2007), pp. 21–26. IEEE (2007)
14. Halmos, P.: The personal service society. Br. J. Soc. **18**, 13–28 (1967). https://doi.org/10.2307/588586
15. Hauder, M., Münch, D., Michel, F., Utz, A., Matthes, F.: Examining adaptive case management to support processes for enterprise architecture management. In: 2014 IEEE 18th International Enterprise Distributed Object Computing Conference Workshops and Demonstrations, pp. 23–32. IEEE (2014)
16. Hauder, M., Pigat, S., Matthes, F.: Research challenges in adaptive case management: a literature review. In: 2014 IEEE 18th International Enterprise Distributed Object Computing Conference Workshops and Demonstrations, pp. 98–107. IEEE (2014)
17. Kurz, M., Herrmann, C.: Adaptive case management—Anwendung des business process management 2.0-Konzepts auf schwach strukturierte Geschäftsprozesse. In: Sinz, E.J., Bartmann, D., Bodendorf, F., Ferstl, O.K. (eds.) Dienstorientierte IT-Systeme für hochflexible Geschäftsprozesse, Report for FLEX-2011-008, pp. 241–265. University of Bamberg Press, Bamberg (2011)
18. Kurz, M., Schmidt, W., Fleischmann, A., Lederer, M.: Leveraging CMMN for ACM: examining the applicability of a new OMG standard for adaptive case management. In: Proceedings of the 7th International Conference on Subject-Oriented Business Process Management, p. 4. ACM (2015)

19. Lantow, B.: Adaptive case management—a review of method support. In: Buchmann, R., Karagiannis, D., Kirikova, M. (eds.) The Practice of Enterprise Modeling. PoEM 2018. Lecture Notes in Business Information Processing, vol 335. Springer, Cham (2018)
20. Liebhold, R., Trinczek, R.: Experteninterview. In: Kühl, S., Strodtholz, P., Taffertshofer, A. (eds.) Handbuch Methoden der Organisationsforschung: Quantitative und Qualitative Methoden, pp. 32–56. VS Verlag für Sozialwissenschaften, Wiesbaden (2009)
21. Lovelock, C.H.: Classifying services to gain strategic marketing insights. J. Market. **47**(3), 9–20 (1983)
22. Maier, R., Hadrich, T.: Knowledge management systems. In: Encyclopedia of Knowledge Management, 2nd edn., pp. 779–790. IGI Global (2011)
23. Mayring, P.: Qualitative Inhaltsanalyse: Grundlagen und Techniken. Beltz Verlag, Weinheim (2015)
24. Michel, F., Matthes, F.: A holistic model-based adaptive case management approach for healthcare. In: 2018 IEEE 22nd International Enterprise Distributed Object Computing Workshop (EDOCW), pp. 17–26. IEEE (2018)
25. Motahari-Nezhad, H.R., Swenson, K.D.: Adaptive case management: overview and research challenges. In: Conference on Business Informatics, pp 264–269. Wien (2013)
26. Mundbrod, N., Reichert, M.: Process-aware task management support for knowledge-intensive business processes: findings, challenges, requirements. In: 2014 IEEE 18th International Enterprise Distributed Object Computing Conference Workshops and Demonstrations, pp. 116–125. IEEE (2014)
27. Nickerson, R.C., Varshney, U., Muntermann, J.: A method for taxonomy development and its application in information systems. Eur. J. Inf. Syst. **22**, 336–359 (2013)
28. Oelerich, G., Otto, H.-U.: Empirische Forschung und Soziale Arbeit: Ein Studienbuch. VS Verlag für Sozialwissenschaften/Springer Fachmedien, Wiesbaden (2011)
29. Rupp, C.: Requirements-Engineering und -Management: Professionelle, iterative Anforderungsanalyse für die Praxis. Carl Hanser Verlag, München (2007)
30. Rupp, C.: Requirements-Engineering und -Management: Aus der Praxis von klassisch bis agil. Carl Hanser Verlag, München (2014)
31. Schröder, C., Schlepphorst, S., Kay, R.: Bedeutung der Digitalisierung im Mittelstand. Institut für Mittelstandsforschung, Bonn (2015)
32. Zeithaml, V.A., Parasuraman, A., Berry, L.L.: Problems and strategies in services marketing. J. Market. **49**(2), 33–46 (1985)

Chapter 8
Service-Dominant Design: A Digital Ideation Method

Jonas Bürkel and Alfred Zimmermann

Abstract The digital transformation is today's dominant business transformation having a strong influence on how digital services and products are designed in a service-dominant way. A popular underlying theory of value creation and economic exchange that is known as the service-dominant (S-D) logic can be connected to many successful digital business models. However, S-D logic by itself is abstract. Companies cannot directly use it as an instrument for business model innovation and design in an easy way. To address this a comprehensive ideation method based on S-D logic is proposed, called *service-dominant design* (*SDD*). SDD is aimed at supporting firms in the transition to a service- and value-oriented perspective. The method provides a simplified way to structure the ideation process based on four model components. Each component consists of practical implications, auxiliary questions and visualization techniques that were derived from a literature review, a use case evaluation of digital mobility and a focus group discussion. SDD represents a first step of having a toolset that can support established companies in the process of service- and value-orientation as part of their digital transformation efforts.

Keywords Service-dominant logic · Ideation method · Value-orientation · Value networks · Digital transformation

8.1 Introduction

Since Adam Smith's analysis of the wealth of nations during the Industrial Revolution [34], economic exchange was predominantly concerned with tangible manufactured goods, efficient utilization of labor and maximizing productivity and output. Creating

J. Bürkel (✉) · A. Zimmermann
Informatics, Reutlingen University, Herman Hollerith Center, Reutlingen, Germany
e-mail: buerkelj@gmail.com

A. Zimmermann
e-mail: alfred.zimmermann@reutlingen-university.de

A. Zimmermann et al. (eds.), *Architecting the Digital Transformation*, Intelligent Systems Reference Library 188,
https://doi.org/10.1007/978-3-030-49640-1_8

value was regarded as an isolated process, without the necessity of consumer involvement as the manufacturer occupied the dominant position of economic exchange [13].

When looking upon today's global economy this perspective has changed. Especially since the second half of the twentieth century, with the increasing advancement of technological capabilities, *value* is instead redefined as being experientially perceived and cocreated [29]. Tangible products are enriched with intangible services as companies are shifting their focus towards being *service providers* [8]. Information technology (IT) enables new business models where consumers are actively involved in their own value creation and where economic exchange is a productive dialogue whose dominant position is mostly on the side of the consumer—the *beneficiary* of a respective service [18]. This change of perspective resulted in established businesses being disrupted by innovative digital competitors. Companies are following the need to act, often by striving to transform themselves into a digital enterprise. However, a successful *digital transformation* is no easy task.

This chapter suggests an approach of identifying and structuring the scope of a digital transformation in a simplified and practical way. The approach, called *service-dominant design* (*SDD*), is heavily based on a theoretical framework that was originally presented as the "service-centered dominant logic of marketing" [38]. Becoming known as the *service-dominant* (*S-D*) *logic* it gained a lot of momentum in the academic literature [19] and was continuously updated over the course of the last 15 years [39, 40]. S-D logic discusses a new economic and societal mindset based on concepts such as value cocreation in networks of actors, the contextual and experiential nature of value and its facilitation leveraging routinized mechanisms called institutions which in turn form larger *service ecosystems* [40].

To connect those abstract building blocks with actionable items that can be integrated into a company's digital strategy and business models a corresponding toolset is needed that provides insights especially regarding the following questions:

- How can potential beneficiaries of novel digital services be identified and categorized in a structured manner?
- How can the digitization of a company result in value added for itself and its specific businesses and stakeholders?
- How can a digital offering be designed such that consumers perceive it as valuable and innovative?

SDD is oriented at being a comprehensible, visual ideation method that supports the business model innovation and design processes. It includes preceding research on S-D logic as a theoretical foundation for an S-D transformation model [7]. A literature review, a use case evaluation of digital mobility and a focus group discussion were conducted as part of this research. Its derived findings imply a high potential of leveraging the practical implications of S-D logic for ideation purposes as well. Therefore, the goal of this chapter is to do just that, to put these findings in the context of business model ideation.

Section 2 discusses the S-D logic theory. Section 3 presents a way of structuring the ideation process by integrating the practical implications that were derived from the preceding literature review. Section 4 then includes the additional findings of the S-D transformation model research and presents the concluding tool including its component structure, visualization techniques, practices, and auxiliary questions. Finally, Sects. 5 and 6 discuss the tool's current limitations and potential subjects for future research as well as draw a conclusion of how the presented ideation method can be leveraged to understand and address the key success factors of digitally transforming a business.

8.2 Service-Dominant Logic

S-D logic was originally proposed in 2004 by Stephen Vargo and Robert Lusch in their article "Evolving to a New Dominant Logic for Marketing". It is presented as an emerging dominant logic that focuses on intangible resources, the cocreation of value and close business relationships. Thornton and Ocasio define such a dominant logic or institutional logic "as the socially constructed [...] assumptions, values, beliefs, and rules by which individuals [...] provide meaning to their social reality" [35, p. 804]. *Service* in the context of S-D logic is defined by Vargo and Lusch "as the application of specialized competences (knowledge and skills) through deeds, processes, and performances for the benefit of another entity or the entity itself" [38, p. 2].

This demonstrates the focus of S-D logic on an emergent understanding of economic exchange that favors service based on specialized competences over traditional manufactured goods. Vargo and Lusch mostly discuss service in general while this chapter will adapt their thoughts to digital service offerings. The specialized competencies that constitute service are defined as so-called *operant resources*. They stand in contrast to traditional *operand resources*, which are resources that describe physical materials that are tangible and finite [3]. S-D logic summarizes its statements in eleven foundational premises (FPs) which are illustrated in Table 8.1 [40, p. 8].

In addition to a shift from operand to operant resources, there is also a shift in the perception of the nature of *value*. In the traditional institutional logic of economy that Vargo and Lusch call the goods-dominant (G-D) logic value is seen as something that is embedded into a product during the manufacturing process. This is called *value-in-exchange* and contrasts their concept of *value-in-context*, value that is defined as something that is cocreated together with the consumer and is derived from the benefit that is achieved by the consumption of a specific service in a specific context [36, 38, 40]. The consequence is that only the individually determined benefit of the consumer defines the value added of any product or service. This supports the introductory statement that the dominant position of any economic exchange today is on the side of the consumer. Skålén and Edvardsson, therefore, describe the shift from G-D logic to S-D logic as a company's move from "What can we do for you?" to "What can you do with us?" [31].

Table 8.1 FPs and axioms of S-D logic

Axiom	FP	
$Axiom_1$	FP_1	Service is the fundamental basis of exchange
	FP_2	Indirect exchange masks the fundamental basis of exchange
	FP_3	Goods are distribution mechanisms for service provision
	FP_4	Operant resources are the fundamental source of strategic benefit
	FP_5	All economies are service economies
$Axiom_2$	FP_6	Value is cocreated by multiple actors, always including the beneficiary
	FP_7	Actors cannot deliver value but can participate in the creation and offering of value propositions
	FP_8	A service-centered view is inherently beneficiary oriented and relational
$Axiom_3$	FP_9	All social and economic actors are resource integrators
$Axiom_4$	FP_{10}	Value is always uniquely and phenomenologically determined by the beneficiary
$Axiom_5$	FP_{11}	Value cocreation is coordinated through actor-generated institutions and institutional arrangements

Service is the Fundamental Basis of Exchange The first axiom of S-D logic states that any economic exchange consists of service provision of some sort. The singular term *service* highlights the process orientation of S-D logic, rather than the plural term *services* which would highlight an output orientation concerned with a manufactured number of units [22]. Vargo and Lusch base their first axiom on the work of Frédéric Bastiat who already claimed in the nineteenth century that "services are exchanged for services" [5, p. 162]. Bastiat also discussed concepts such as the distinction between *use-value* and *exchange-value* and stated that every actor's economic activity consists of rendering and receiving services—both points define the essence of S-D logic.

Value is Cocreated by Multiple Actors, Always Including the Beneficiary The second axiom of S-D logic states that the beneficiary of a service "is always a party to its own value creation" [40, p. 9]. This process of *value cocreation* represents a positive statement that is described as "the actions of multiple actors, often unaware of each other, that contribute to each other's wellbeing" [40, p. 8]. A concept of multi-actor value networks is introduced that replaces the traditional producer-consumer interaction. Any exchange influences different categories of stakeholders that are not necessarily customers. In addition to value cocreation, Plé and Chumpitaz Cáceres introduce the term *value codestruction* and define it as "as an interactional process between [actors] that results in a decline in at least one of the [actors]' well-being" [27, p. 431]. Value networks can therefore both positively and negatively influence its participants.

All Social and Economic Actors are Resource Integrators The third axiom of S-D logic states that the fundamental effort of any social or economic activity is the integration and application of operand and operant resources to serve another entity or oneself [21, 36]. This axiom emphasizes several of the non-axiom FPs of

S-D logic. Resource integration as the basis of service provision highlights that all economies are service economies (FP_5). It can also be connected to the primacy of operant resources that is defined in FP_4 and the inclusion of traditional goods as a mechanism of service provision (FP_3). Lusch and Nambisan conclude that resource integration also embodies the basis for service innovation which they define as the "rebundling of diverse resources that create novel resources that are beneficial (i.e. value experiencing) to some actors in a given context" [18, p. 161].

Value is Always Uniquely and Phenomenologically Determined by the Beneficiary The fourth axiom of S-D logic was discussed at the beginning of this section. The value of service provision is always determined by the beneficiary based on its uniquely perceived *value-in-context* [39]. The full value added is then derived from a function of collective wellbeing as value is cocreated in value networks of multiple actors [37]. A related FP is FP_7 that explicitly states that any actor (for example a company) can only propose that a service is of potential value to a beneficiary, but "cannot unilaterally create and/or deliver value" [39, p. 8].

Value Cocreation is Coordinated Through Actor-Generated Institutions and Institutional Arrangements The last axiom is the most abstract one and refers to the latest extension of S-D logic: *service ecosystems*. Lusch and Vargo define a service ecosystem as "a relatively self-contained, self-adjusting system of resource-integrating actors that are connected by shared institutional logics and mutual value creation through service exchange" [20, p. 161]. Within this service ecosystem, an institution is defined as a routinized mechanism to facilitate resource integration and service exchange through the coordination of actors. An institutional arrangement is an assemblage of interdependent institutions [40, p. 7]. All these concepts are summarized in the S-D logic's value cocreation cycle that is shown in Fig. 8.1.

Finally, while S-D logic could appear as an abstract academic theory a closer investigation with respect to its practical implications reveals parallels to today's disruptive digital services and successful business models. These implications will be discussed in more detail after defining a general structure for combining S-D logic with identifying an effective digital strategy in Sect. 3.

8.3 Digital Ideation

The process of digitally transforming a company is complex, which is a challenge when defining a tool that is aimed at supporting it. Because of this, the focus of SDD is not on completeness but on comprehensibility and simplicity. Its goals are to ask the right questions and to emphasize considering service- and value-orientation over plain digitization. SDD provides an ideation framework consisting of four components. Each component is comprised of practical implications, visualization techniques, and auxiliary questions. In combination, they provide a uniform approach for identifying a company's desirable digital services and business models that directly

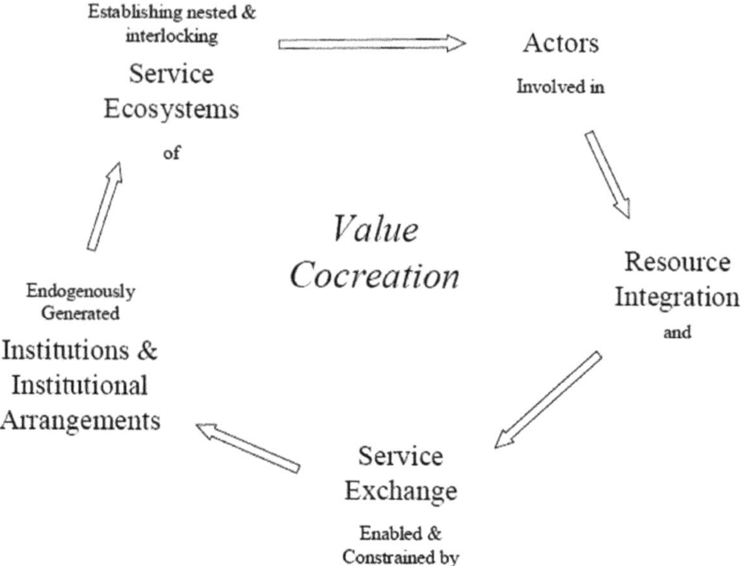

Fig. 8.1 Narrative and process of S-D logic

correspond to cocreating superior stakeholder value. Based on literature definitions of the relevant aspects of *strategy* [2, p. 52, 14, p. 98, 28, p. 285] and the layers of abstraction of S-D logic the four model components are identified hierarchically. They are first discussed in theory and are then connected to their respective practical implications. For each component, up to three overall implications are presented that were derived from a literature review. The literature review analyzed 22 journal articles which discuss S-D logic both in theory and in practice [7].

Corporate Identity The first building block of SDD is a company's corporate identity which is defined as "the mix of elements which gives organizations their distinctiveness" [4, p. 254]. This component describes a company's organizational behavior, its personality and its purpose *why* it is doing business. S-D logic proposes a new mindset of service- and value-orientation that must be integrated into this overall purpose. It must be commonly understood that value is collaboratively cocreated by serving each other. The overall goal of a company must be to center on personalized cocreation experiences that are truly unique to everyone [29]. In total the three overall practical implications of S-D logic with respect to corporate identity are:

- Define and enact a service-centered company philosophy and values
- Ensure a high transformation commitment
- Establish service-centered relationships and practices between employees and managers.

In addition to the company philosophy, relevant values to incorporate are continuous improvement, a ubiquitous customer focus, proactivity and actively assisting the customer in its value creation efforts [11, 31]. Considering environmental and societal impact as well as establishing strong relationships with a firm's extended customer base help improving brand value, collaboration and value cocreation in large networks of actors [22]. They encourage stakeholders to voluntarily become involved in value cocreation activities. Examples include contributions to open source software projects, online reviews and recommendations or encyclopedias. The second implication goes with all large company transformations. For a transformation to be successful a strong overall commitment is required. This includes top-management as much as regular employees.

Finally, there's the roles and the relationship between employees and managers within a firm. FP_4 states that operant resources i.e. competencies and skills are the fundamental source of strategic benefit. The employee is therefore regarded as the single most essential source of innovation, organizational knowledge and value [11, 31]. Managers need to be servant-leaders that focus on coaching and on empowering employees in the delivery of world-class service. The relationship between employees and managers is coined by frequent conversation and dialogue. Additionally, S-D logic encourages practices such as information symmetry, the active rejection of internal competition, active knowledge sharing, 360-degree feedback, and relationships that are built on trust, open communication and solidarity.

Stakeholder Network The second building block of SDD covers a company's stakeholder network, where and how it is positioned within its service ecosystem. Based on the company's identity and its operant resources the relevant stakeholders to interact with can be identified, classified and analyzed with respect to their individual *pains* and *gains*. This component connects a company's purpose with its businesses to pursue and therefore describes *how* it is creating value. While the identification of stakeholders can follow any ideation process it is recommended to follow the *six markets model* for a subsequent classification [9]. This model encourages considering multiple categories of non-customer stakeholders in addition to traditional customers namely (1) referral markets, (2) supplier and alliance markets, (3) influence markets, (4) recruitment markets and (5) internal markets [12]. Finally, it is recommended to visualize the classified stakeholders using the customer profile of the popular *value proposition canvas* [25]. The overall practical implications of the stakeholder network component are:

- Perform a comprehensive stakeholder analysis and categorization
- Take advantage of supply network management
- Leverage technological capabilities that suit the overall company purpose and goals.

As part of the first implication the literature highlights the importance of considering formal organizations such as governments, universities and communities to ensure a diversity of stakeholders to interact with [6]. This is to improve value cocreation capabilities as each stakeholder brings in its individual competencies that can be leveraged by partnering and collaboration. The second implication summarizes recommendations and values that are needed in more complex value networks. Similarly to the corporate values discussed above the literature points out the benefits of close collaboration, mutual access to information as well as knowledge and training sessions [1] that ensure that a network of value cocreating actors can reliably deliver service, provide at least a minimum level of quality and a high value-in-context [17]. Finally, this component is used to identify the overall technological direction a company should take. Based on its stakeholders' life projects and goals that are reflected by their pains and gains instruments represented by groups of technologies can be identified that are of high value for the majority of the stakeholders. A popular instrument that facilitates value cocreation and results in valuable network effects is the *digital platform* where examples include Twitter, Kickstarter or Airbnb.

Value Proposition The third building block of SDD encapsulates a company's value propositions. Based on the identified stakeholder network this component is used to define how to address each stakeholder's individual pains and gains. This is visualized using the value map of the *value proposition canvas* [25]. Separating the pain relievers and gain creators from their corresponding pains and gains into a second component helps focussing on the creation of unprejudiced profiles in the first step. Value propositions can then be formulated more neutrally which increases the degree of *fit* between customer profile and value map and therefore the cocreated value-in-context. The goal of SDD is not the justification of an existing product and service portfolio but the identification of novel digital services that best correspond to the actual needs of the network [3, 32]. Value propositions are also related to the question of *how* a company is delivering value. They represent the basis for the definition of adequate business models that best address the ecosystem's opportunities. The overall practical implications of the value proposition component are:

- Provide a meaningful service-centered mission statement
- Define and adhere to value proposition guidelines.

The preceding focus group discussion highlighted the need to explicitly connect the identified value propositions with a company's corporate identity [7]. Therefore, one major implication is the definition of a stakeholder-independent overall value proposition that represents the company's purpose and goals. It should resemble a mission statement that is considered during the definition of each value proposition. The goal is to ensure an alignment between purpose and value cocreation activities which is needed for differentiation in the marketplace by integrating a company's character and philosophy with its actions. A second implication in the literature is the discussion how a meaningful value proposition is defined. In general, value propositions must be defined from the perspective of the customer and should be directed at its life-projects and goals. They are oriented at creating unique value and at maximizing the customer's value-in-context [8, 12, 15]. If possible, the definition of this component includes feedback loops with relevant actors to improve the resulting statements and the degree of *fit* between the customer profiles and value maps.

Business Model Finally, after the identification of a set of value propositions that best address the customer and non-customer stakeholders' pains and gains the fourth component of SDD is used to identify and define the *realization* of these value propositions in suitable business models. This component is directed at answering the question of *what* a company is doing to cocreate value. Each business model corresponds to one value proposition and contains everything that is needed to deliver the proposed value. It is recommended to visualize this building block using the *business model canvas* presented by Osterwalder and Pigneur [26]. This canvas is intended as a "shared language that allows [one] to easily describe and manipulate business models to create new strategic alternatives" [26, p. 15]. Two fields of the canvas are already included in previous SDD components: (1) the value proposition and (2) the customer segments. This helps integrating this component into the overall ideation method. Furthermore, the business model canvas in general is highly compatible with S-D logic, because of dedicated fields for customer relationships (see FP_8), key activities, key (operand and operant) resources, and key partners, i.e. additional actors in the value network that are needed for the cocreation of value. The overall implications of the business model component that occur in the literature are:

- Enact practices and tools for an effective customer relationship management
- Evaluate the degree of control shifting
- Evaluate the degree of standardization versus individualization.

Regarding the nature and the attributes of relationships Le Meunier-FitzHugh et al. emphasize the importance of trust and a long-term focus of relationship building [17]. Actively managed long-term customer relationships must center around a two-way productive dialogue. Stakeholders should be involved holistically. Mutual transparency, collaborative risk assessments, and gain sharing, as well as risk-based pricing, improve collaboration and the resulting value-in-context. The customer is always an active audience and feedback loops are therefore essential for successful value cocreation. Contextual knowledge is collected, meaning that stakeholder-specific information is actively leveraged to improve service provision. Mutual access to goals and needs as well as the collection and management of this shared contextual knowledge is a priority. The goal of a company should be to learn the client's business and then solve the client's problem in the client's context [33]. A second implication of defining service- and value-oriented business models is a shift of control to the consumer. While the necessity of this is implied by the S-D logic theory, Moeller argues that the degree of this control shifting should be explicitly evaluated [23]. Too much control on the side of other stakeholders might negatively affect autonomy, risk, efficiency, and quality. Arnould et al. on the other hand argue that the higher the degree of control of the stakeholders, the higher is the resulting value-in-context as the beneficiary has more influence on the service that is provided. Plé and Chumpitaz Cáceres' thoughts on value codestruction [27] support Moeller in the aspect that while shifting control can be beneficial, a company should always consider potential negative impacts as well. The third implication discusses the ratio between standardization and individualization with respect to a company's product and service portfolio. Cestino and Berndt recommend offering a high degree of individualization [8] which is supported by Skålén and Edvardsson who claim that portfolio diversification improves the ability to fulfill the customer's *higher-order needs* [31]. On

Table 8.2 SDD component summary

Component	Deliverables	Auxiliary questions
Corporate Identity	Overall practices	What is the overall company philosophy? What is its essence from a service-centered perspective? Which practices need to be enacted to enable service-centricity on a corporate level?
Stakeholder Network	Stakeholder network diagram, stakeholder profiles	Who are the strategic customer stakeholders and non-customer stakeholders? How might they interact with each other? How can they be classified and what are their individual life projects and goals that could be addressed?
Value Proposition	Overall value proposition, value maps	Which value proposition would best fit to each stakeholder's pains and gains? What is needed to cocreate superior value-in-context?
Business Model	Business model canvases	How are stakeholder relationships managed? Which tools and processes are in use to enable information sharing and contextual knowledge management? Which business models are needed to best realize each stakeholder's value proposition? What are the resulting products and services?

the other hand, Moeller states that more standardization reduces risk while more individualization improves value-in-context [23]. There are industries where a high degree of individualization is not economically viable, where a service-centered business model still heavily relies on large quantities of physical goods as a distribution mechanism of service provision.

All four SDD components are summarized by their main deliverables and auxiliary questions in Table 8.2.

8.4 Concluding Methodology

Section 3 presented the four building blocks of SDD as one way to structure an ideation process towards service- and value-orientation. Starting with the corporate identity down its hierarchical structure, answering and visualizing the auxiliary questions shown in Table 8.2 provides support with respect to:

- Clearly defining the personality and purpose of the company *why* it is doing business, defining service-centered company practices and values
- Identifying strategically relevant customer and especially non-customer stakeholders supported by the six markets model, understanding their individual goals, pains, and gains
- Clearly defining, based on the company purpose, *how* these stakeholders can best be addressed by individualized value propositions
- Identifying novel (digital) business models that are tailored to those value propositions and describe concise approaches of *what* must be delivered such that it is highly valuable for the respective stakeholder.

This process must be conducted open-minded without restrictions to a company's current offerings. A corresponding workshop would first discuss the principles and axioms of S-D logic followed by an iterative collaboration on the components' deliverables following their hierarchical order. The preceding focus group discussion highlighted the parallels of this approach with Simon Sinek's *Golden Circle* [30, p. 37]. SDD encourages discussing the *why* before focusing on the *how* and *what*, which follows the S-D logic principles as well as Sinek's attributes of successful companies. All value propositions and business models that are derived from applying SDD can communicate the overall purpose and the personality of the company.

Ultimately, the essence of SDD is a structured methodology based on academic research that helps identifying opportunities for digitalization that are directly related to addressing customer needs and at creating value. This is to prevent digital transformations where large-scale changes and investments do not result in the expected outcomes as they are not accepted by the consumer or not compatible with the company's external perception in its ecosystem. After applying SDD, the three to five most promising business models are selected and implemented. Radical change should be treated with caution, promising reasonable steps towards digital improvements and new services are favored [10]. If possible, existing portfolio items and digitization projects are incorporated into these transformation efforts. So how does this look like in practice? SDD primarily relies on two popular visualization techniques for its application: Osterwalder's value proposition canvas and business model canvas whose original versions are shown in Figs. 8.2 and 8.3, respectively [25, 26].

SDD is using the *business model canvas* to describe the digital business model in a uniform way, to enable an initial conceptual modeling of digital products and services as well as to model the performance promise of a company or product, its infrastructure, customers and finances. The template consists of nine building blocks for value proposition, customers, activities, partners, resources, customer channels, customer relationships, costs, and revenue. The central aspect of the business model canvas is the description of offerings that revolve around value propositions. The canvas was originally proposed by Osterwalder and Pigneur [26] based on the *business model ontology* as presented in Osterwalder's dissertation [24]. The *value proposition canvas* [25] expands the business model canvas whose performance promise is projected onto the value proposition canvas and divided into three perspectives: Products and services, profit generators and painkillers associated with the aspects of the

Fig. 8.2 Business model canvas

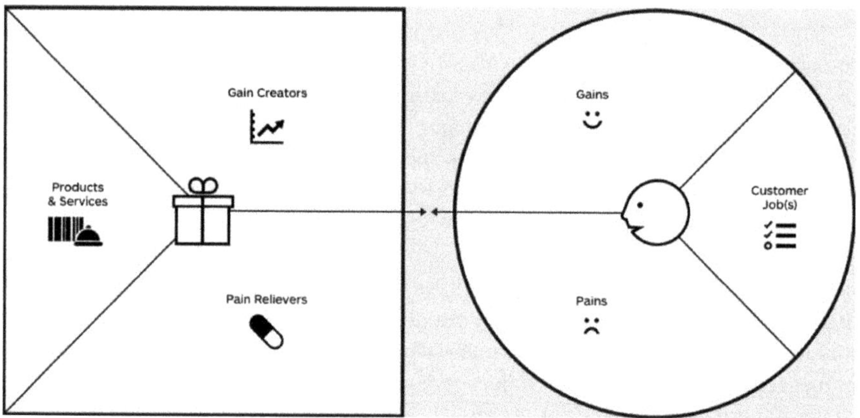

Fig. 8.3 Value proposition canvas

customer segment, such as customer jobs, profits, and pains. Both canvases are highly compatible with S-D logic, there are building blocks for its key concepts such as value cocreation, resource integration and relationship management. However, for SDD, additional emphasis is put on *value* instead of revenues, *expenses* instead of costs and (customer and non-customer) *stakeholders* instead of customers. SDD recommends the utilization of the value proposition canvas, separated into its customer profile and value map, before the definition of a business model. Therefore, the business model canvas' customer segment and value proposition are already defined beforehand which allows for a better integration of the model components.

In order to demonstrate the benefits of using SDD in practice this chapter evaluates the simplified digital transformation of a small mechanical engineering company. Its main product line are industrial water jet cutters for which the company is also providing maintenance services. When conducting an ideation workshop to identify new value streams the first optimizations that come to mind are often those that are popular and in active discussion at that time. Current examples would include flexible pricing based on subscriptions or pay-per-use models or adding machine learning combined with Open Platform Communications Unified Architecture (OPC UA) interfaces to enable fleet-wide predictive maintenance and intelligent optimization of the manufacturing process. However, when considering the structure of SDD these ideas correspond to its very last component, the business model component which is oriented at *what* to do to improve value-in-context. The *why* and *how* are neglected as much as getting familiar with the potential of leveraging the stakeholder ecosystem.

Using SDD the workshop would first concentrate on constructing a service-oriented corporate identity and philosophy. Following S-D logic's FP_3 goods are distribution mechanisms of service provision. An industrial water jet cutter is purchased because it provides high-precision cutting services for various materials and shapes [38]. With this shared understanding the company could redefine its overall purpose to become a *digital service provider of high-value industrial cutting solutions*. The goal is to maximize the customer's phenomenologically experienced value-in-context when receiving this service, by leveraging technological capabilities, stakeholder network effects as well as identified contextual knowledge [33] and higher-order needs. The personality of the company could be shifted towards being an innovative digital facilitator of modern supply chains that aims to reduce complexity and risk for its customers. Commitment is the key success factor of this building block. It is important that both managers and employees support the new philosophy and personality. Therefore, its outcome should not be the perfect vision based on global standards but a reasonable extension and enhancement of familiar company culture aspects and values.

Based on SDD's stakeholder network component a resulting set of stakeholders to interact with is displayed in Table 8.3. Each stakeholder is assigned to at most two markets of the *six markets model* [12, pp. 226–227] and has a set of pains and gains to potentially address by value propositions.

Both customer and non-customer stakeholders are considered. This is a simplified view, in an SDD workshop all stakeholders would be identified, classified and evaluated regarding their interrelationships. Then for each relevant stakeholder a customer profile is constructed consisting of its *customer jobs*, *pains* and *gains*. The goal is to get an in depth understanding of the current ecosystem state. Who are our competitors and potential partners? Are there non-customer parties that share our interests? Which patterns can be identified in the ecosystem's pains and gains that our corporate purpose and philosophy could address?

Subsequently, individually tailored value propositions can be defined for each relevant stakeholder. Figure 8.4 shows one exemplary customer profile on the right and one corresponding value map that results from the value proposition component of SDD on the left for the *manufacturing company* stakeholder. Following Sect. 3,

Table 8.3 Stakeholder network actors

Stakeholder	Markets	Pains	Gains
Manufacturing company	Customer, referral	High investments for new machines, complex supply chains	Outsource responsibilities, simplify manufacturing process, reduce risk
University	Alliance, recruitment	Accessibility of real-world research projects e.g. on machine learning	Research on prestigious new business models
Government	Alliance, influence	Global digitization and open markets threaten traditional economy	Innovative economy, create trust based on technology, security and reliability
Basic materials supplier	Supplier	Complex customer network and logistics	

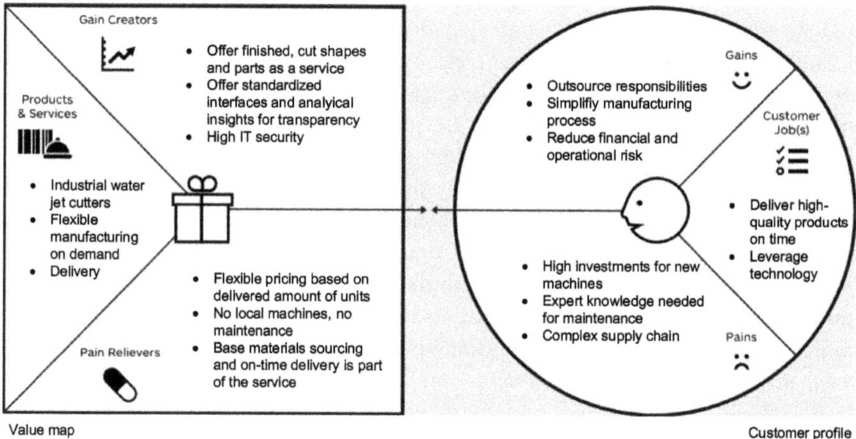

Fig. 8.4 Visualization example of manufacturing company

an SDD workshop would also include the definition of an overall value proposition that is aligned to the company purpose ("a *digital service provider of high-value industrial cutting solutions*") as well as the definition of guidelines and criteria all value propositions must adhere to. The above pain relievers and gain creators noticeably contain some of the popular ideas that were mentioned upfront such as flexible pricing or standardized interfaces. However, at this stage they can be directly related to a stakeholder pain or gain that is being addressed and they are backed by a corporate philosophy and values that clearly communicate *why* the company follows these ideas. This results in integrity and alignment which is even more important, the more stakeholders and value propositions a company is targeting.

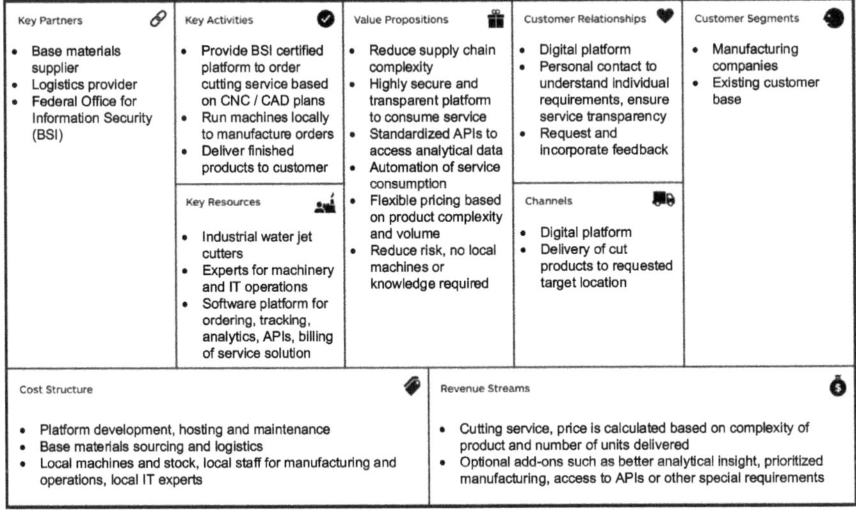

Fig. 8.5 Visualization example of manufacturing company business model

Finally, the business model component of SDD is applied to define suitable business models for each value proposition, describing a concise approach of *what* needs to be done in order to best address an opportunity for value cocreation, based on the identified gain creators and pain relievers. The resulting business model targeting the *manufacturing company* stakeholder is shown in Fig. 8.5.

All stakeholder pains and gains are in some way addressed, resulting in a potentially high value-in-context for the involved parties. Regarding the discussed areas of conflict (1) of shifting control to the consumer and (2) of standardization versus individualization this business model favors both *individualization* and *shifting a lot of control* to the consumer by allowing it to provide its own computer-aided design (CAD) plans and computer numerical control (CNC) instructions and by allowing an automation of service consumption using the digital platform provided by the mechanical engineering company. This maximizes the service's value added but also increases risk and could negatively impact autonomy and quality. So those factors need to be kept in mind when designing the platform's software solution. In the end the key to a successful S-D logic business model is a deep understanding of the stakeholder network. The exemplary model allows the mechanical engineering company, by analyzing its own data, to get very detailed insights into its consumers' priorities and needs. Based on this knowledge and based on explicit stakeholder feedback the above model can be iteratively improved to enhance the experienced value-in-context, which directly corresponds to the company's overall goal of becoming a digital service provider of high-value industrial cutting solutions.

Regarding the SDD methodology in its entirety there is no explicitly recommended visualization technique for the corporate identity component. However, especially with respect to corporate branding the *brand identity prism* is helpful to illustrate the relevant aspects of brand value and brand value cocreation [16, p. 183] similarly to the six markets model with respect to the stakeholder network analysis. As SDD is applied in the context of more use cases and further research, different ways of visualizing each component might emerge depending on the respective situation.

Initially, the building blocks of SDD were validated in the context of an exemplary digital mobility use case [7]. This use case was based on an electric car manufacturer that wants to transform its business to become a provider of digital mobility services. Applying SDD for this use case highlighted the importance of focus and prioritization during the ideation workshop. Especially the identification of potential stakeholders and their respective classification according to the six markets model should be restricted to a reasonable amount of entities. Ultimately, opportunities were identified in the areas of (1) launching a collaborative research and investment platform for electromobility to cocreate innovation within the partner ecosystem and (2) creating a marketplace for third-party extensions to modern digital displays and infotainment systems where external parties can become engaged in the process of developing a state-of-the-art infotainment experience. The analysis also resulted in a brand identity prism highlighting areas where a change in the brand's external perception was needed to best support the novel company purpose. Furthermore, a list of values and practices to enact was derived affecting organization, management, and strategy. These last steps were omitted in this chapter to keep the example as simple as possible.

8.5 Limitations and Future Research

The current state of the SDD ideation method is based on preceding research consisting of a literature review covering 22 journal articles, one use case in one industry and one focus group discussion [7]. Therefore, it only represents an initial concept, an explorative collection of hypotheses that serve as a starting point for the goals that are defined throughout this chapter. Both regarding its scope as well as its scientific maturity SDD is not generally applicable in practice at this point. Both its qualitative and quantitative validations are limited. The focus group pointed out several potential extensions which would improve the value added of an SDD ideation workshop and the application of the methodology in general. In summary, a more extensive validation incorporating the opinions of S-D logic experts and practitioners as well as an extension of the method's building blocks are required. Future research should encourage the continuous application and the theoretical and practical discussion of the ideation method to continually improve it step by step and to enrich it with more experience resulting from digital transformation projects. A collection of potential research streams is visualized as an SDD research roadmap in Fig. 8.6.

Achieving scientific maturity includes integrating additional academic sources and conducting additional focus group discussions and interviews on the qualitative side while applying the model in a broader range of industries in ideation workshops on the quantitative side. In a second step, research should then focus on extending

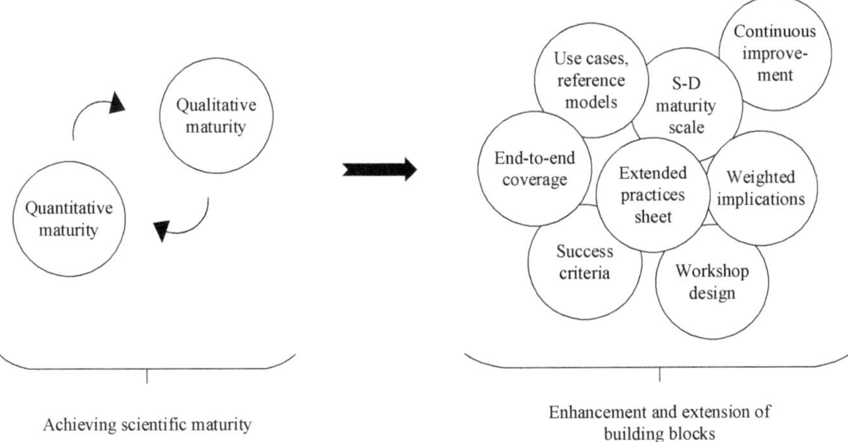

Fig. 8.6 SDD research roadmap

SDD based on the experiences made at this point. Examples include the creation of an S-D logic maturity scale, a detailed science-based workshop design or the definition of specific, measurable success criteria.

8.6 Conclusion

Ultimately, the pursued vision of all research streams beyond the scope of this chapter is to establish a mature and generally applicable methodology that supports the identification of novel digital value streams and business models by fostering service- and value orientation in the context of the digital transformation. All transformation efforts must be interconnected with a company's overall purpose and directly relate to a specific set of stakeholder needs that are addressed in order to cocreate a high phenomenologically perceived value-in-context. This chapter outlines a way of structuring this abstract vision into four components that in turn consist of practical implications and visualization techniques. In summary, you need to (1) define who you are and why you are doing business, (2) understand your position in your service ecosystem and identify the relevant customer and non-customer stakeholders to interact with, (3) define individually tailored value propositions for each stakeholder that best address its higher-order-needs and (4) describe and implement (digital) business models that best enable you to address the identified opportunities for value cocreation.

From a managerial perspective SDD improves the *effectiveness* of business model ideation, as shown in Sect. 4. The methodology ensures alignment and integrity while redefining a company's *why*, *how* and *what* [30], thereby improving its external authenticity and acceptance in the market. It ensures that participants of a workshop understand their overall company purpose from a service- and value-oriented perspective and structure their stakeholder knowledge and value propositions accordingly before discussing the actual *what*—which avoids the collection of popular technological topics that provide no real value added for the company's strategy. Therefore, the resulting business models, when launched as part of a digital transformation program, have a higher chance of succeeding while ideas that do not correspond to the ecosystem's needs can be discarded early on.

As Davenport and Westerman state companies know by now that large digital transformation projects combined with high financial investments and radical change often fail [10]. SDD enables companies to tackle the need to become digitalized progressively in reasonable steps with smaller investment and risk—by sequentially launching its resulting business models that best address the identified stakeholder ecosystems' potential for value cocreation. As they state: "Instead of ramping up quickly, only to ramp down painfully, it would be much better if companies can make steady progress toward the right end state without making such costly mistakes" [10]. The SDD ideation method extends this statement by providing a corresponding methodology and by encouraging companies to view digitization from a service-centered perspective without just following trends or market pressure. If we think about a *service-dominant transformation* first and then extend our digital architecture to comply with the requirements of our service ecosystem we can best leverage and benefit from today's predominant logic of economic exchange.

References

1. Altuntas Vural, C.: Service-dominant logic and supply chain management: a systematic literature review. J. Bus. Ind. Market. **32**(8), 1109–1124 (2017). https://doi.org/10.1108/JBIM-06-2015-0121
2. Andrews, K.R.: The concept of corporate strategy. In: Foss, N.J. (ed.) Resources, Firms, and Strategies: A Reader in the Resource-Based Perspective. Oxford University Press, Oxford (1997)
3. Arnould, E.J., Price, L.L., Malshe, A.: Toward a cultural resource-based theory of the customer. In: Lusch, R.F., Vargo, S.L. (eds.) The Service-dominant Logic of Marketing: Dialog, Debate, and Directions, pp. 91–104. M.E. Sharpe, Armonk, NY (2006)
4. Balmer, J.M.T.: Corporate identity, corporate branding and corporate marketing—seeing through the fog. Eur. J. Mark. **35**(3/4), 248–291 (2001). https://doi.org/10.1108/03090560110694763
5. Bastiat, F.: Selected essays on political economy. In: de Huszar, G.B. (ed.) Irvington-on-Hudson: Foundation for Economic Education (trans: Cain, S.) (1995) (Original work published 1848)
6. Ben Letaifa, S., Edvardsson, B., Tronvoll, B.: The role of social platforms in transforming service ecosystems. J. Bus. Res. **69**(5), 1933–1938 (2016). https://doi.org/10.1016/j.jbusres.2015.10.083
7. Bürkel, J.: Strategy innovation: a service-dominant transformation model. Master's dissertation. ESB Business School, Reutlingen (2018)

8. Cestino, J., Berndt, A.: Institutional limits to service dominant logic and servitisation in innovation efforts in newspapers. J. Media Bus. Stud. **14**(3), 188–216 (2017). https://doi.org/10.1080/16522354.2018.1445163
9. Christopher, M., Payne, A., Ballantyne, D.: Relationship Marketing: Bringing Quality, Customer Service and Marketing Together. Butterworth-Heinemann, Oxford (1991)
10. Davenport, T.H., Westerman, G.: Why So Many High-Profile Digital Transformations Fail. Harvard Business Review. Retrieved from https://hbr.org/2018/03/why-so-many-high-profile-digital-transformations-fail (2018)
11. Davey, J., Alsemgeest, R., O'Reilly-Schwass, S., Davey, H., FitzPatrick, M.: Visualizing intellectual capital using service-dominant logic. Int. J. Contemp. Hosp. Manag. **29**(6), 1745–1768 (2017). https://doi.org/10.1108/IJCHM-12-2015-0733
12. Frow, P., Payne, A.: A stakeholder perspective of the value proposition concept. Eur. J. Mark. **45**(1/2), 223–240 (2011). https://doi.org/10.1108/03090561111095676
13. Grönroos, C., Voima, P.: Critical service logic: making sense of value creation and co-creation. J. Acad. Mark. Sci. **41**(2), 133–150 (2013). https://doi.org/10.1007/s11747-012-0308-3
14. Henderson, J.C., Venkatraman, N.: Strategic alignment: a model for organizational transformation through information technology. In: Kochan, T.A., Useem, M. (eds.) Transforming Organizations. Oxford University Press, New York (1992)
15. Hilton, T., Hughes, T.: Co-production and self-service: the application of service-dominant logic. J. Market. Manag. **29**(7–8), 861–881 (2013). https://doi.org/10.1080/0267257X.2012.729071
16. Kapferer, J.-N.: The new strategic brand management: advanced insights and strategic thinking, 5th edn. Kogan Page, London, PA (2012)
17. Le Meunier-FitzHugh, K., Baumann, J., Palmer, R., Wilson, H.: The implications of service-dominant logic and integrated solutions on the sales function. J. Market. Theory Pract. **19**(4), 423–440 (2011). https://doi.org/10.2753/MTP1069-6679190405
18. Lusch, R.F., Nambisan, S.: Service innovation: a service-dominant logic perspective. MIS Quart. **39**(1), 155–175 (2015). https://doi.org/10.25300/MISQ/2015/39.1.07
19. Lusch, R.F., Vargo, S.L.: Service-dominant logic: reactions, reflections and refinements. Market. Theory **6**(3), 281–288 (2006). https://doi.org/10.1177/1470593106066781
20. Lusch, R.F., Vargo, S.L.: Service-dominant Logic: Premises, Perspectives, Possibilities. Cambridge University Press, Cambridge (2014)
21. Lusch, R.F., Vargo, S.L., O'Brien, M.: Competing through service: Insights from service-dominant logic. J. Retail. **83**(1), 5–18 (2007). https://doi.org/10.1016/j.jretai.2006.10.002
22. Merz, M.A., He, Y., Vargo, S.L.: The evolving brand logic: a service-dominant logic perspective. J. Acad. Mark. Sci. **37**(3), 328–344 (2009). https://doi.org/10.1007/s11747-009-0143-3
23. Moeller, S.: Customer Integration—A Key to an Implementation Perspective of Service Provision. J. Serv. Res. **11**(2), 197–210 (2008). https://doi.org/10.1177/1094670508324677
24. Osterwalder, A.: The business model ontology—a proposition in a design science approach. Dissertation. Université de Lausanne, Lausanne (2004)
25. Osterwalder, A.: Achieve product-market fit with our brand-new value proposition designer canvas. Retrieved from http://businessmodelalchemist.com/blog/2012/08/achieve-product-market-fit-with-our-brand-new-value-proposition-designer.html (2012)
26. Osterwalder, A., Pigneur, Y.: Business Model Generation: A Handbook for Visionaries, Game Changers, and Challengers. Wiley, Hoboken, NJ (2010)
27. Plé, L., Chumpitaz Cáceres, R.: Not always co-creation: introducing interactional co-destruction of value in service-dominant logic. J. Serv. Mark. **24**(6), 430–437 (2010). https://doi.org/10.1108/08876041011072546
28. Porter, M.E.: From competitive advantage to corporate strategy. In: Goold, M., Luchs, K.S. (eds.) Managing the Multibusiness Company: Strategic Issues for Diversified Groups. Routledge, London, NY (1996)
29. Prahalad, C.K., Ramaswamy, V.: Co-creating unique value with customers. Strat. Leadersh. **32**(3), 4–9 (2004). https://doi.org/10.1108/10878570410699249

30. Sinek, S.: Start With Why: How Great Leaders Inspire Everyone to Take Action. Portfolio/Penguin, London (2009)
31. Skålén, P., Edvardsson, B.: Transforming from the goods to the service-dominant logic. Market. Theory **16**(1), 101–121 (2016). https://doi.org/10.1177/1470593115596061
32. Skålén, P., Gummerus, J., von Koskull, C., Magnusson, P.R.: Exploring value propositions and service innovation: a service-dominant logic study. J. Acad. Mark. Sci. **43**(2), 137–158 (2015). https://doi.org/10.1007/s11747-013-0365-2
33. Skjølsvik, T.: The impact of client-professional relationships in ex ante value creation: a service-dominant logic perspective. J. Bus. Bus. Market. **24**(3), 183–199 (2017). https://doi.org/10.1080/1051712X.2017.1345259
34. Smith, A.: An inquiry into the nature and causes of the wealth of nations. Reprint, W. Strahan and T. Cadell, London (1904) (Original work published 1776)
35. Thornton, P.H., Ocasio, W.: Institutional logics and the historical contingency of power in organizations: executive succession in the Higher Education Publishing Industry, 1958–1990. Am. J. Sociol. **105**(3), 801–843 (1999). https://doi.org/10.1086/210361
36. Vargo, S.L.: Customer integration and value creation. J. Serv. Res. **11**(2), 211–215 (2008). https://doi.org/10.1177/1094670508324260
37. Vargo, S.L., Akaka, M.A., Vaughan, C.M.: Conceptualizing value: a service-ecosystem view. J. Creat. Value **3**(2), 117–124 (2017). https://doi.org/10.1177/2394964317732861
38. Vargo, S.L., Lusch, R.F.: Evolving to a new dominant logic for marketing. J. Market. **68**(1), 1–17 (2004). https://doi.org/10.1509/jmkg.68.1.1.24036
39. Vargo, S.L., Lusch, R.F.: Service-dominant logic: continuing the evolution. J. Acad. Mark. Sci. **36**(1), 1–10 (2008). https://doi.org/10.1007/s11747-007-0069-6
40. Vargo, S.L., Lusch, R.F.: Institutions and axioms: an extension and update of service-dominant logic. J. Acad. Mark. Sci. **44**(1), 5–23 (2016). https://doi.org/10.1007/s11747-015-0456-3

Chapter 9
A Semantically Integrated Conceptual Modelling Method for Business Process Reengineering

Remigijus Gustas and Prima Gustienė

Abstract Information system methodologies project different modelling aspects into disparate dimensions and it creates difficulties to achieve integrity in various diagram types. Concepts cannot be analysed in isolation from interactions and state changes, which is the case in current traditional system development methods. The lack of an integrated graphical modelling approach creates problems in rethinking and changing business processes. In this chapter, we present Semantically Integrated Conceptual Modeling (SICM) method and demonstrate how the method can be applied for business process re-engineering. A well-known case study is analyzed to demonstrate how service interactions among enterprise organizational components can be represented to detect the essential similarities and differences of business processes. Service architecture is graphically described in terms of interactions among organizational and technical enterprise system components. One of the advantages of SICM method is based on a single type of diagram, which represents interactive, transitional and the structural aspects of concepts. State changes result from processes that are represented by creation and termination events of various concepts.

Keywords Digital Business · Service interactions · Reengineering · Decomposition · Integration · Actions · Transitions from one class to another

9.1 Introduction

Business process re-engineering is the fundamental rethinking and radical redesign [3] of processes for achieving necessary improvements. The conceptual model of activities designed to produce a specified output can represent a business process. It

R. Gustas (✉) · P. Gustienė
Information Systems Department, Karlstad University, 651 88 Karlstad, Sweden
e-mail: Remigijus.gustas@kau.se

P. Gustienė
e-mail: Prima.gustiene@kau.se

A. Zimmermann et al. (eds.), *Architecting the Digital Transformation*,
Intelligent Systems Reference Library 188,
https://doi.org/10.1007/978-3-030-49640-1_9

implies a strong emphasis on how work is performed within an organization. Despite the widespread adoption of Business Process Reengineering (BPR), it has failed in many cases to deliver promised results. The lack of integrated method is one of the reasons behind BPR failures [1].

Process mapping should provide models and method for identifying a current business process, which is used to specify an expected business process. It is a critical link that can be applied to significantly improve a process. In what order the processes need to be changed is a million dollar question. No company reengineers all processes simultaneously. Choices need to be made that can be based on the different criteria of the efficiency, importance or feasibility of an expected business processes. Having identified the potential improvements, the development of different conceptual models of processes has to be done. Gunasekaran and Kobu [8] reviewed the literature on modelling and distinguished many different BPR techniques, which include conceptual models, simulation models, object-oriented models, integration definition models, network models, and knowledge based models. All these modelling techniques define different aspects of a system using various types of diagrams. Integration principles of various dimensions of such diagrams are quite unclear. The lack of the graphical conceptual modelling method that helps to detect semantic integrity between static and dynamic aspects of process specification is a reason for BPR project failures.

Enterprise business process models may have many modeling projections, which are typically described by using different types of diagrams. It is common to all industrial versions of system analysis and design methods to distinguish disparate views or dimensions of an enterprise architecture. The Zachman Framework [16] can be viewed as a taxonomy for understanding these diagrams that are used for representation of organizational and technical system architecture. They define different dimensions of business application and data architecture such as Why, What, How, Who, Where and When. Detection of inconsistency among different architecture views is a fundamental problem in most Information Systems (IS) methodologies.

Inter-model consistency and integrity is hard to achieve for non-integrated model collections [7]. Modeling techniques that are realized as set of models are difficult to comprehend for business experts, who are responsible for reengineering organizations and for communicating essential changes of enterprise architectures. There is often semantic discontinuity and overlapping in specifications, because static and dynamic constructs do not fit perfectly. Inter-model consistency is difficult to achieve in this situation. Most conceptual modeling languages are plagued by the paradigm mismatch between static and dynamic constructs of meta-models [7]. Nearly all enterprise information system graphical models are collections of the static and dynamic aspects of specifications.

This paper is organized as follows. The next section, presents notation of semantically integrated conceptual modelling method. The third section demonstrates applicability of the method for Ford Motor Company. Three different situations are analyzed in this section to demonstrate how service architectures are visualized before and after radical change. In the fourth section, similarities and differences of two situations are discussed. The concluding section outlines the perspectives of business process reengineering in the context of graphical Semantically Integrated Conceptual Modelling (SICM) method.

9.2 Semantically Integrated Conceptual Modelling

Actors, actions and exchange flows are elements that are necessary for the demonstration of value exchange [11, 13]. Economic resources are special types of concepts, which represent moving things. Rectangles with shaded background are used for denotation of moving resources and dotted line boxes are used to represent data flows. Actors are represented by square rectangles. Actions, which are performed by actors, are represented by ellipses. Actions are necessary for transferring resource or information flows from one actor to another. Data stores are concepts, which represent static aspects of a system. Concepts and actors can be in various states. Graphical notation of SICM method is presented in Fig. 9.1.

Interaction dependencies can be used to conceptualize services between various enterprise system actors. Since actors are implemented as organizational and technical system components, they can use each other according to the prescribed service interaction patterns to achieve their goals. The simplest service interaction loop can be represented by two interaction dependencies into opposite directions between service requester and service provider. Interaction loop between two actors indicates that they depend on each other for some specific actions. Service providers are actors that typically receive service requests, over which they have no direct control, and initiate service responses that are sent to service requesters. Service requests and service responses can be used to define more complex interactions. In this way, designers are able to construct a blueprint of interacting components, which are represented by different actors. An enterprise system is typically defined as a set of interacting loosely connected actors, able to perform specific services on request. Two actors and transfer of two resource flows into opposite directions are illustrated in Fig. 9.2.

Figure 9.2 shows actions Deliver and Pay, which can be triggered at any time. It is not stated in which situation a certain action will be initiated. Vendor initiates delivery

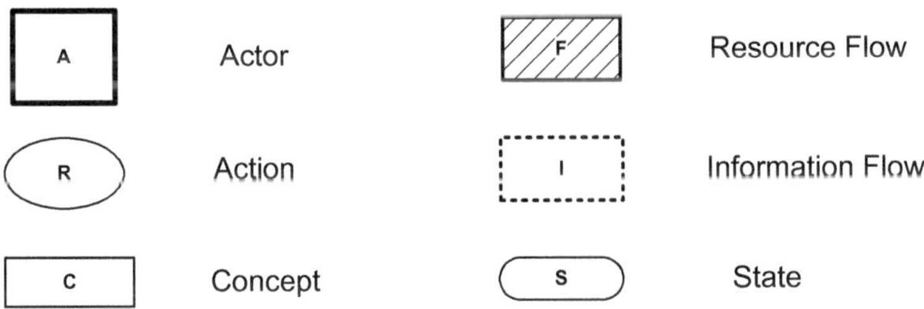

Fig. 9.1 Graphical notation of SICM

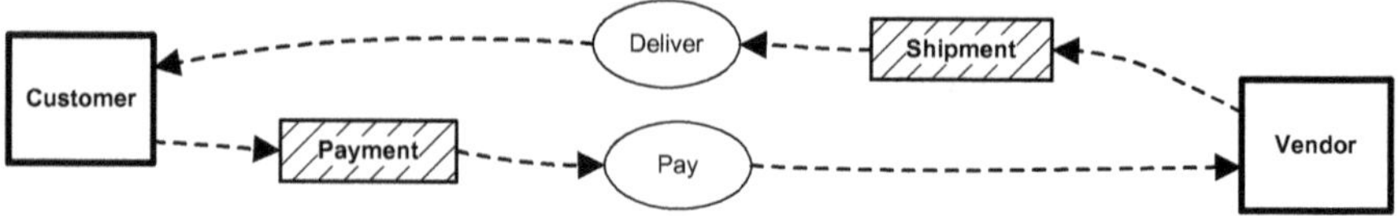

Fig. 9.2 Value flow exchange

action, because Shipment's moving direction is from Vendor to Customer. Customer receives a shipment flow. Conversely, Customer triggers the payment by the action of pay. Pay action and Deliver action represents the increment and decrement events of this process. The process of paying is essentially an exchange of a Shipment for a Payment from the point of view of both actors. For a Vendor, the paying action is an increment event and the delivery action is a decrement event, because it decreases the value of a resource. For a Customer, the delivery action is an increment event and the paying action is a decrement event. The terms of increment and decrement actions depend on the actor, which is the focus of this model.

Our buying and selling example focuses on the core phenomenon. Most customers pay in advance of shipment, but some customers may pay later when they receive the delivered products. If we consider the case of on line sales, then customers provide credit card details before the product items are delivered. Some customers may receive an invoice later and pay for all their purchases in a certain period. All these cases are covered by the same service interaction loop, which is illustrated in Fig. 9.2. When the shipment is delivered, then the delivery fact is registered in a system by a newly created object (Shipment) with all its mandatory properties. The transition arrow (⟶) points from the class Order to Shipment. This transition is shown in Fig. 9.3.

Action of Pay initiated by Customer creates an object of Payment. The moving resource flow (Payment) is consumed by Vendor. If we want to represent that the Deliver action precedes the Pay action, then the created Shipment should be linked by the transition arrow with the Pay action, which indicates the creation of the next

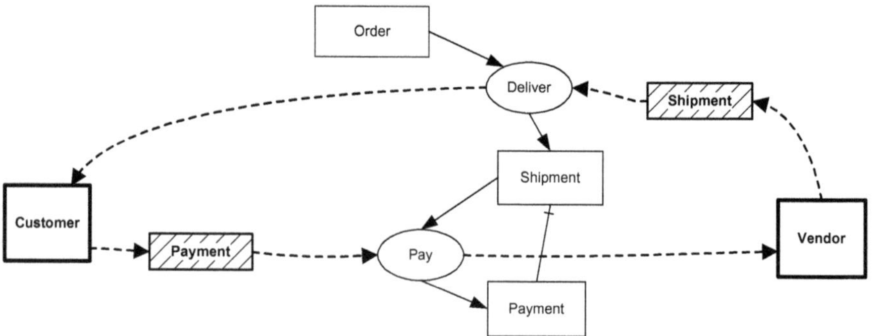

Fig. 9.3 Interaction loop, where the delivery action precedes the paying action

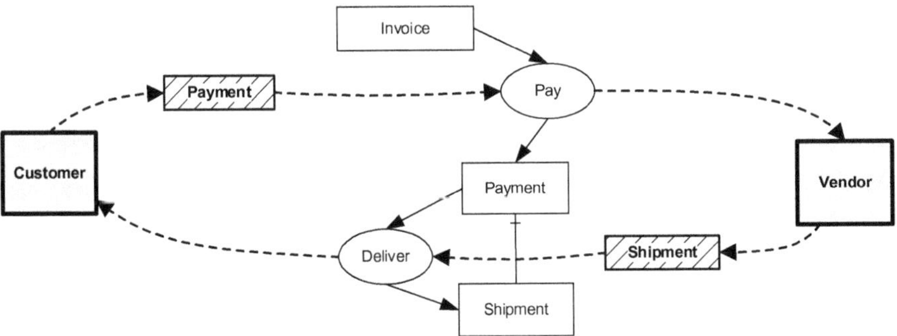

Fig. 9.4 Interaction loop, where pay action precedes deliver action

object of Payment. It creates the Payment from Shipment, which is the property of Payment (see dependency from Payment to Shipment). We may want to ask for the payment in advance to the shipment. In this case, we can show first the action of pay, which is designed to transform the concept of invoice to payment. Transition arrow from payment to shipment connects the second action of delivery. In this way, the payment flow will be exchanged to the shipment. This process is represented in Fig. 9.4.

Actors represent physical subsystems and structural changes of concepts represent data stores of a system. In such a way, the diagram illustrates actions, which result in changes of the attribute values. All actions are used to show the legal ways in which the actors interact with each other. Structural changes of objects can be manifested via static properties of objects. They are represented by the mandatory attributes. The mandatory attributes can be linked to a class by the single-valued (—+) or by multi-valued (—+<) attribute dependencies. Semantics of static dependencies are defined by cardinalities, which represent a minimum and maximum number of objects in one class (B) that can be associated with the objects in another class (A). Single-valued dependency is defined by the following cardinalities: (0, 1; 1, 1), (0, *; 1, 1) and (1, 1; 1, 1). Multi-valued dependency denotes either (0, 1; 1, *) or (1, 1; 1, *) cardinality.

Structural changes of objects are manifested via object properties. Properties can be understood as mandatory attribute values. If diagrams are used to communicate semantic details of a conceptualized system unambiguously, then optional properties should be proscribed. The decline in cognitive processing performance that occurs when optional attributes and relationships are used appears to be substantial [2]. If —+or —+<, then concept A is viewed as a class and concept B—as a property of A. One significant difference of the SICM is that the association ends cannot be used to denote mappings in two opposite directions, because these concepts have to be represented by data stores. The justification of this way of modelling can be found in the other papers [9, 10]. The main reason for introducing nameless attribute dependencies is to improve the stability of conceptualizations. The semantic links are defined by cardinalities, which represent a minimum and maximum number of

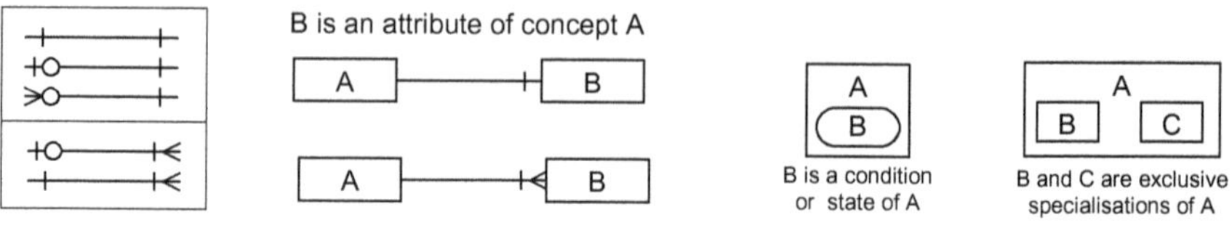

Fig. 9.5 Notation of attribute dependencies

objects in one class (B) that can be linked to objects in another class (A). Graphical notation of an attribute dependency between A and B is represented in Fig. 9.5.

Any data store is a passive concept that can be defined as an exclusive complete generalization of two or more specialized concepts. Passive concepts can also be characterized by state [6] or condition. The notation of exclusive generalization and the notation of state are represented in Fig. 9.5 as well. Static dependencies are very important, because they can help to reason about sequences of creation events of various passive concepts (see Figs. 9.3 and 9.4). For instance, a shipment is a single-valued dependency from payment, which is represented in Fig. 9.3. The same payment is dependent on shipment in the next figure (Fig. 9.4). In this way, every interaction is able to produce new facts, which are represented by different classes of objects [5].

9.3 Three Situations in Ford Motor Company

To understand the internal service interaction effects between service requesters and service providers, graphical representations of processes should be integrated with data models. The dynamic aspects of service interactions can be expressed by defining structural changes of passive concepts. Passive concepts are ordinary or derived classes, which are characterized by mandatory attributes. Values of such attributes are created or consumed in service requests and service responses. Requests, responses and passive classes are fundamental for understanding various aspects of integrated modelling. From an ontological perspective, when two subsystems interact one may affect the state of another. The changes of objects are characterized by object properties. Interaction dependency R(A---▶B)between two active concepts A and B indicates that subsystem A can perform action R on one or more subsystems B. Actions are able to manipulate object properties. Property changes may trigger objects transitions from one class to another.

To demonstrate how interactions can be used to analyze service architectures, we take the well-known Ford Case as an example [14]. The enterprise system architecture before radical change is described as follows: '*When Ford's purchasing department wrote a purchase order to a vendor, it also sent a copy of this order to accounts payable. Later, when material control received the goods from the vendor, it sent a copy of the receiving document to accounts payable. Meanwhile, the vendor sent an invoice to Accounts Payable. It was up to Accounts Payable, then, to match the*

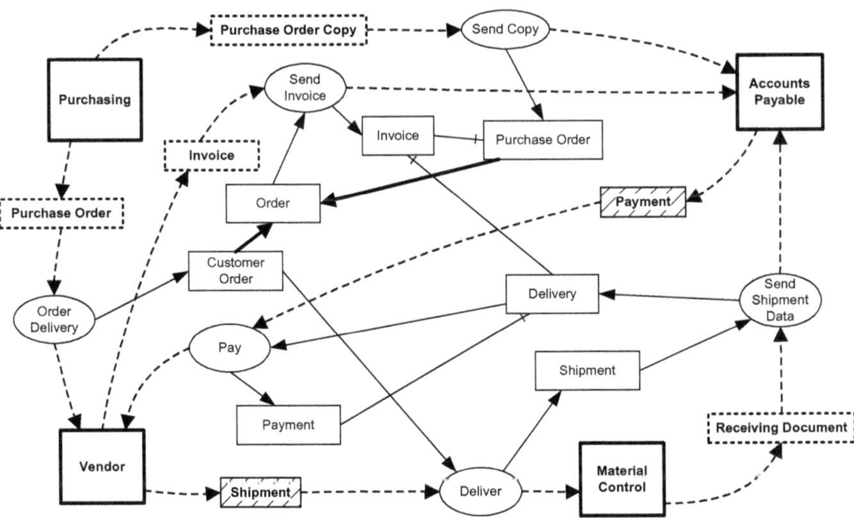

Fig. 9.6 Service architecture before radical change

purchase order against the delivery and the invoice. If they matched, the department issued payment. The department spent most of its time on mismatches, instances where the purchase order, receiving document, and invoice disagreed. In these cases, an Accounts Payable clerk would investigate the discrepancy, holding up the payment'.

The effects of described actions are expressed by using passive classes, which are represented in Fig. 9.6.

The diagram in Fig. 9.6 represents the conceptualization of two service loops (ordering-shipment and invoicing-paying). Purchasing division is entitled to order goods by sending a purchase order flow to a Vendor. If it is accepted, then an order is archived as a Customer Order. Vendor is obliged to Send Invoice and Deliver Shipment for every outstanding Customer Order. Send Invoice action is reclassifying Order to Invoice, which should match one Purchase Order. As we can see, creation of an Order object is necessary for generating a second service request, which will trigger Send Invoice action. A Customer Order object is used to trigger the Deliver action from Vendor to Material Control that is necessary for the creation of a Shipment object. When Material Control receives the shipment flow, it sends the Receiving Document to Accounts Payable. This is done with the help of Send Shipment Data action, which is designed to remove a Shipment object and to create a Delivery object. The Delivery object requires existence of one Invoice object. It is necessary for triggering the Pay action, which would create a Payment object. There is an explicit post-condition requirement for creating every Payment. The Payment must be linked to one Delivery (see mandatory single-valued dependency). If this dependency cannot be established, then this situation violates the consistency of the model. In this case, the Pay action cannot go forward and the payment flow to Vendor cannot be delivered. It is a precondition to start an investigation and prove that something is wrong with this payment. The alternative of investigation is missing in the presented diagram.

We can show how the situation was improved in Ford Motor Company by visualizing the service architecture after the radical change. The reengineering of the enterprise system in Ford was possible in two ways. These two alternatives are described by Hammer [14]:

> One way to improve the situation might have been to help the Accounts Payable clerk investigate mismatches between Delivery and Invoice or between Invoice and Purchase Order more efficiently. But a better alternative was to prevent these mismatches. Ford Motor Company introduced invoice-less paying. When the purchasing department initiated a purchase order, it entered the information into an online database (DB). This order is not sent to anyone. When the shipment arrives, then the receiving clerk checks the database to see if it corresponds to an outstanding purchase order.

If so, the receiving clerk accepts Shipment and makes Payment. If the receiving clerk cannot find a database entry for the shipment, it simply returns the order. Computation-neutral graphical description of the new situation, after radical change, is represented in Fig. 9.7.

The situation after radical change is much simpler. It represents an alternative to the Pay action, which is triggered by a Return action. This alternative was not included in the service architecture before radical change. Purchasing department takes care of Order creation, which is entered to the shared database (DB). DB is an active concept, which is connected by a composition link to Vendor (DB —▸—Vendor). The Purchase Order flow is sent to the shared DB by the Order Delivery action. When Vendor delivers the Shipment flow to a Receiving Clerk, then Delivery is created. Receiving Clerk is an active concept, which belongs to the Purchasing (Receiving Clerk —▸—Purchasing). Newly created instance of Delivery has a link to exactly one Order. Delivery is a passive concept, which denotes data store. If order delivery

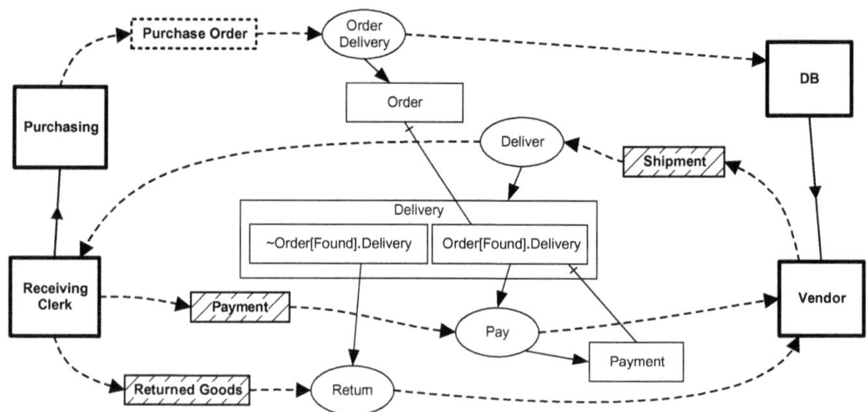

Fig. 9.7 Service interactions after radical change

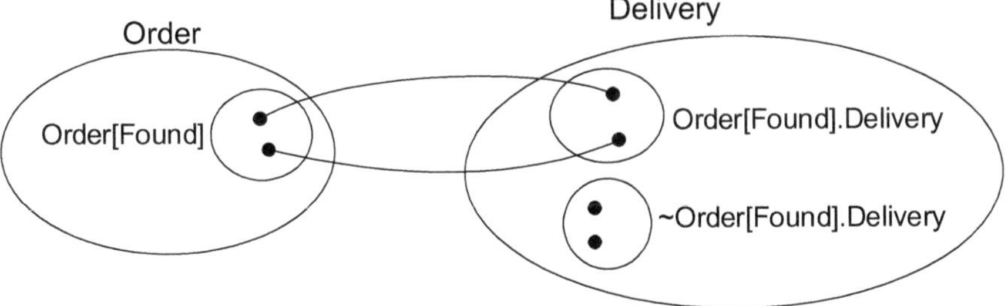

Fig. 9.8 Graphical example of prefixing and negation operations

is found (represented by derived Order[Found].Delivery object), then it can be paid by a Receiving Clerk. Triggering Pay action sends Payment information flow to the Vendor that would create the payment object. If the delivery order is not found (~Order[Found].Delivery), then Receiving Clerk returns goods flow to the Vendor. Returned Goods is the same Shipment flow, which was sent in the opposite direction. The return of this flow is initiated, because the order of this delivery was not found.

Expression Order[Found].Delivery represents a specialized concept of Delivery. '.' represents a prefixing operation on Delivery concept in the context of Order[Found] concept. The prefixing operation A.B can take place if concept B is dependent on concept A by any kind of static dependency (see Fig. 9.8). The derived concept A.B defines the subset of concept B objects, which are associated with at least one object from concept A.

Expression ~Order[Found].Delivery represents the negation of concept Order[Found].Delivery. It can be used in the following situation:

$$\sim A.B = A - \sim A.B, \sim A.B \Rightarrow B.$$

$\Rightarrow$ is an inheritance link between two concepts.

Examples of negation and prefixing operations are presented in Fig. 9.8.

Fundamentally, two kinds of changes occur during a reclassification action: removal of an object from a precondition class and creation of an object in a post-condition class. Sometimes, the postcondition class may require preservation of the precondition class. So, static dependencies have a higher priority as compared to object removal request. Examples of such preserved classes of objects are Order[Found].Delivery and Payment. Preservation of Order[Found].Delivery object is required by the postcondition of Pay action (see Fig. 9.7). The creation of Payment object requires preservation of the Order[Found].Delivery object, because these two classes are linked by a single valued dependency. Quite often objects are not be preserved. These objects pass several classes and then are removed. Interaction dependencies represent creation (Order Delivery and Deliver action), termination (Return action) or reclassification (Pay action) events from Order to Payment (see Fig. 9.7). When reclassification action is triggered, then creation and termination events are performed at the same time.

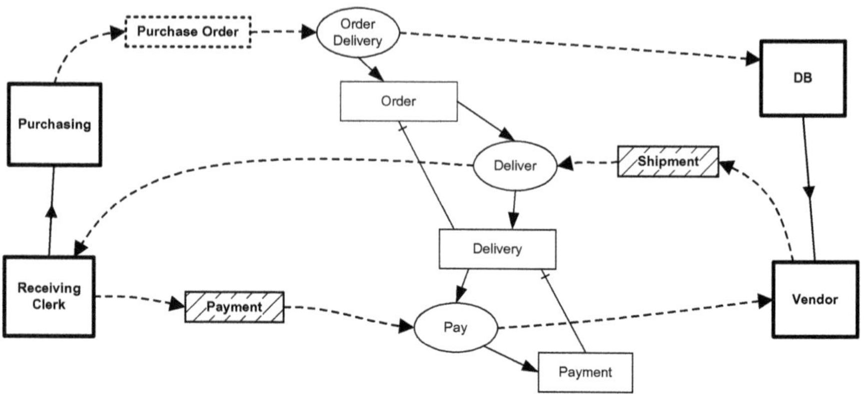

Fig. 9.9 Simplified description after radical change

Conceptual model of the service architecture after radical change, which is presented in Fig. 9.7, can be even more simplified, if creation of deliveries without corresponding orders are not allowed. It means that an Order is always necessary for the reclassification to a Delivery. In this case, if Delivery is not linked to any Order, then it cannot be recorded in the system. Consequently, it cannot be delivered and cannot be returned. Only deliveries, that are linked to orders, can be reclassified to Payment by the Receiving Clerk (Pay action). This simplified situation is shown in Fig. 9.9.

Service system after radical change, which is represented in Fig. 9.9, is even more efficient as compared to the first situation. Since, deliveries that are not found can be eliminated. These deliveries are not expected to be returned to the Vendor. In other words, there is no need to return what is not delivered by Ford Motor Company.

9.4 Similarities and Differences of Two Situations

Service interactions are important for understanding similarities of business processes, but their comprehension is not sufficient for studying the essential semantic differences of processes when reengineering organizations. The structural and behavioral aspects need to be analyzed in terms of transition effects and mandatory static attribute dependencies between concepts. Essential similarities and differences of service architectures can be identified by analyzing graphical descriptions of two situations in Ford Motor Company. They illustrate not just essential interactions, but also some related behavioral and structural aspects of service architectures. Fundamental similarities of service interaction are not difficult to identify with the help of the inference rules, which are based on the hierarchical structure of a system. This structure is presented in Fig. 9.10.

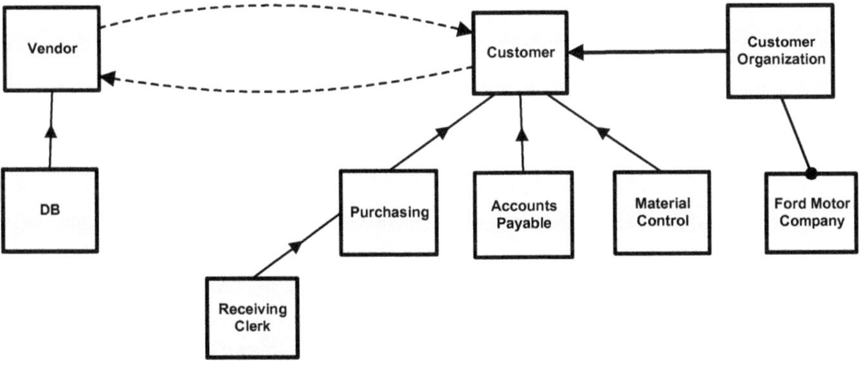

Fig. 9.10 Composition and generalization hierarchy of Ford Motor Company

This hierarchy shows composition, generalization, interaction and classification links between subsystems. Composition dependency can be used for the detection of inconsistently specified interaction dependencies between actors. Interaction dependency is characterized by the following inference rule [12]:

If A —▶— B and R(A ---▶ C) then R(B ---▶ C).

For instance, if (DB —▶—Vendor) and Order Delivery(Customer ---▶ DB) then Order Delivery(Customer ---▶ Vendor).

After application of the inference rules, similarities of two different situations (Figs. 9.7 and 9.9) after radical change can be graphically expressed by the diagram, which is represented in Fig. 9.11.

More specific actors inherit interaction dependencies from the more generic actors. Interaction dependency is inherited according to the following inference rule [12].

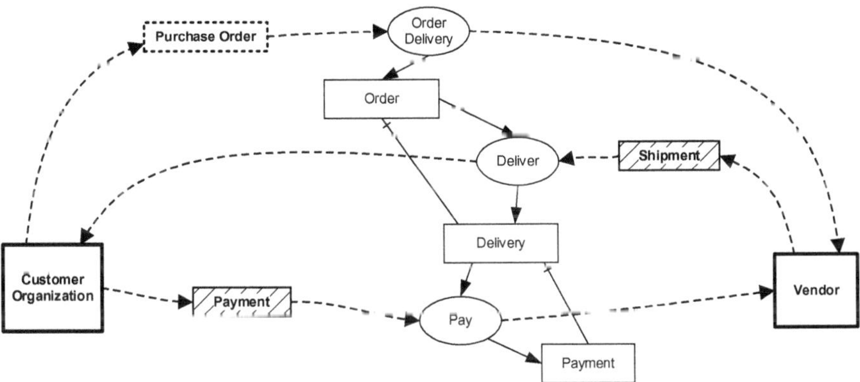

Fig. 9.11 Essential business process after radical change

If A ➡ B and R(B ---▶ C) then R(A ---▶ C).

If Customer Organization ➡ Customer and Order Delivery (Customer ---▶ Vendor) then Order Delivery (Customer Organization ---▶ Vendor).

We can see that this derived organizational system architecture is the same in two different situations, which are analyzed in this paper. This diagram characterizes in a concise form the basic inter-organizational structure between Customer Organization and Vendor, which is defined in terms of service requests and service provision. Overlapping interactions of two different situations demonstrate that the generic business process structure significantly overlaps. The analysis of the same case with the DEMO approach confirms stability of the essential model as well [5]. However, our service-oriented diagrams demonstrate some important differences between the situation before radical change and the generic situation after radical change. It is not difficult to understand that one service interaction loop before radical change is different. Pay action follows Deliver action in all three diagrams (see Figs. 9.6, 9.7 and 9.9), but in the first diagram an Invoice object cannot be created directly by Deliver action until Send Invoice action is triggered (since Delivery and Invoice are linked by a single valued dependency). This Invoice object (related via Delivery) is required as a precondition for triggering the Pay action. The Send Invoice action has to be triggered by Vendor and this is unnecessary in the architecture after radical change. The essential interactions before radical change are represented in Fig. 9.12.

Service interactions before radical change can be expressed as follows:

1. if Order Delivery(Customer ---▶ Vendor)
   then Send Invoice(Vendor ---▶ Customer),
2. if Order Delivery(Customer ---▶ Vendor)

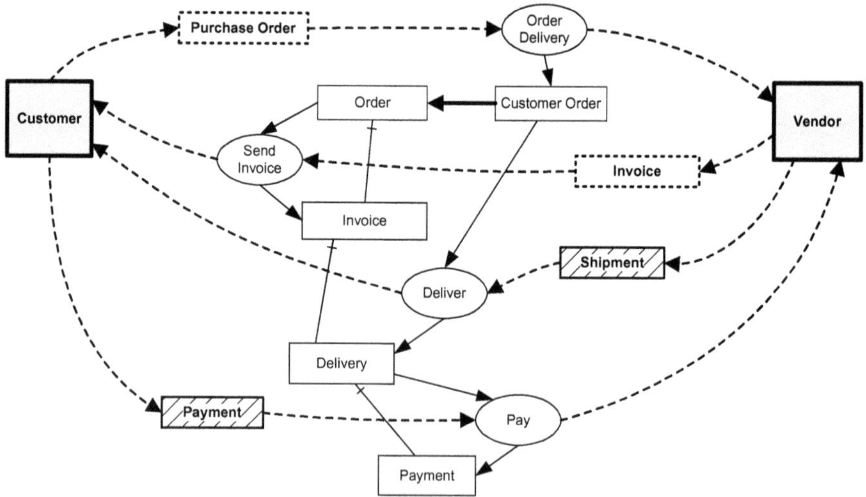

Fig. 9.12 Essential business process before radical change

then Deliver(Vendor ---▶ Customer),
3. if Deliver(Vendor ---▶ Customer)
then Pay(Customer ---▶ Vendor).

The second and the third service interaction loop is exactly the same in all three situations. Only the first service interaction loop is quite different, which is a characteristic feature of the situation before radical change. There is one more essential semantic difference in the service architecture after radical change. It clearly reveals an alternative service interaction loop, which is missing in the previous conceptualization. The third interaction loop contains an alternative action and flow (see Fig. 9.7) of Returned Goods. This service loop can be expressed as follows:

if Deliver(Vendor ---▶ Customer)
then Pay(Customer ---▶ Vendor),
otherwise Return(Customer ---▶ Vendor).

The alternative interaction prescribes what happens if Pay action fails. Please note that Return action is missing in the service architecture before radical change, because the conditions on returning goods were unclear.

When analyzing business processes, we often need to explicitly show graphical differences before and after radical change. Most information system design methodologies are 'blind' in expressing such alternatives, because they fail to identify conditions on which these interactions are dependent. For instance, the Returned Goods flow is not represented in any way by the DEMO approach [4].

9.5 Conclusions

Business Process Reengineering tries to improve the redesign of organizational business processes to achieve better productivity. BPR involves the examination and redesign of business processes in an organization. Most BPR methods do not use conceptual models, because they do not integrate structural, behavioral and interactive aspects of service architectures. The failure to visualize interaction and object transition effects in one model makes it difficult to understand essential differences of business process architectures when reengineering enterprise systems. In this paper, we demonstrated three different alternatives of business processes in Ford Motor Company. We have shown how to represent similarities and differences of these business processes. Similarities have demonstrated that even if the business process was dramatically changed, the basic structure significantly overlaps. The differences of business processes were represented by the SICM method. Traditional conceptual modelling methods cannot express such semantic differences, because static structures are not clearly linked to various actions.

Most BPR methodologies are not integrating structural, interactive and behavioral aspects of service architectures. Basically, business process is the way we perform our work and BPR is the process of changing the way we do our work so that we can do it better. Business process can be represented by conceptual model of activities designed to produce a specified output. Three different situations in our example demonstrate how interactions should be linked together with structural changes of passive concepts. Transition dependencies between passive concepts help designers to reason about alternatives or iterations of service interaction loops.

The visual analysis of semantic differences between three conceptualizations of business process reveals that the service architecture before radical change is more complicated in many ways. The Deliver action alone (see the service architecture after radical change) is not sufficient for triggering the Pay action by Accounts Payable. Triggering this action before radical change require matching of an Invoice flow, which is sent by a Vendor. It is obvious from the presented diagram that Pay action will fail in the following cases:

1. Failure to trigger Send Copy action by Purchasing or to receive Purchase Order Copy by Accounts Payable,
2. Failure to trigger Send Invoice action by Vendor, which will create an Invoice (see mandatory dependency from Invoice to Purchase Order),
3. Failure to trigger Deliver action by Vendor or to trigger Send Shipment Data action by Material Control, which is matching a Delivery with an Invoice.

According to the presented diagram, these three situations cannot take place in the service architecture after radical change, because the business process is more straightforward. It is unclear how these semantic differences can be represented using DEMO method, because the method does not express the conditions on which these interactions are dependent. To the best of our knowledge, expressing parallel, alternative and iterative behavior of service interactions, in combination with structural properties, is a challenge of most business process modeling methods.

DEMO approach [5], as well as the enterprise modeling language Archimate [15], puts active concepts into the foreground of modelling. This tradition is quite successful for the analysis of the external behavior of enterprise system components. In contrast, the object-oriented paradigm is based on modeling the static and dynamic aspects of passive concepts, which are represented by various classes of objects. The majority of the textbooks in the area of systems analysis and design recommend to concentrate first on domain modeling, which is based on analyzing classes, attributes, relationships, state transitions and related effects. Most conventional system analysis and design approaches put forward modeling of passive concepts. Only a few emerging approaches try to express deep interplay [6] between active and passive structures of concepts. Conventional information system analysis and design methods project structural, interaction and state-transition aspects using different types of diagrams. This makes difficulties while trying to integrate static and dynamic aspects of enterprise IS architectures.

References

1. Al-Mashari, M., Irani, Z., Zairi, M.: Business process reengineering: a survey of international experience. Bus. Process. Manag. J. **7**(5), 437–455 (2001)
2. Bodart, F., Patel, A., Sim, M., Weber, R.: Should optional properties be used in conceptual modelling? A theory and three empirical tests. Inf. Syst. Res. **12**(4), 384–405 (2001)
3. Dennis, A.R., Carte, T.A., Kelly, G.G.: Breaking the rules: success and failure in groupware-supported business process reengineering. Decis. Support Syst. **36**, 31–47 (2003)
4. Dietz, J.L.G.: The deep structure of business processes. Commun. ACM **9**(5), 59–64 (2006)
5. Dietz, J.L.G.: Enterprise Ontology: Theory and Methodology. Springer, Berlin (2006)
6. Dori, D.: Object-Process Methodology: A Holistic System Paradigm. Springer, Berlin (2002)
7. Glinz, M.: Problems and deficiencies of UML as a requirements specification language. In: Proceedings of the 10th International Workshop on Software Specification and Design, pp. 11–22. San Diego (2000)
8. Gunasekaran, A., Kobu, B.: Modelling and analysis of business process reengineering. Int. J. Prod. Res. **40**(11), 2521–2546 (2002)
9. Gustas, R.: A look behind conceptual modeling constructs in information system analysis and design. Int. J. Inf. Syst. Model. Des. **1**(1), 78–107 (2010)
10. Gustas, R., Gustiene, P.: Conceptual modeling method for separation of concerns and integration of structure and behaviour. Int. J. Inf. Syst. Model. Des. **3**(1), 48–77 (2012)
11. Gustas, R., Gustiene, P.: Conceptual modeling patterns of business processes. Int. J. Adv. Intell. Syst. **7**(4), 716–727 (2014)
12. Gustas, R., Gustiene, P.: Principles of semantically integrated conceptual modelling method. In: Business Modeling and Software Design, pp. 1–26. Springer International Publishing (2017)
13. Gustiene, P., Gustas, R.: Modelling patterns for business processes. In: Proceedings of the 25th International Conference ICIST 2019, pp. 161–172. Springer Nature, Switzerland (2019)
14. Hammer, M.: Reengineering work: don't automate, obliterate. Harv. Bus. Rev., 104–112 (1990)
15. Lankhorst, M.: Enterprise Architecture at Work: Modelling, Communication and Analysis. Springer, Berlin (2005)
16. Zachman, J.A.: A framework for information system architecture. IBM Syst. J. **26**(3) (1987)

Part III
Digital Architecture

Chapter 10
Architecting Digital Products and Services

Alfred Zimmermann, Rainer Schmidt, Kurt Sandkuhl, Dierk Jugel, Christian Schweda, and Justus Bogner

Abstract Enterprises are currently transforming their strategy, processes, and their information systems to extend their degree of digitalization. The potential of the Internet and related digital technologies, like Internet of Things, services computing, cloud computing, artificial intelligence, big data with analytics, mobile systems, collaboration networks, and cyber physical systems both drives and enables new business designs. Digitalization deeply disrupts existing businesses, technologies and economies and fosters the architecture of digital environments with many rather small and distributed structures. This has a strong impact for new value producing opportunities and architecting digital services and products guiding their design through exploiting a Service-Dominant Logic. The main result of the book chapter extends methods for integral digital strategies with value-oriented models for digital products and services which are defined in the framework of a multi-perspective digital enterprise architecture reference model.

Keywords Digital transformation · Digital products and services · Service-dominant logic · Value modeling · Digital architecture

10.1 Introduction

Data, information and knowledge are fundamental core concepts of our everyday activities and are driving the digital transformation of today's global society [12]. New services and smart connected products expand physical components by adding

A. Zimmermann (✉) · D. Jugel · C. Schweda · J. Bogner
Herman Hollerith Center, Informatics, Reutlingen University, Reutlingen, Germany
e-mail: alfred.zimmermann@reutlingen-university.de

R. Schmidt
Informatics and Mathematics, Munich University, Munich, Germany
e-mail: rainer.schmidt@hm.edu

K. Sandkuhl
Informatics and Electrical Engineering, University of Rostock, Rostock, Germany
e-mail: kurt.sandkuhl@uni-rostock.de

A. Zimmermann et al. (eds.), *Architecting the Digital Transformation*, Intelligent Systems Reference Library 188,
https://doi.org/10.1007/978-3-030-49640-1_10

information and connectivity services using the Internet. Influenced by the transition to digitalization many enterprises are presently transforming their strategy, culture, processes, and their information systems.

Digitalization [24] defines the process of digital transformation, which is promoted by important technological trends: Internet of Things, artificial intelligence, cloud, edge and fog computing, services computing, big data, mobile systems, and social networks. The disruptive change of current business interacts with all information systems that are important business enablers for the digital transformation. Digital services and products amplify the basic value and capabilities, which offer exponentially expanding opportunities. Digitalization enables human beings and autonomous objects to collaborate beyond their local context using digital technologies. The exchange of information enables better decisions of humans, as well as of intelligent objects. Furthermore, social networks, smart devices, and intelligent cars are part of a wave of digital economy with digital products, services, and processes, which call for an information-driven vision [6].

The digital transformation deeply disrupts existing enterprises and economies. Digitalization fosters the development of IT systems with many, globally available and diverse rather small and distributed structures, like Internet of Things or mobile systems. Since years a lot of new business opportunities appeared using the potential of the Internet and related digital technologies, like Internet of Things, services computing, cloud computing, big data with analytics, mobile systems, collaboration networks, and cyber physical systems. This has a strong impact for architecting digital services and products integrating high distributed systems and services.

Unfortunately, the current state of art in research and practice of enterprise architecture lacks an integral understanding of an integral value and service perspective of design models connecting a digital strategy with a digital enterprise architecture for the digitalization of products and services. Our goal is to extend previous quite static approaches of enterprise architecture to fit for a flexible and adaptive digitalization of intelligent products and services. This goal shall be achieved by introducing new mechanisms for an integral design of digital products and services, considering a service-dominant logic perspective, for a value-oriented modeling approach as part of an effective digital architecture.

Our chapter focuses on a fundamental research question:

> How can a value-oriented digital architecture for digital products and services be modeled to support the open-world integration and management for a huge amount of dynamically growing micro-granular intelligent systems and services?

We will proceed as follows. First, we are setting the fundamental architectural context for digital transformation introducing digital products and services. We integrate fundamental perspectives and principles of the service-dominant logic for architecting digital products and services. Then we present our digital modeling approach for systematically defining value-oriented digital products and services in order to map digital business operating to digital composition models and digital architectures. Based on the target of digital architectures we are focusing on modeling micro-granular digital systems and services and provide insides to our methods and mechanisms for architectural decision management for multi-perspective digital architectures. Finally, we conclude in the last section our research findings mentioning also our future work.

10.2 Digital Transformation

New high-performance IT technologies are strategic drivers of the digital transformation. Digital transformation [21, 25] primarily concerns the formulation of a digital strategy and new digital business models geared to disruption. The most important transformations concern the transition from analog to digital platforms in a way that is focused on customer value.

For years we have been experiencing a hype about Digitalization, in which the terms digitization, digitalization and digital transformation are often confusingly used. The origin of the term digitalization is the concept of digitization. According to [8], there are four stages of digitization (Fig. 10.1). At the basic level of substitution, analog media are replaced by digital media, which is called digitization. Digitization doesn't replace the original analog information, but provides a digital copy of it. This first transformation level essentially represents the digital substitution of original analog representations, such as paperwork, in computer-based media in order to

Digitalization Level	Description	Transformation Type	Example
1. Substitution	Technology as a direct tool substitute, with no functional change	Digital Enhancement (1)	Scientific paper as pdf file
2. Augmentation	Technology as a direct tool substitute, with functional improvements	Digital Enhancement (2)	Enhanced pdf file with direct connectors to processes / tools
3. Modification	Technology allows significant task redesign	Digital Transformation (1)	Paper submission automatically triggers the subsequent review process
4. Redefinition	Technology allows creation of new tasks, previously inconceivable	Digital Transformation (2)	Digital platform and ecosystem of living scientific conferences, journals, and other assets with co-creating people and intelligent services

Fig. 10.1 Levels of digitalization: digital enhancement versus digital transformation

store, process and transmit them digitally. In this way, only the static information is digitized without functional change, while the process in this first transformation step remains the same as in the analog world.

The next stage of digital improvement is augmentation, where the digitized information consists of a combination of basic information and is complemented by functional improvements such as menus to activate associated processing steps. In a further transformative step, the modification of the processes opens up new possibilities for exchange and communication. The aspect of digital transformation by modification is based on functional extensions, which enables significant transformations of processing, e.g. process automation with digital technologies and reduced user interaction.

Digital transformation is broader than digitalization and usually involves a digital strategy at the beginning. Digitalization of business leads to digital business. This transformation of business is an important topic of digital change management enabling the digital transformation. Digitalization is the use of digital technologies and data to enable digital value propositions for customers and to improve business while expanding the revenue.

Finally, at the level of the highest redefinition (digitalization) completely new forms of interaction and structures become possible. At the highest level, the transformative redefinition enables new digital perspectives, e.g. through digital platforms and ecosystems, which were previously not possible. Many domains of social life are redefined by digital communication and social media platforms. Digitalization enables changes to new digital business models with expanded value perspectives for customers and improved revenue opportunities. Digitalization is therefore more about shifting processes to attractive high automated digital business operations and not just communication by using the Internet. The digital redefinition usually causes disruptive effects on business. Going beyond the value-oriented perspective of digitalization, digital business requires a careful adoption of human, ethical and social principles.

Considering close related concepts of digitization, digitalization and digital transformation [5] we conclude: Digitization and digitalization is about digital technology, while digital transformation is about the changing role of digital customers and the digital change process. We digitize information, we digitalize processes and roles for extended platform-based business operations, and we digitally transform the business by promoting a digital strategy, customer-centric and value-oriented digital business models, and an architecture-driven digital change.

Digital technologies change the way we communicate and collaborate for value co-creation with customers and other stakeholders, even with competitors. Digital technologies have changed our view on how to analyze and understand a magnitude of real-time accessible data from multiple perspectives. Digital transformation has also changed our understanding on how to innovate in global processes to architect and develop digital products and services faster than ever approaching for the best available digital technology and quality. Digitalization force us to look differently on value creation for and together with customers.

Five strategic domains define the focus of a digital transformation in [21]: customers, competition, data, innovation, and value. Most important strategical changes of customers in a digital business are: customers as dynamic network, two-way communication for co-creation, customers are key influencers, marketing to inspire purchase and loyalty, reciprocal value flows, and economies of value. Strategic changes also affect competitors, as: competition across fluid industries, blurred differentiation between partners and competitors, competitors cooperate in key areas (coopetition), key assets reside in outside networks, platforms and ecosystems with partners who exchange value, and winner-takes-all due to network effects. The strategic perspective on data in digital business can be described as: data is generated and processed in real-time, data provides valuable information, unstructured data is increasingly processable and valuable, value of data results from connected data across silos, data is a key asset for value creation. Digital innovation is promoted by: decisions based on testing and validating, testing of ideas is cheap and easy, experiments are conducted constantly by everyone, challenge of innovation is to solve the right problem, early and cheaply failures, focus on minimum viable prototype and fast iterations. Finally, the improved digital value perspective results from: value proposition defined by changing customer needs, uncover the next opportunity for customer value, evolve before you must to stay ahead, judge change by how it could create your next business, and stay always active and not looking for complacency.

10.3 Digital Products and Services

The digital transformation is the current dominant type of business transformation having IT both as a technology enabler and as a strategic driver. Digitized services and associated products are software-intensive [24] and therefore malleable and usually service-oriented [35]. Digital products are able to increase their capabilities via accessing cloud-services and change their current behavior.

In the beginning, digitization was considered a primarily technical term [32]. Thus, a number of technologies is often associated with digitization [35]: cloud computing, big data often combined with advanced analytics, social software, and the Internet of Things. New technologies such as deep learning are strategic enablers and strong related with advances in digitalization. They allow computing to be applied to activities that were considered as exclusive to human beings. Therefore, the present emphasis on digitalization become an important area of research.

New ways of interaction with the customer are enabled [14] by combining a product consisting of hardware and software with cloud-provided services. Current research suggests that different customers will use such devices for different use cases enabling new ways of triggering and interaction with business processes. An example is Amazon Alexa, in [30] and Fig. 10.2, that consists of a physical device with microphone and speaker e.g. Echo Dot, and services, called “Alexa skills”. The

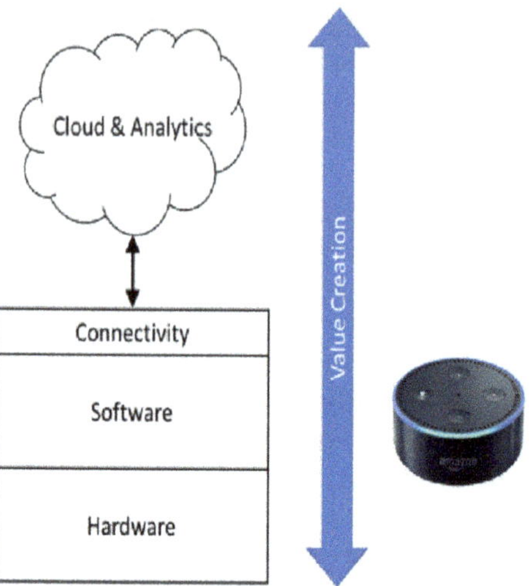

Fig. 10.2 Hybrid value creation for digital products

set of Alexa skills is dynamic and can be tailored to the customer's requirements during run-time. The lifecycle of digitized products is extended by the acquisition and decommissioning of services.

Our thesis is, that digitalization embraces both a product and a value-creation [24] perspective. Digitized products and services support the co-creation of value together with the customer and other stakeholders in different ways. First, there is a permanent feedback to the provider of the product. The internet connection of the digitized product allows to collect permanently data on the usage of the product by the customer. Second, the data provided by a large number of digitized products are able to provide new insights, which are not possible with data from a single device. Current research argues that digital products and services are offering disruptive opportunities [12] for new business solutions, having new smart connected functionalities.

Classical industrial products are static [24]. You can only change them to a limited extent, if at all. On the contrary, digitized products are dynamic. They contain both hardware, software and (cloud-)services. They can be upgraded via network connections. In addition, their functionality can be extended or adapted using external services. Therefore, the functionality of products is dynamic and can be adapted to changing requirements and hitherto unknown customer needs. In particular, it is possible to create digitized products and services step-by-step or provide temporarily unlockable functionalities. So, customers whose requirements are changing can add and modify service functionality without hardware modification.

Digital products [12] are able to capture their own state and submit this information into linked contexts. This is the basis for the so called servitization of products. Not a physical product, but a service is sold to the customer. The service usage is measured and lays the foundation for usage-based billing models. The provider can remotely determine, whether the product is still functional and trigger, where appropriate, maintenance and repairs. Evaluation of status information and analysis of the history of use of the product can be predicted when a malfunction of the product is probable. A maintenance or replacement of the product is performed before predicted data of failure. The data collected also provide information for a repair on the spot, so that a high first-time solution rate can be achieved. At the same time, storage can be improved in this way of spare parts. By this means, preventive maintenance can be implemented. Unscheduled stoppages can this way be significantly reduced.

Digital products also enable network effects [12] that grow exponentially with the number of participating devices. An increase in the number of digitized products increases the incentive for providers of add-on services and complementary skills. At the same time this increase the attractiveness for further digitized products. In summary, an exponential growth can be achieved. Therefore, significant first-mover advantages exist. Network effects emerge not only for the functionality but also for the analytical exploitation of data collected by the digitized products. These effects are called network intelligence [12]. By bringing together data from many devices and not only single devices, trends can be detected much earlier and more accurately. Further improvements can be achieved by linking data from different sources, also external ones. In this way, it is possible to establish correlations that would not have been possible considering data from a single device. This effect increases with the number of devices.

Digital products and services [24, 14] become part of an information system, which accelerates the learning and knowledge processes across all products. The manufacturer can win genuine information about the use of the product. Important information for the development of new products can be obtained in this way. Therefore, a number of other beneficial effects can be achieved as network optimization, maintenance optimization, improved restore capabilities, and additional evidence against the consideration of individual systems.

Traditional products were created with a tayloristic view in mind, that emphasized the separation of production and consumer in order to enable centralized production and thus scaling effects. Now, the co-creation approach of service-dominant logic [28, 29] can be implemented because of a persisting continuous connectivity of digital products with the manufacturer. The consumer converts dynamically to be co-producer. Platforms are complementary to products, which cooperate via standardized interfaces.

10.4 Principles of Service-Dominant Design

The Service-Dominant (S-D) logic [28, 29] is a fundamental service-centered approach and to some extend opposite to the traditional goods-centered paradigm for large parts of the traditional business. The principal idea is that all economic exchanges can be defined as service-to-service exchanges considering also associated real or digital products. The origin of the service-dominant logic relies on ten fundamental axioms [28] for defining service businesses, including digital services and products. The origin of service-dominant logic was slightly extended through modifications and additional premises [29] to a body of five axioms and eleven foundational premises.

Fundamental premises of the S-D logic [28, 29] promote a clear service perspective and above all the exchange of services based on a network-centered actor for actor generalization, predicating the first axiom: a service is the fundamental basis of exchange, and deriving the indirect exchange as fundamental basis of exchange. S-D logic relies on decoupling of information from related physical forms. Goals provide just distribution mechanisms for service provisioning, which implies that all economies are service economies. Operant resources, like knowledge and people, are the fundamental source of strategic benefit.

The second axiom is that the value is co-created by several actors, including the beneficiary, e.g. the customer, but also the service provider. The next fundamental understanding of S-D logic in the third axiom is that actors are resource integrators, but cannot deliver value by themselves, but can participate in the creation of value propositions, and promoting that a service-centered view should be beneficiary-oriented. All social and economic actors are resource integrators, not only by providing a service but also integrating various even external resources.

The last two axioms postulate that value is always determined by the beneficiary, and value co-creation is coordinated through actor-generated institutions and institutional arrangements. Institutions are in this sense human-defined rules, norms, and beliefs that enable and constrain actions.

Through exploiting the base of service-dominant logic and by means of design research a focused set of four design principles for business-model-based management methods was elaborated in [3]. This service-dominant perspective of inspiring design principles motivates and sets the base for our digital service-oriented design methodology. These design principles not only apply to capture the general guidance and methods for designing digital services and products, but also to illustrate their organizational implications.

The first principle defines the proactive base for an ecosystem-oriented management by positioning the orchestration tasks for specific actors in a service ecosystem, defining an organization's role as focal orchestrator in the service ecosystem, and for sharing the risks, costs, and revenues among multiple actors.

The second principle about a technology-based management defines responsibilities for using digital infrastructures, for decoupling informational assets from products and facilitate product exchange, and for driving value creation through digital channels.

Principle three about mobilization-oriented management postulate the mobilization of operand resources, like knowledge and capabilities, which are the fundamental source of strategic benefit, and further uncovering and utilizing internal knowledge.

The last principle about co-creation-oriented management demands for customer involvement, to reflect on co-creation through customer journey as dynamic interaction, and for recalibrating service bundles to optimize customer's experience.

In addition to methods for digital business modelling, we also cover theoretical principles for the design of a digital infrastructure, which are part of a theoretical framework [4]: structural integrity, elasticity, ambidexterity, connectivity, generativity, and modularity.

Structural integrity of the digital infrastructure is the fundamental basis for control and stability. The digital infrastructure aims at stabilization in order to use the actor-to-actor network. Elasticity of a digital infrastructure supports autonomy and change. The digital infrastructure is looking for dynamism to open new ways of resource integration that adapt to external stimuli. Different ways of organizing actors should be suitable for innovation opportunities.

The ambidexterity of the digital infrastructure is the basis for reconciling integrity and elasticity, seeking a balance between structural integrity and elasticity. A digital infrastructure should enable both control and stability as well as autonomy and change. The connectivity base is the fundamental mechanism of the structural integrity of the digital infrastructure. The connectivity of the digital infrastructure reduces communication costs and increases its speed and reach. Generativity is the basic mechanism for the elasticity of the infrastructure, generating new results, structures and behaviors. Modularity is the basic mechanism of ambidexterity of the digital infrastructure to increase the separation and recombination of digital components.

10.5 Value-Oriented Architecture

The business and technological impact of digitalization [12] has multiple aspects, which directly affect digital architectures of service-dominant digital products and services. Unfortunately, our current modeling approach for designing proper digital service and product models suffers from having many uncontrolled diverse modeling approaches and structures, where value-orientation of integral composed services and systems is only partially fulfilled. High quality digital models should follow a clear value and service perspective.

But today, we currently have no sound value relationship from digital strategies, to the resulting digital business modeling, and subsequently to a value-oriented enterprise architecture, which today often has seldom proper aligned service and product model representations. The core idea of the present contribution and current paper is to present and discuss a new introduced integral value-oriented model composition approach by linking digital strategies with digital business models for digital services and close aligned products by means of an extended multi-perspective digital enterprise architecture model.

Value is commonly associated with worth and aggregates potentially required categories like worth, importance, desirability and usefulness. The concept of value is important in designing adequate digital services with their associated digital products, and to align their digital business models with value-oriented enterprise architectures. From a financial perspective the value of the integrated resources and the price defines the main parts of the monetary worth.

A current conceptualization of value as a service-based view is offered by [27, 18] considering a conceptual framework of service-dominant (S-D) logic [28, 29] and its service-ecosystem perspective. The distinction between the concepts of value-in-use and value-in-exchange dates back to the antiquity and continue to influence our today's value view. Since the work of Adam Smith and the development of economic science the value-in-exchange as a measure for price a person is willing to pay for a service or a product moved to the forefront. Smith recognized the value-in-use as the real value and value-in-exchange as the nominal value.

We present our view of an integrated value perspective combined with a service perspective in Fig. 10.3. Today, we are experiencing a starting set of now not well consolidated digital strategy frameworks, like in [6, 23] which are loosely associated with traditional strategy frameworks.

The digital marketing discipline nowadays shifted to a nominal use of the value perspective [27] considering customer experience and customer satisfaction as important value-related concepts. Characteristics of value modeling for a service ecosystem were elaborated by [3]. Value has important characteristics: value is phenomenological, co-created, multidimensional, and emergent. Value is phenomenological means that value is perceived experimentally and differently by various stakeholders in the varying context within a service ecosystem. Value is co-created though the integration and exchange of resources between multiple stakeholders and related organizations.

Value is also multidimensional, which means that value is aggregated up of individual, social, technological and cultural components. Value results as emergent value from specific manifestations of relationships between resources and resource combinations. Therefore, the resulting real value cannot be determined ex-ante. Value propositions are value promises for a typical, but not exactly known customer at design time and should be realized later when using these digital services and associated products.

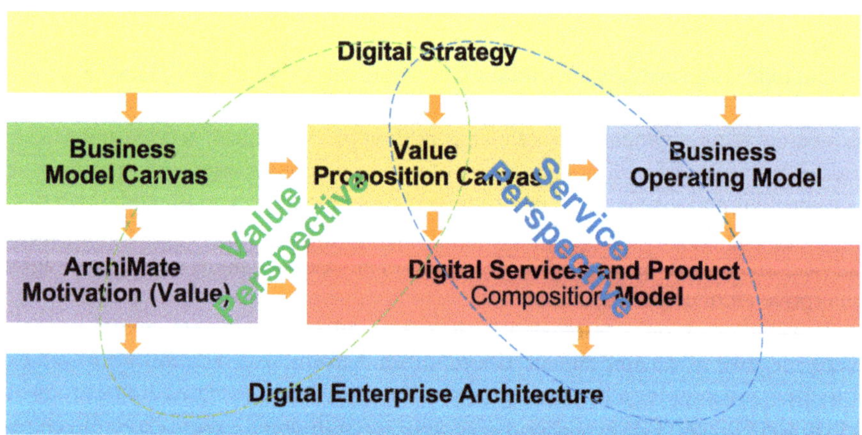

Fig. 10.3 Value and service perspective of service-dominant logic

Our starting point is a model of the digital strategy, which gives the digital modeling direction and sets the base and a value-oriented framing for the business definition models, with the business model canvas [19], and the value proposition canvas [20]. Having the base models for a value-oriented digital business we map these base service and product models to a digital business operating model. An operating model [22] strategically defines the necessary level of business integration and standardization for delivering services and products to customers. From the value perspective of the business model canvas [19] results suitable mappings to enterprise architecture value models [13] with ArchiMate [17]. Finally, we are setting the frame for the systematic definition of digital services and associated products by modeling digital services and product compositions following semantically related composite patterns.

We have integrated the Business Model Canvas [19] into our comprehensive digital modeling framework to uniformly describe the digital business model respectively to enable a first conceptual modeling of digital products and services as well as a firm's or product's value proposition, infrastructure, customers, and finances. The Business Model Canvas was initially proposed by Alexander Osterwalder [19] and is based on his PhD-Thesis on Business Model Ontology. Central part of the Business Model Canvas [19] is the description of offerings based on value propositions. First, the Business Model Canvas indirectly describes offerings giving priority to the value propositions to meet the needs of its customers. A company's value proposition is what distinguishes it from its competitors.

The value proposition provides value through various elements such as newness, performance, customization, "getting the job done", design, brand/status, price, cost reduction, risk reduction, accessibility, and convenience/usability. The value propositions may contain demanded or supposed business requirements which can be related to qualitative or quantitative descriptions. The Value Proposition Canvas [20] extends the Business Model Canvas [19] to conceptualize and exactly define digital services and products for customers which are connected to the perceived value of digital products or services, and the potential product/market context. The value proposition from the Business Model Canvas [19] is projected to the Value Proposition Canvas [20] and divided into three perspectives as products and services, creators of profit and painkillers, which are associated to the customer segment aspects, as customers jobs, profits, and pain.

The primary motivation of successful organizations is to provide value to one or more stakeholders, typically considering value for clients at first. This includes the modeling of value creation, capturing, and value delivery by using discrete value producing tasks. Classical concepts of value chains and value networks are seminal for lean value streams and for applying the current fundamental TOGAF [16] series guide on value streams [18]. Value chain modeling focuses on an economic perspective while value networks primarily shows participants involved in creating value.

Value streams, as in [18], model an end-to-end value view of value-adding activities as value stream stages from the customer's or stakeholder's perspective. Therefore, value streams enable digital business models which are closer to the definition and not the implementation of organizational core activities. Value streams are defined as compositions of value stages from the value-perspective for the addressed stakeholders.

From using value stream models and mappings we can summarize important benefits. Value stream models are the base for decision making helping to envision and prioritize the impact from strategic plans, for managing the stakeholders' engagement, and supporting the deployment of new business solutions. Business capabilities enable value stages and value streams, which are focused to the viewpoint of customers. Value streams provide a framework for better requirement analysis, case management, and supports modeling of digital services. Finally, value streams are focused on how business value is achieved for specific stakeholders, particularly for customers.

10.6 Digital Enterprise Architecture

Enterprise Architecture Management [11], as today defined by several standards like [16] and [17] uses a quite large set of different views and perspectives for managing current IT. An effective architecture management approach for digital enterprises should additionally support the digitalization of products and services [24] and be both holistic and easily adaptable [35]. Furthermore, a digital architecture sets the base for the digital transformation enabling new digital business models and technologies that are based on a large number of micro-structured digital systems with their own micro-granular architectures [34], like IoT [26, 1, 33] mobile devices, or with Microservices [2, 15].

We are extending our service-oriented enterprise architecture reference model for the context of digital transformation with micro-granular structures and considering associated multi-perspective architectural decision-making [9] models, which are supported by viewpoints and functions of an architecture management cockpit. DEA - Digital Enterprise Architecture Reference Cube provides an architectural reference model [35] for bottom-up integrating dynamically composed micro-granular architectural models [34] (Fig. 10.4). DEA for architecting digital products and services is more specific than existing architectural standards of architecture management, like in [16, 17]. The bottom-up composition of living architectural models fundamentally extends existing quite static frameworks from practice.

DEA provides today with our current research ten integral architectural domains for a holistic architectural classification model, which is well aligned to embed also mirco-granular architectures for different digital services and products. DEA abstracts from a concrete business scenario or technologies, because it is applicable for concrete architectural instantiations to support digital transformations independent of different domains. The Open Group Architecture Framework TOGAF

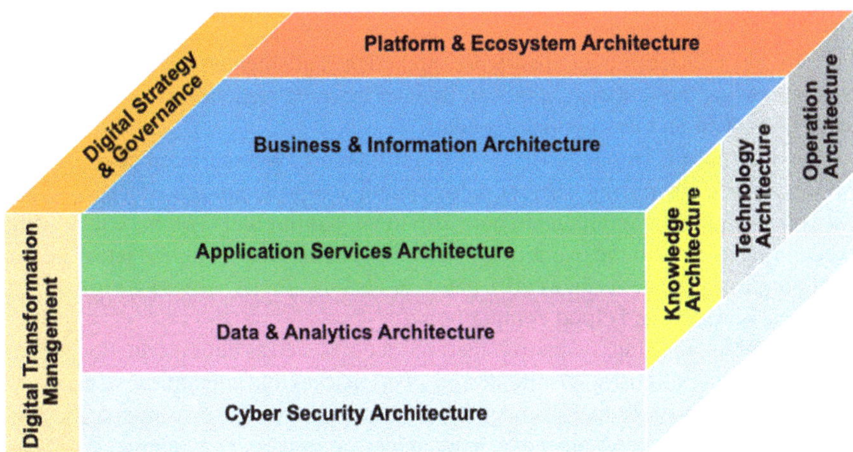

Fig. 10.4 Digital enterprise architecture reference cube

[16] provides the basic blueprint and structure for extended service-oriented enterprise architecture domains. Metamodel extensions are additionally provided by considering and integrating ArchiMate Layer models from [17].

DEA extends by a holistic view the metamodel-based extraction and bottom-up integration for micro-granular viewpoints, models, standards, frameworks and tools of a digital enterprise architecture model. DEA frames these multiple elements of a digital architecture into integral configurations of a digital architecture by providing an ordered base of architectural artifacts for associated multi-perspective decision processes.

Metamodels and their architectural data are the core part of the enterprise architecture. Enterprise architecture metamodels [11, 17] should enable decision making [10] as well as the strategic and IT/business alignment. Three quality perspectives are important for an adequate IT/business alignment and are differentiated as: (i) IT system qualities: performance, interoperability, availability, usability, accuracy, maintainability, and suitability; (ii) business qualities: flexibility, efficiency, effectiveness, integration and coordination, decision support, control and follow up, and organizational culture; and finally (iii) governance qualities: plan and organize, acquire and implement deliver and support, monitor and evaluate (e.g., [31])

Digitalization promotes massively distributed systems, which are based on the development of IT systems with many rather small and distributed structures, like Internet of Things, mobile systems, cyber physical systems, etc. Additionally, we have to support digitalization by a dense and diverse amount of different service types, like microservices, REST services, etc. and put them in a close relationship with distributed systems, like Internet of Things. The change from a closed-world modeling perspective to more flexible open-world composition and evolution of system architectures defines the moving context for adaptable systems, which are essential to enable the digital transformation. This has a strong impact for architecting digital services and products. The implication of architecting micro-granular systems and services considering an open-world approach fundamentally changes modeling contexts, which are classical and well defined by quite static closed-world and all-times consistent and less complex models.

The Internet of Things (IoT) [1, 26, 33] connects a large number of physical devices to each other using wireless data communication and interaction based on the Internet as a global communication environment. Additionally, we have to consider challenging aspects of the overall software and systems architecture to integrate base technologies and systems, like cyber-physical systems, social networks, big data with analytics, services, and cloud computing.

The Internet of Things, supports smart products as well as their production enables enterprises to create customer-oriented products in a flexible manner. Devices, as well as human and software agents, interact and transmit data to perform specific tasks as parts of sophisticated business or technical processes. The Internet of Things embraces not only a things-oriented vision [1] but also an Internet-oriented and a Semantic-oriented one. A cloud-centric vision for architectural thinking of a ubiquitous sensing environment is provided by [26].

Microservices and Microservices Architectures (MSA), as in [2, 15], is considered to be an important enabler for the digital enterprise and the digital transformation. Software developing enterprises have switched to integrate Microservice architectures to handle the increase velocity. Therefore, applications built this way consist of several fine-grained services that are independently scalable and deployable. The fast-moving process of digitalization demands flexibility to adapt to rapidly changing business requirements and newly emerging business opportunities.

Microservices addresses our second fundamental use-case for micro-granular architectures, which are developed and operated in an open-world. The fundamental concept of architecture is defined as structure of components, their inter-relationships, together with principles and guidelines for governing their design and evolution. The open-world approach fundamentally changes the rules of engineering and management by following a high distributed and globally metaphor for the new setting of a digital business operating model. This bottom-up tailored digital operating model changes the perspective of a classical top-down oriented enterprise architecture.

Microservices should be designed to be self-contained by integrating with specific needed platform and infrastructural elements. Microservices does not require a large pre-existing infrastructure. As exemplified by DevOps [15], Microservices support processes of Continuous Development (CD) in small environments and Continuous Integration (CI). Additionally, Microservices should also naturally support resiliency and scalability in both cloud and on-premise environments.

Architecture governance, as in [31], defines the base for well aligned management practices through specifying management activities: plan, define, enable, measure, and control. Digital governance should additionally set the frame for digital strategies, digital innovation management, and Design Thinking methodologies. The second aim of governance is to set rules for a value-oriented architectural compliance based on internal and external standards, as well as regulations and laws. Architecture governance for digital transformation changes some of the fundamental laws of

Fig. 10.5 Example: enterprise architecture cockpit

traditional governance models to be able to manage and openly integrate a plenty of diverse micro-granular structures, like Internet of Things or Microservices.

To effectively support architecture managers for navigation, interaction and decision support during knowledge intensive management activities in a complex architectural space we have adopted the idea of a Decision Support System (DSS) [10], as in Fig. 10.5, which gives the idea of managing decisions by navigating using an enterprise architecture cockpit [9, 7].

A cockpit presents a facility or device via which multiple viewpoints on the system under consideration can be consulted simultaneously. Each stakeholder who attends a cockpit meeting can utilize a viewpoint that displays relevant information. Stakeholders can leverage views that fit the particular role, like application architect, process owner or infrastructure architect.

The viewpoints are applied simultaneously and are linked to each other showing multi-perspective architectural dependencies and the impact of change. Changes in one view are pointing to updates and dependencies in other views as well.

Jugel et al. [10] have elaborated a collaborative approach for decision-making for EA management. They identify decision making in such architectural exploration environments as a complex knowledge-intensive process, which strongly depends on the participating stakeholders. Therefore, the collaborative approach presented is based on the methods and techniques of adaptive case management (ACM).

10.7 Conclusion

Based on our fundamental research question we have first set the context proceeding from digital transformation to a systematic value-oriented digital product and service modeling according the service-dominant logic. To be able to support the dynamics of digital transformation with flexible software and systems compositions we have leveraged an adaptive architecture approach for open-world integrations of globally accessed systems and services with their local architecture models.

We contribute to the literature in different ways. Looking to our results, we have identified the need and solution mechanisms for a value-oriented integration of digital strategy models through suitable digital business models up to models for service-dominant products as part of a value-based digital enterprise architecture. To integrate micro-granular architecture models from an open-world we have extended traditional enterprise architecture reference models and enhanced them with state of art elements from agile architectural engineering to support the digitalization of products, services, and processes. This is a major extension of our seminal work on reference enterprise architectures, to be able to openly integrate through a continuously bottom-up approach a huge amount of global available and heterogeneous micro-granular systems, having their own local architectures.

Strength of our research results from our novel approach of integrating a digital strategy with digital business operating models for architecting flexible compositions of digital products and services. Limits are still resulting from an ongoing validation of our research and open issues in managing inconsistencies and semantic dependencies.

Future research will cover mechanisms for flexible and adaptable integration of digital architectures. We are working to extend human-controlled dashboard-based decision making by AI-based intelligent systems for decision support, as well as for architectural data and model analytics with semantic representations combined with deep learning mechanisms.

References

1. Atzori, L., Iera, A., Morabito, G.: The internet of things: a survey. J. Comput. Netw. **54**, 2787–2805 (2010)
2. Balakrushnan, S., Mamnoon, O., Bell, J., Currier, B., Harrington, E., Helstrom, B., Maloney, P., Martins, M.: Microservices Architecture. White Paper, The Open Group (2016)
3. Blaschke, M., Haki, M.K., Riss, U., Aier, S.: Design principles for business-model-based management methods-A service-dominant logic perspective. In: Maedche, A. et al. (Eds.) DESRIST 2017, pp. 179–198. Springer (2017)
4. Blaschke, M., Haki, M.K., Riss, U., Winter, R., Aier, S.: Digital infrastructure-A service-dominant logic perspective. In: Thirty Seventh International Conference on Information Systems, Dublin (2016)
5. Bloomberg, J.: Digitization, Digitalization, And Digital Transformation: Confuse Them At Your Peril. Forbes (2018)
6. Bones, C., Hammersley, J., Shaw, N.: Optimizing Digital Strategy—How to Make Informed, Tactical Decisions That Deliver Growth. Kogan Page (2019)
7. Emery, D., Hilliard, R.: Every Architecture Description needs a Framework: Expressing Architecture Frameworks Using ISO/IEC 42010. IEEE/IFIP WICSA/ECSA, pp. 31–39 (2009)
8. Hamilton, E. R., Rosenberg, J.M., Akcaoglu, M.: Examining the substitution augmentation modification redefinition (SAMR) model for technology integration. Tech Trends, vol. 60, pp. 433–441. Springer (2016)
9. Jugel, D., Schweda, C. M. (2014) Interactive functions of a cockpit for enterprise architecture planning. In: International Enterprise Distributed Object Computing Conference Workshops and Demonstrations (EDOCW), Ulm, Germany, 2014, pp. 33–40. IEEE

10. Jugel, D., Schweda, C.M., Zimmermann, A.: Modeling decisions for collaborative enterprise architecture engineering. In: 10th Workshop Trends in Enterprise Architecture Research (TEAR), held on CAISE 2015, Stockholm, Sweden, pp. 351–362. Springer (2015)
11. Lankhorst, M.: Enterprise Architecture at Work: Modelling, Communication and Analysis. Springer (2017)
12. McAfee, A., Brynjolfsson, E.: Machine, Platform, Crowd. Harnessing Our Digital Future. W. W. Norton & Company (2017)
13. Meertens, L. O., Iacob, M. E., Nieuwenhuis, L.J.M., van Sinderen, M.J., Jonkers, H., Quertel, D.: Mapping the Business Model canvas to ArchiMate. SAC 2012, pp. 1694–1701. ACM(2012)
14. Möhring, M., Keller, B., Schmidt, R., Pietzsch, L., Karich, L., Berhalter, C.: Using Smart Edge Devices to Integrate Consumers into Digitized Processes: The Case of Amazon Dash-Button. BPM, Workshops, LNBIP, pp 374–383. Springer (2018)
15. Newman, S.: Building Microservices Designing Fine-Grained Systems. O'Reilly (2015)
16. Open Group: TOGAF Version 9.2. The Open Group (2018)
17. Open Group: ArchiMate 3.0 Specification. The Open Group (2016)
18. Open Group: Value Streams. The Open Group (2017)
19. Osterwalder, A., Pigneur, Y. Business Model Generation. Wiley (2010)
20. Osterwalder, A., Pigneur, Y., Bernarda, G., Smith, A., Papadokos, T.: Value Proposition Design. Wiley (2014)
21. Rogers, D.L.: The Digital Transformation Playbook. Columbia University Press (2016)
22. Ross, J. W., Weill, P., Robertson, D.: Enterprise Architecture as Strategy—Creating a Foundation for Business Execution. Harvard Business School Press, Harvard Business School Press (2006)
23. Ross, J.W., Sebastian, I.M., Beath, C., Mocker, M., Moloney, K.G., Fonstad, N.O.: Designing and Executing Digital Strategies. ICIS 2016, Dublin(2016)
24. Schmidt, R., Zimmermann, A., Möhring, M., Nurcan, S., Keller, B., Bär, F.: Digitization–Perspectives for Conceptualization. In Advances in Service-Oriented and Cloud Computing, pp. 263–275. Springer International Publishing (2016)
25. Sebastian, I. M., Mocker, M., Ross, J. W., Moloney, K. G., Beath, C., Fonstad, N. O.: How Big Old Companies Navigate Digital Transformation. MIS Quarterly Executive, September 2017, pp. 197–213
26. Uckelmann, D., Harrison, M., Michahelles, F. (2011) Architecting the Internet of Things. Springer
27. Vargo, S.L., Akaka, M.A., Vaughan, C.M.: Conceptualizing value: a service-ecosystem view. J. Creat. Value **3**(2), 1–8 (2017)
28. Vargo, S.L., Lusch, R.F.: Service-dominant logic: continuing the evolution. J. Acad. Mark. Sci. **36**(1), 1–10 (2008)
29. Vargo, S.L., Lusch, R.F.: Institutions and axioms: an extension and update of service-dominant logic. J. Acad. Mark. Sci. **44**(4), 5–23 (2016)
30. Warren, A.: Amazon Echo: The Ultimate Amazon Echo User Guide 2016 Become an Alexa and Echo Expert Now!. CreateSpace Independent Publishing Platform, USA (2016)
31. Weill, P., Ross, J.W.: IT Governance: How Top Performers Manage It Decision Rights for Superior Results. Harvard Business School Press (2004)
32. Weill, P., Woerner, S.: Thriving in an increasingly digital ecosystem. MIT Sloan Manag. Rev. (2015)
33. WSO2: A Reference Architecture for the Internet of Things. Version 0.9.0 (2015). https://wso2.com. Accessed 22 March 2019
34. Zimmermann, A., Schmidt, R., Sandkuhl, K., Wißotzki, M., Jugel, D., Möhring, M. Digital Enterprise Architecture—Decision Management for Micro-Granular Digital Architecture. In Hallé, S., Dijkman, R.., Lapalme, J. (eds.) EDOC 2017 with SoEA4EE, pp. 29–38. IEEE Computer Society (2017)
35. Zimmermann, A., Schmidt, R., Sandkuhl, K., Jugel, D., Bogner, J., Möhring, M. (2018) Evolution of enterprise architecture for digital transformation. In: Proceedings of EDOC 2017/SoEA4EE, pp. 87–96, IEEE Computer Society

Chapter 11
Adaptive and Resilient Business Architecture for the Digital Age

Václav Řepa and Oleg Svatoš

Abstract The process of enterprise digitalization has become in recent years unstoppable. In the global competition, enterprises have to use all the means available in order to be able to respond to competitors' innovations or to be able to implement innovations that would take them ahead of the competition. With the rising complexity of the used business and the information technologies, usage of enterprise architecture becomes inevitable so that the business-IT alignment is maintained. In this chapter we describe a method how to develop a resilient and adaptive minimal business architecture so that business requirements can be clearly defined, their support by the functionality of applications evaluated and the missing support resolved. The method is illustrated with the generally accepted EA standards TOGAF/ArchiMate, UML and BPMN and its benefits are discussed. We also provide the reader with a meta-model that specifies how these standards in the discussed areas match together.

Keywords Digitalization · Enterprise architecture · Business process

11.1 Introduction

There are significant changes being made in the businesses which undergo the process of digital transformation. The reason is that the substance of digitalization is not the inherent automation, but an application of new technologies connected with a radical redesign of current business processes so that the services and products, the businesses provide to their customers, deliver much higher comfort and value. The digital transformation is a very expensive process and therefore changes and also the

V. Řepa (✉) · O. Svatoš
Faculty of Informatics and Statistics, University of Economics in Prague, Prague, Czech Republic
e-mail: repa@vse.cz

O. Svatoš
e-mail: svatoso@vse.cz

A. Zimmermann et al. (eds.), *Architecting the Digital Transformation*, Intelligent Systems Reference Library 188,
https://doi.org/10.1007/978-3-030-49640-1_11

outcomes have to be significant otherwise it does not pay off. Simple automation of current practices or cost-saving project is not just enough [1].

The substance of the digital transformation is then in creating something that does not exist yet. The general approach to such a project is to start with a model, which allows one to test one's ideas and form something that at least works on a paper. Only when one elaborates one's ideas in a relevant formal model(s), it makes sense to start the digital transformation process, which not only includes implementation of a particular application but and mainly, the change of the business itself: business processes, products, services as well as required competences [2]. For this purpose, there have been developed modeling standards that allow one to capture the business system itself and also the information system (IS) which would support it. But the individual modeling standards are not usually enough. The model has to represent architecture built on several views (diagrams) at the modeled business system, each putting emphasis on different aspects of reality. This usually requires several standards to be used. When creating a "vision of future reality" on paper, the mutual consistency of different diagrams, which picture the same future reality from different perspectives, is the main tool how to get to an agreement that the suggested digital transformation is feasible and makes sense at least on a paper. At the stage of analysis, this is probably the only way how to get to this conclusion. The required speed, with which the digital transformation has to be implemented and deployed, is in the digital age relatively high [3] and there is usually not enough time to develop business architecture in such detail that there would be some simulation possible. It would cost too much time, money and effort. Nevertheless, the completeness and correctness of the business architecture is an important issue, as any significant change in the process of implementation or even in the deployment, would endanger project delivery. The analytical stage is the place at which one should deal with it primarily. We do not suggest neglecting the detail. We just suggest that the model of the business should focus on a consistent architecture at a conceptual level and a particular detail can be elaborated in the process of implementation through, for instance, prototyping.

Unfortunately although the commonly used standards for business architecture such as TOGAF/ArchiMate, BPMN or UML provide analysts and architects with wide variety of diagrams that cover all kinds of perspectives, they are not specifying precisely what is the minimum of diagram types one has to use for a proper business architecture, which one should start with, how exactly the diagrams from different but related standards relate to each other (for instance ArchiMate and BPMN) and how their consistency can be verified.

The main goal of this chapter is to outline the minimum of mutually complementary perspectives which should be incorporated in a business architecture for digital transformation, what diagrams should be used for that and what are the consistency rules among them. Further on we illustrate how to derive requirements on IS functionality from such business architecture so that the IS functionality is fully aligned with the business architecture making so the digital transformation possible. The content of the minimal business architecture for digital transformation, which we here propose, stays at the conceptual level as we want to provide the readers with a general approach to the business system modeling, which is alterable for the individual needs of individual businesses. We do not get into an implementation detail. First, it is specific for each enterprise and its current business, application and technological architecture and second, there have been a number of methodologies already introduced for this, either traditional [4–6] or agile [7].

For the conceptual level, there are many modeling standards and frameworks available yet none of them covers the required detail of business architecture for the digital transformation completely. We base our approach on the most popular standard for enterprise architecture i.e., TOGAF, which we link together with the UML and BPMN. We take into consideration also the ArchiMate as its meta-model extends TOGAF concepts and makes them more tangible. Both TOGAF and ArchiMate standards consider the UML and BPMN as a valid extension for further architecture detail elaboration, but they do not elaborate their relation in detail. Neither does BPMN or UML. We fill these blank spaces based on the Methodology for Modelling and Analysis of Business Processes (MMABP). It not only maps the relevant relations among these standards together but also suggests a minimum of specific diagrams from these standards which should be used and put together so that one gets the complete view at the modeled business system and is able to evaluate business architecture's completeness, correctness, and consistency.

A practical point of view at the business system modeling is also considered. The creation of the presented minimal business architecture would not be possible without equally capable computer-aided software engineering (CASE) tool. It has to be able to not only support the individual modeling standards used but it also has to be able to capture the relations between entities of different diagrams of different standards. In our case, we use the open-source tool Modelio and its TOGAF module, which is based on [8] and covers fully all the diagrams specified in TOGAF. The interconnection between TOGAF, UML, and BPMN is in this module also included. Thanks to it all the diagrams and models, presented in this chapter, it is possible to create, manage and have interlinked and consistent at one place in one tool. Modelio has also an enterprise edition that meets with the standards an enterprise may require.

This chapter consists of five main sections. First, we present core principles the proposed minimal business architecture is built on. In the second section, we go over the global analysis and the diagrams that should be elaborated during this phase so that in the following section we can dive into the detailed analysis of particular aspects identified in the global analysis. In the fourth section, we put all the introduced diagrams together and based on the presented meta-model we discuss the general consistency rules these diagrams have to comply with. In the fifth section, we illustrate how to derive function requirements on IS based on the business architecture so that the technological architecture is fully aligned with the business.

11.2 Core Principles of Minimal Business Architecture

As mentioned in the previous paragraph the approach presented in this chapter is based on the MMABP methodology. MMABP is a general methodology for modeling business systems. We understand the concept of the business system as any system created and constantly developed by people, the aim of which is achieving so-called business goals. In this conception, the business system consists of mutually collaborating business processes that are altogether focused on achieving particular business goals. Achieving of business goals is generally determined by the general rules of the environment in which the processes operate. We call the collection of those rules *business rules*. MMABP also pays special attention to the IS as an integral part of the business system even if it itself is a model of the business system.

Unlike the modeling standards, which specify a great number of types of diagrams each applicable for a specific case, the MMABP proposes a minimal number of types of diagrams that have to be elaborated in order to have a compact and consistent business architecture which is ready for digital transformation.

There are two basic kinds of models of the business (Real World) system:

- **Real World ontology** represented by models expressing the important features of the business environment—modality and causality. These models are in informatics also called structural models as they are focused on the structure of the business system—which objects it consists of and how they can interact. For the ontology models MMABP uses standard modeling language UML [9] which contains diagrams for sufficient modeling of the Real World modality as well as its related causality:

  - ***Class Diagram*** is used for the static model of the basic modality of the Real World in terms of the conceptual model. The conceptual model represents the so-called system model as it describes the whole system of mutually related business objects. This kind of description principally does not allow capturing the temporal structural aspects of the Real World (i.e., its causality) as it is focused on the common aspects of the whole system while temporal aspects are principally located just to particular elements of the system.
  - ***State Chart*** is used for the model of the causality connected with the Real World object in terms of the so-called object life cycle. This kind of description completes the system model with temporal aspects (i.e., causality). Each model is focused on a single object and describes the causality of the Real World relevant for the given object in the form of its life cycle.
  - UML defines some basic relationships between both diagrams which are also a part of the MMABP rules for modeling the Real World ontology.

- **Real World behavior** represented by models expressing the relevant behavior of business actors as a system of business processes. Unlike in the case of the ontology, there is no integrated language in informatics for the complete description of both—the system of processes and the temporal aspects of a single process.

MMABP uses the standard business process modeling language **BPMN** [10] for the model of the run (i.e., temporal aspects) of the process complemented with the **Process Map** for the description of the process system. Even though the Process Map is a necessary and often used type of diagram, it is not present in the BPMN standard and has to be complemented from another resource. In this chapter, we use the TOGAF Event diagram [11] for this purpose. Looking from the TOGAF perspective, the MMABP extends the TOGAF/ArchiMate business architecture by missing detail necessary for digital transformation represented by BPMN.

The above described four basic diagrams for modeling the business system the MMABP completes with the traditional diagram for the description of the needed functionality of the IS—**Data Flow Diagram (DFD)** [5]. This diagram should not be regarded absolutely as a model of the Real World as the functionality of the IS itself is actually a model of the Real World. Nevertheless, DFD models just the conceptual contents of the IS (its functionality) which also plays an important role in the business system. Therefore the DFD has to be also regarded as an integral part of the set of Business System model and its conceptual relationships to other models have to be defined in the methodology (see Fig. 11.1).

In both kinds of Real World models MMABP distinguishes between two basic types of models:

- **Global models** are focused on the global characteristics of the system, therefore we also call them **system models**. To keep the orientation on the whole system such a model has to abstract from the temporal Real World aspects. These models thus represent the naturally *static view*. MMABP uses the UML Class Diagram

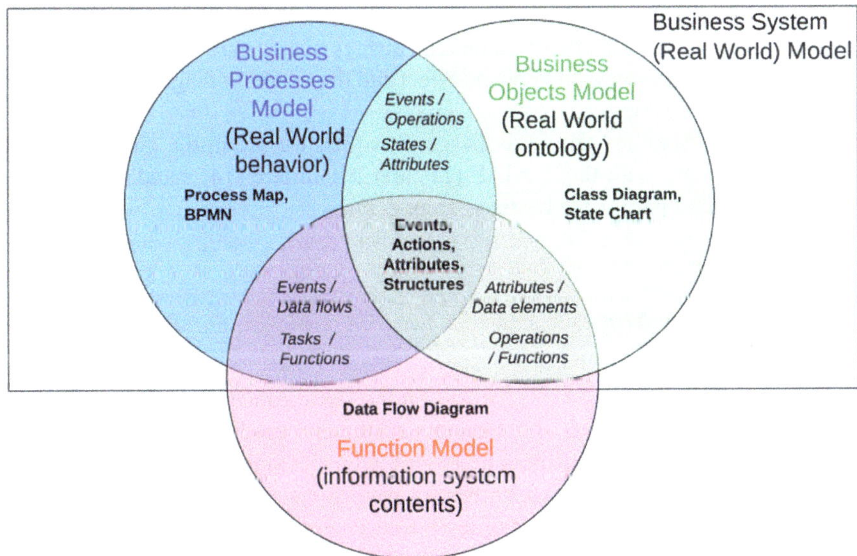

Fig. 11.1 MMABP models

for the global ontological model and the Process Map for the global behavioral model (see above).
- **Detailed models** are focused on details of the Real World. These models are focused on *temporal aspects* of the Real World from both perspectives: ontological and behavioral. In order to be able to express the temporal aspects, each detailed model has to be always focused just on that part of the whole modeled system which is temporally unambiguous, i.e., it can be described as a single algorithm. Therefore we call these models detailed. MMABP uses UML State Chart for the detailed ontological model and BPMN for the detailed behavioral model (see above).

For a more detailed explanation of the above-described kinds of models see [12] and for a detailed description of how to proceed when modeling them see [13].

11.3 Business System Global Analysis

The global analysis, as described above, consists of two diagrams: the Process Map and the Class Diagram. In this section, we describe both these diagrams and illustrate them in a simple use case.

Throughout the following sections, we will use a simple online bookstore example to illustrate the method. The example is shared by all the following sections so that we can demonstrate the multidimensionality of the method and also its compactness and its focus on consistency. We focus only on core elements for minimal business architecture. The other elements provided by the standards used, which enrich the diagrams with additional context, we leave up to an analyst's specific needs and consideration. Additional context should be always treated carefully as adding additional detail may also have an opposite effect and the resulting diagram may become unclear and unreadable.

The Process Map and the Class Diagram, which we use in the global analysis, are fully compliant with the TOGAF [11] and ArchiMate [14] standards and they represent standard part of the business architecture.

11.3.1 Process Map

The Process Map (see Fig. 11.2) focuses on the significance of the business for its customers through capturing what business customers' needs the business is able to satisfy. It maps system of business processes, which business function they belong to and causal relations among them: who or what starts a business process and who or what waits for the target output of the business process.

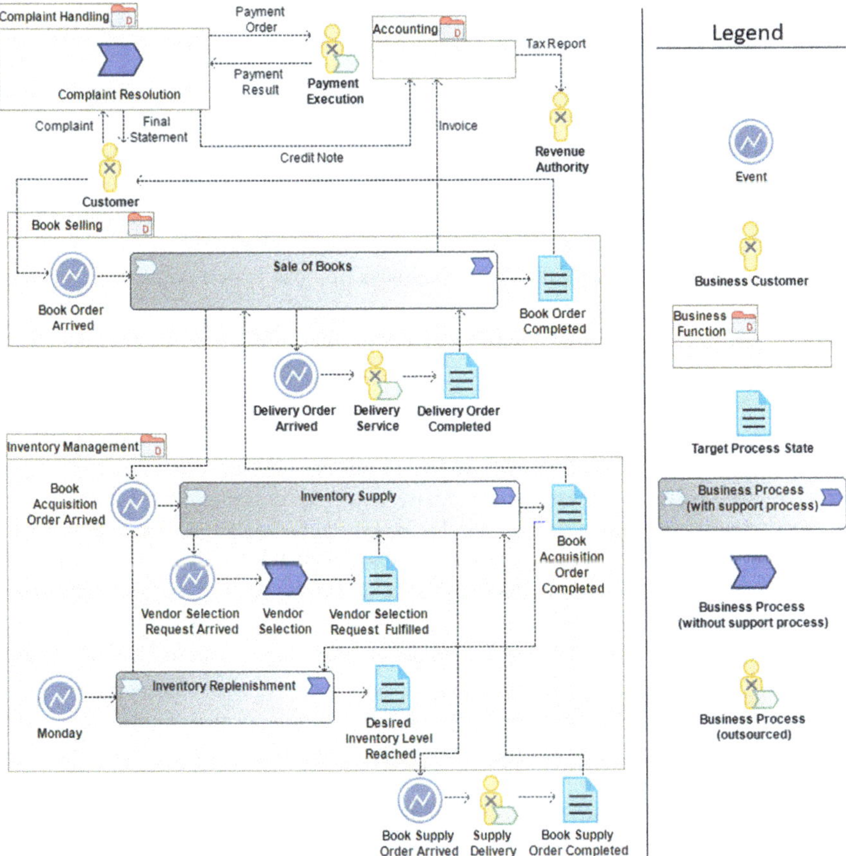

Fig. 11.2 Example process map (TOGAF event diagram)

The Process Map is based on the TOGAF Event diagram and the notation used is from [8]. Definitions of used diagram elements are also compliant with ArchiMate [15], which extends the TOGAF general definitions and specifies the business functions and processes in a more precise way.

A business process is a series of activities of the business deliberately making efforts to create such target output (product or service) that is essential for the satisfaction of the specific need of the customer of the process. A business process covers the entire business cycle of pursuing have satisfaction—from an expression of the customer's need (initiating event) to its satisfaction by the service or product (target process state). Expression of customer's need may not always be explicit (e.g., order), but it may also be implicit (e.g., an incident whose solution is defined within the SLA).

The by default end-to-end character of the business process proposed in this definition leads to a specific view at the business processes. Business processes no longer can acquire an "inferior" form of what is called a sub-process. There is only one kind of business process, which has to always fit the definition. Nevertheless, they may perform in two roles: key or support. A key process is a process directly linked to the customer of the business function and covers the whole business cycle from the expression of its customer's need to its satisfaction. A support process is a process which supports other processes of the business with its target output which is essential for the satisfaction of the specific need of the customer of the key process.

A business function is a grouping of business processes of an enterprise into such a whole, which represents one of the capabilities of the enterprise that enable the enterprise to meet its goals. Every business process is part of just one business function.

The Process Map is developed iteratively as the individual business functions, business processes, business customers, and causal relationships between them are gradually identified during the whole analysis of the business system. The main principle, how the diagram is set up, is the goal decomposition. This principle builds on the premise that the process model maps intentional goal-oriented behavior represented by the business processes that span from the expression of the customer's need to its satisfaction by the produced service or product. One should start the Process Map creation with the identification of customers of the business and their needs that then lead one to the key business processes and their target states. These target process states can be then broken down to milestones that represent target states of their support processes. This way the support processes are identified and thanks to this step-by-step goal decomposition, all business processes are captured in the diagram and there are no "inferior" sub-processes hidden in the detail.

The Process Map presented here has several unique advantages over the value chain maps, which are commonly used for capturing this level of process abstraction. First of all, the goals of business processes, the Process Map is built on, do not change frequently (otherwise the business would be in chaos) this makes the Process Map relatively stable. The changes happen inside the processes. Second, in the Process Map, one does not have to deal with capturing sequences of business processes and how they change over time. It is the role of the key processes to coordinate their support processes and this coordination is captured in their detailed diagrams where it belongs. Adding or removing support processes from the Process Map is then a simple task as there is no remapping of process sequences necessary. So is the change in insourcing or outsourcing of a process.

The example diagram (see Fig. 11.2) also shows that the Process Map does not necessarily have to have all business functions or business processes elaborated in detail. The Process Map allows an analyst to focus only on particular business functions and processes that are of one's interest.

11.3.2 Class Diagram

The Class Diagram is a model of a structure of the business system that describes the modeled business system at the object class level, including their properties

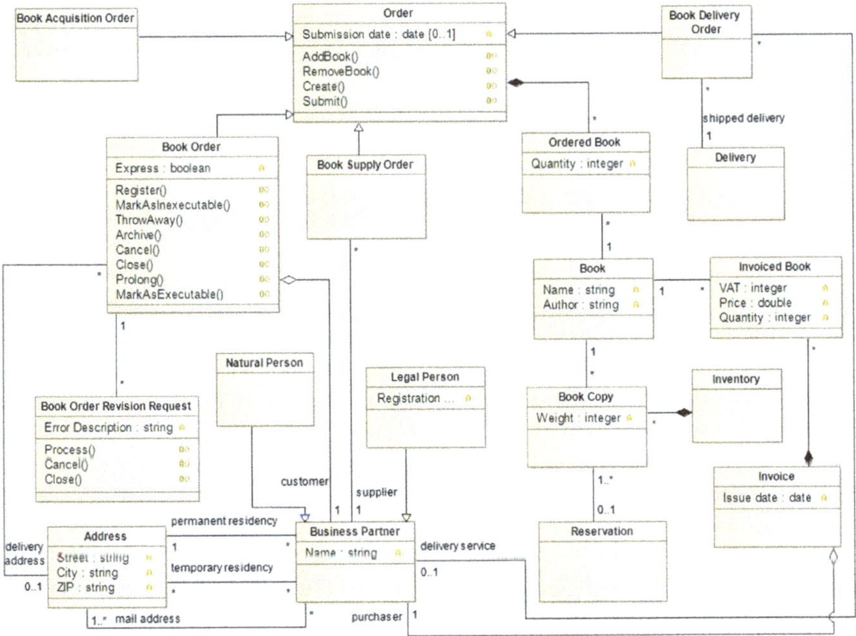

Fig. 11.3 Example class diagram

(attributes and operations) and relations between them. As the notation, we use the standard UML Class Diagram (see Fig. 11.3).

The Class Diagram captures the structure of reality with a focus on its substance, leaving aside any technological or implementation details. The purpose of this model is to understand the reality of the way the business understands it and to formally capture it in terms the business understands and uses. One can look at it as a more formal form of a so-called business dictionary. The Class Diagram has to be modeled the object-oriented way and not the "database way" as many people confuse class diagrams with ER diagrams and also not the "object design way" as this model is made at a conceptual level and not at the implementation level. This means, as Fig. 11.3 shows, usage of core concepts for capturing abstraction: a generalization, an aggregation and association roles. Only then the Class Diagram can be relatively resilient and easy to change at the same time. If done properly, changes of objects and their properties do not have to be replicated—they are done only at the one place where they are necessary and the changes in technology or implementation have no impact as the diagram is captured at a conceptual level. The Class Diagram should reflect all the concepts we find in the other diagrams, especially in the process diagrams so that it is clear what they mean and how they relate to each other. The process of creation is the continual as the analysis proceeds. The identification of operations is not subject to global analysis—operations of each class of objects are filled in after there are captured their object life cycles in the specific diagrams (the State Charts).

11.4 Business System Detailed Analysis

The purpose of the detailed analysis is to take the business processes and business objects identified in the global analysis and elaborate them into reasonable detail. The detailed diagrams provide us with a temporal overview of how the processes deliver their value and with what *business rules* are the objects bound. This detail is what interconnects the abstract business architecture with the specific application functions (see Fig. 11.7) and only based on it one can specify properly the business requirements on the functionality of the IS. Without it, the possibilities of business and application architectures alignment are very limited.

The diagrams for detailed processes are not covered by TOGAF/ArchiMate specification, for this case that it refers to BPMN, which we eventually use. The object life cycle diagrams are not completely unknown to TOGAF as it works with the *Data Lifecycle Diagram* [11]. Since capturing of object life cycles is not clearly defined in the TOGAF specification, we use the UML State Machine diagram which suits this purpose well.

11.4.1 Detailed Process Diagram

The detailed process diagrams elaborate the individual business processes from the Process Map in detail. The diagrams capture deliberate behavior pursuing the goal of a particular business process. They should be created at two levels of detail: process steps and tasks. These two levels allow an analyst to differentiate two perspectives one can look at the process detail: from an external point of view (interaction of the process with its surroundings) and internal (what actual work has to be done).

A diagram at the process step level (see Fig. 11.4) focuses on the description of how a business process is synchronizing its run with its environment. It clearly shows points (process states) at which the process execution has to stop and wait and what expected feedback from the environment lets the process execution continue and which processing way. Diagrams at the task level (see Fig. 11.5) draw the detail of process steps and focus on descriptions of chains of particular tasks that need to be performed.

The detailed processes not only capture the possible processing paths from the process initiating event to the process target state (process customer's view captured by the Process Map) but also the steps that the business has to take after the need of the customer has been satisfied so that run of a business process is finalized also from the business point of view. This is for instance reflected in the diagram (see Fig. 11.4)

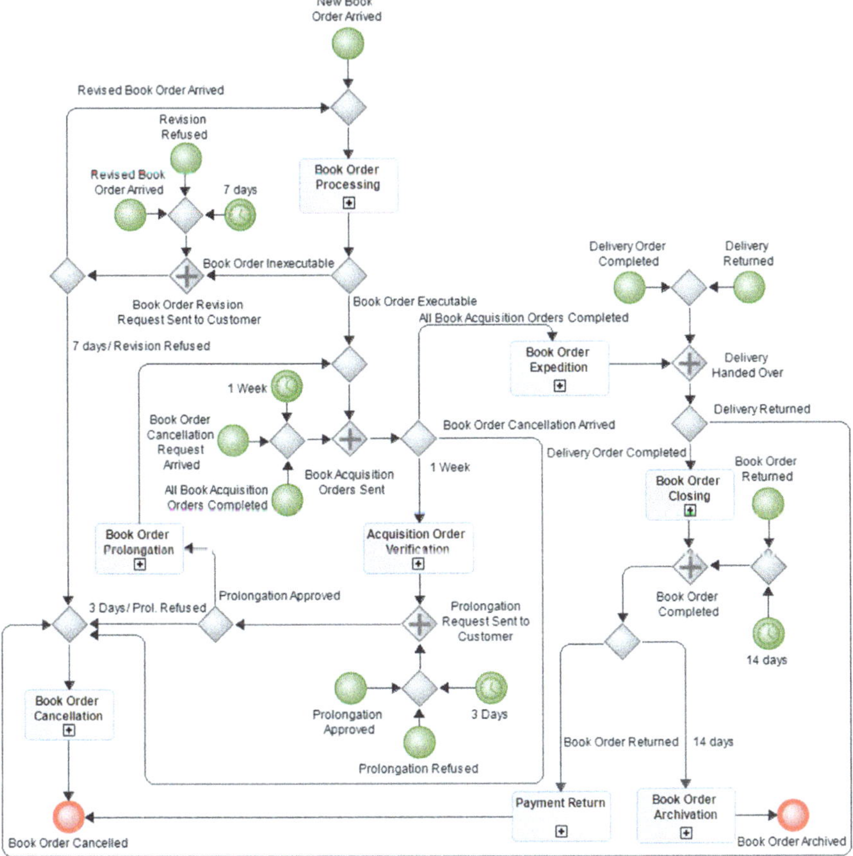

Fig. 11.4 Example process steps of business process sale of books

such a way that even the target process state is reached (Book Order Completed) there are further steps taken which lead to the final states of the business process (Book Order Canceled or Archived).

When creating the diagrams, one should be aware of the fact that a process task is the highest detail a business process model gets into and therefore the definition, where the detail stops, is important. Modeling standards leave this up to an analyst, but that causes only troubles and misunderstandings in our opinion. As we understand it, a task is a unit of work performance of which changes a state of one significant object to another according to the possibilities given by the object's life cycle (see linked object states from Fig. 11.5). The focus on states from objects' life cycles is essential as they define how the people of the business think about their business and what they find important. This is reflected in life cycles of individual business objects, as the life cycle states represent moments in an object life, the business recognizes and therefore it has a name and meaning for them. Tasks bound to object states then

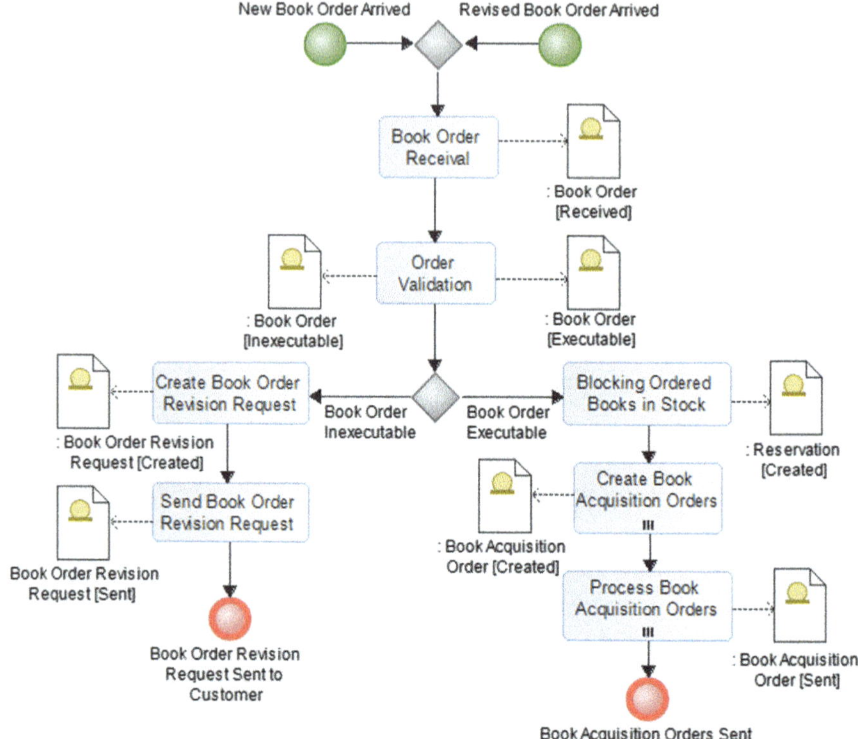

Fig. 11.5 Example tasks of book order processing process step

represent such work done which output also has meaning for the business. This is essential for the model's resilience as this rule implies that all changes to a business process, which do not have an impact on object states, occur inside the individual tasks and have no impact on a process diagram.

Aggregations of chains of tasks of a business process that can be performed completely without interruption (see Fig. 11.5) are process steps (see Fig. 11.4) and the interruptions between them are captured in form of process states (the named AND gates from Fig. 11.4). A process state then represents a point in a process execution where, with the exception of the final state, process execution stops and waits for one of the expected events which would determine in which direction the process execution will proceed. Usually, a process state represents a state of a significant object(s), the setting of which is the output of the preceding activity or the process step.

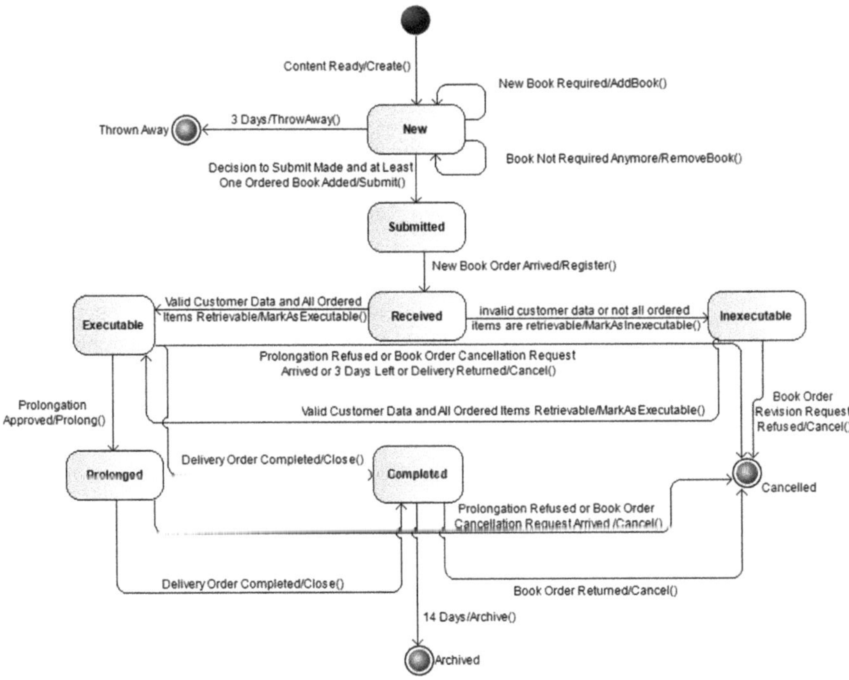

Fig. 11.6 Example object life cycle diagram of a book order

11.4.2 Object Life Cycles

An object's life cycle is a description of a system of business rules capturing essential states of an object for the particular business and reasons for transitions among them (see Fig. 11.6). Unlike the business process, an object's life cycle is a description of common rules given by nature, not of deliberate behavior towards a goal.

Although we are talking about a life cycle of an object, a life cycle as a generalization is always tied to a particular class of objects. A life cycle of an object is an abstraction of all possible instances of the lives of objects belonging to one class and it has to be ensured that the lives of all instances of one class are within its limits and rules.

When creating such a diagram, one should not overfill the diagram with all kinds of object states that come to one's mind but one always has to consider its essentiality for the business. An object state represents a milestone in an object's life which the business recognizes and has a name and meaning for it.

An object always stays in one of its states until the reason for the transition from the current state to another occurs. A special case is a final state in which the life of the object ends. An object may have several life cycles, as one object can perform different roles at the same time (e.g., Business Partner from Fig. 11.3). Each role may mean for an object different life cycle (for instance life in the role of customer vs life in the role of supplier), but it is always valid that in the scope of one life cycle an object may be in just one elementary state.

It is not reasonable to model life cycles for all classes from the Class Diagram. One should focus on significant classes of objects which are essential for the modeled business system and it makes sense from the business point of view to capture their life cycles. At a minimum, these are the classes, state of which is a result of a task (see Fig. 11.5) of a process or those, state of which affects which process tasks will be performed (classes associated with events and split conditions in the business process models). When capturing an object's life cycle, one has to bear in mind that the diagram should capture an object's life completely: from its very beginning to its very end. Life of an object is not limited to one business process or function, but it can span over many of them (for instance a customer).

Life cycles are resilient by their nature, unlike the business processes, as they represent common rules given by nature which are fairly universal.

11.5 How It All Fits Together

The above-presented diagrams provide us with four different but complementing views at the same reality, the same business system. To be true, the individual diagrams must be consistent with each other so that they make together one compact and consistent business architecture. As there is nothing tangible yet at the analytical stage of digital transformation, only the diagrams forming one future business architecture together, it is the consistency of individual diagrams one has available as the main tool for the architecture verification. It is upon the analyst to check the consistency conflicts, clarify them and improve the business architecture so that it fits the expected reality from all modeled perspectives. Only then one can say that at least on a paper the suggested business architecture is feasible. The MMABP defines the needed rules for consistency of the models by means of a set of meta-models. MMABP Business System Meta-Model [16] consists of three basic related models:

- *Business Substance Meta-Model* which defines basic concepts and their basic relationships used for modeling the domain of business objects in terms of the ontology/conceptual modeling. This meta-model is based on the UML Core definition [9] extended with special concepts for the definition of additional modeling aspects mainly the aspects of the life cycles modeling.
- *Business Process Meta-Model* which defines basic concepts and their basic relationships used for modeling the domain of business processes.
- *Business Models Consistency Model* which defines basic relationships among the concepts from both the above-mentioned meta-models defining this way the required consistency of ontology/conceptual and business process models describing the same business system.

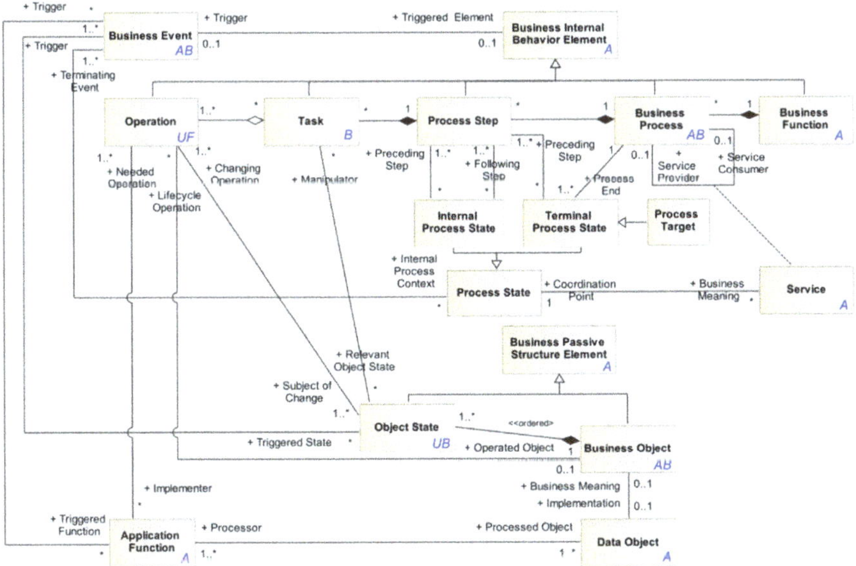

Fig. 11.7 MMABP extended ArchiMate meta-model

Business Models Consistency Model uses the concepts from both meta-models extending them with several special concepts for the precise definition of their relations. Besides the basic *existential* consistency which is interested in the existence of relationships among given concepts MMABP also defines so-called *structural consistency*. Structural consistency deals with the types of these relationships (so-called *structures*) anchoring the correctness of models in the general logic. A more detailed explanation of the structural consistency can be found in [17].

Model from Fig. 11.7 is based on the ArchiMate [14] meta-model, which we have modified and extended in order to contain also the basic information from the MMABP meta-model. All concepts in the model are defined in MMABP. Concepts, which are defined also in other related standards, are labeled in the lower right corner by the letter expressing the given standard (*A*—TOGAF/ArchiMate, *B*—BPMN, *F*—Data Flow Diagram (DFD) and *U*—Unified Modeling Language (UML)). This way the relationships among these standards and MMABP are expressed. The model also contains two concepts original just to MMABP: a *Process State* and a *Process Step* which are crucial for the definition of essential relationships among the technical and business meanings of used concepts as well as the essential relationships between the process (behavioral) and the object (ontological) models through their common concept *Business Event*. The concept *Business Function* is a compound of *Business Processes*, each business process is a compound of *Process Steps* and each process step consists of *Tasks*. This way the meta-model reflects the MMABP four levels of abstraction of the process model: two system levels—*Business Function* and *Process*

Map, and two detailed levels—*Process Steps* model and *Process Tasks* model. For a more detailed explanation of the MMABP levels of process abstraction, see [18].

From the meta-model one can also derive the following consistency rules:

The Process Map and the detailed process diagrams share the starting *event* and target *process state* of each business process and these have to be the same in both diagrams. In addition, the interdependencies between the business processes have to be identical in both diagrams too. In the case of our book shop example, the initiating event and the target state of the key process Sale of Books (see Fig. 11.2) are consistent with the detailed process diagram (see Fig. 11.4). So are the interconnections with the support processes. For instance, the synchronization with Inventory Supply process is represented by the process state Book Acquisitions Orders Sent (see Fig. 11.4), which corresponds to the support process's initiating event (see Fig. 11.2) and the expected events map all the possible eventualities, the synchronization with the support process may end up with, including the support process target state, which again corresponds to the target state defined in the Process Map (see Fig. 11.2). These rules follow the meta-model (see Fig. 11.7) where the concept *Business Event* is defined with all its relationships to other concepts regarding their detailed as well as global meanings. The same goes for the concept of *Process State.*

Further key consistency rules can be found between object life cycles (see Fig. 11.6) and process tasks (see Fig. 11.5). Both diagrams represent a temporal view at the same business but from different perspectives. Combing these two views and checking their consistency gives an analyst a strong tool for model consolidation. Generally, it should be valid that the object states referenced in the process diagrams are referring to the relevant states from the objects' life cycle diagrams. If they are not, they are either missing in relevant object life cycles or they are not essential for the business and then they are redundant in the process diagram (see the relationship between the concepts *Task* and *Object State* from Fig. 11.7). Moreover, there should be valid that object states which are alternative in their life cycles have to be reflected in corresponding alternatives also in the detailed process diagrams. Similarly, repeating object states in an object life cycle invoke the repeating pattern in the process structure. In other words, processes cannot violate the objects' life cycles. The objects can change their states only in those directions the state transitions in their life cycles allow them. This logical dependency is not visible in the meta-model from Fig. 11.7. It is defined in the Business Models Consistency Model as a structural consistency (see [16]).

All classes of objects, their attributes or states referenced in events, process states, conditions, etc. in the diagrams have to have clear formal definition i.e., they have to be reflected in the Class Diagram if they represent an entity or its attribute, or in an object life cycle diagram if they represent state of an object. Object life cycles have to be life cycles of existing classes of objects and if there exists more than one life cycle of one class they have to correspond with the roles the object may perform in.

The Class Diagram provides us with another powerful consistency check which does not need to be applied when modeling the business system only. But when the purpose of the model is to proceed from the business architecture to IS function definition, then it is a must. Relations and attributes of a class specified in the Class

Diagram have to be reflected in its life cycle(s). Transitions in object life cycles should cover at least creation, change, and termination of these relations and attributes so that it is clear under which condition they can be modified and even that they can be modified at all—more on this in the following section.

11.6 Identification of Requirements on Information System Functionality

The identification of requirements on IS functionality based on business architecture stands on the premise that the here presented minimal business architecture of that part of business, which one wants to support by an IS, is consistent and complete. Only then it makes sense to dive into much higher detail than the business architecture has.

The detail of business architecture stops at the level of a process task which execution changes an essential state of an object to another one. Looking at how the object life cycles are being captured for the business architecture one can see that the analytical process outlined above leads one only to identification of only those operations which perform state transitions. For business architecture, this is enough but for the definition of functionality of an IS we have to take into account also those operations which execution has no impact on the object state. For instance, operations like add or remove book (class Order from Fig. 11.3) do not change a state of an order (see Fig. 11.6) and therefore these operations we do not find reflected in the process model in form of corresponding tasks. They are part of the task which covers the whole order setup procedure which output state is the sent/submitted Order (Process Book Acquisition Orders task from Fig. 11.8). This means that a task performance may consist of execution of several operations, but the operations

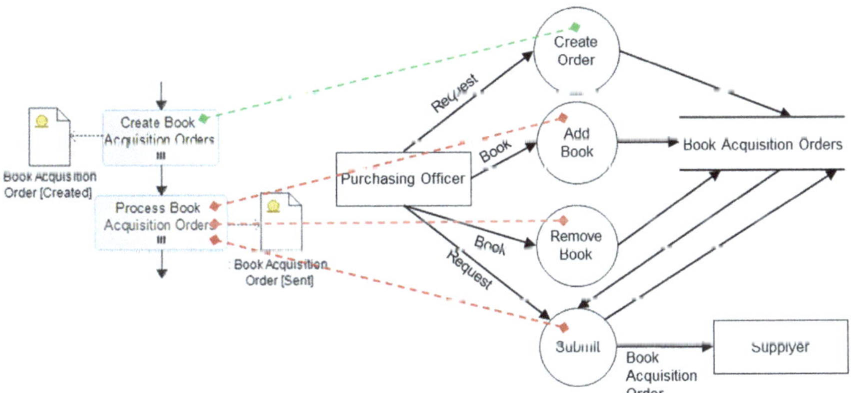

Fig. 11.8 Identification of information system functions based on process tasks

themselves are not the next level of process abstraction as they do not represent the intentional behavior anymore, only processing functions with inputs and an output. The intentional behavior is hidden in the capabilities of the task performer (the actor) who, using one's skills and knowledge, combines the operations in such way which fits the actual situation and keeps the course of processing in the direction of the task's goal (the associated object state).

Tasks and their flow are the leading string of the requirement analysis and the gradually identified operations are the substance of an IS which leads us to its particular functions required for the performance of each task. For such analysis, the UML would use the traditional use case but this would neglect the work done in the business model and also the use case is a rather spontaneous description than a model. Instead, we suggest using two other diagrams that are available—UML class diagram and data flow diagram (DFD).

Before one starts the identification of required IS functionality task by task, it is really helpful to complete the Class diagram from the business architecture by reviewing classes' operations using the consistency rule pointed out in the previous section. The relations and attributes of a class should be reflected in class's operations (i.e., their adding, removing and update) and during the task by task operation identification, one should always consider these operations as they have to be used in some task at least once.

After the enrichment of the Class diagram, one can start the task by task identification of the individual operations which one performs when carrying out the analyzed tasks. Very helpful for this analysis is the DFD. It allows one to note down the information processing operations in form of functions of the IS and the DFD's rules, which specify allowed information flows, guide one through the analysis in a much precise way then the descriptive UML Use Case diagram would. For each task we capture not only list of high-level functions, like we would in the UML Use Case, but also: the information flows which execute the functions, sources and targets of these flows and data stores the functions store their results in and other functions load input data from (see Fig. 11.8). This way the identified information functions are not only derived from the discussions with the business but also cross-checked by the formal rules of the DFD and the class and process diagrams. Moreover, one is able to check the reasonability of the model of information functions using the map of information flows which helps one to review whether the model as a whole makes sense or not and whether all functions are identified so that no information transformation is missing in a complete end-to-end information flow.

The resulting DFD, which now contains all required information functions, can be then in accordance with its rules [5] split and hierarchically organized and used together with the minimal business architecture as a base for specification for particular application architecture/implementation.

11.7 Conclusions

This chapter is about adaptive and resilient business architecture as we believe it is a basic condition for meaningful digital transformation. As discussed in the section Introduction, the digital transformation is not just inherent automation but also an application of new technologies connected with a radical redesign of the current business system. Therefore, it must be started with a precise conceptual model of the business itself in the form of business architecture.

We understand the *adaptivity* of business architecture as its ability to absorb needed changes with as low as possible effort and *resilience* as its resistance to as much as possible changes in the business (Real World) system without the need to be changed. These two qualities obviously support each other. Theoretically absolutely resilient architecture, which does not require any change even if the business system changes any way, is totally adaptive as its adaptation requires the null effort. Nevertheless, these qualities have to be combined also with other qualities like the *coverage* (an extent and depth of the business system features covered by the architecture). We do not encourage building architectures which are so general or so limited that they can automatically absorb or avoid any change in the business; the practical value of such architecture tends to zero. Instead, we propose *minimal business architecture* meaning the architecture maximally adaptive and resilient due to the minimal mode of the reflection of the business system but also complete and consistent (i.e., correctly covering all important aspects of the business).

Use of the *conceptual Class Diagram* allows an exact definition of the proper scope of the needed analysis including the exact border with following actions like design and implementation of the system. Conceptual Class Diagram works as a "business dictionary" defining those object classes and their mutual relationships which are important for the given business and abstracting the technical as well as implementation aspects of the system since they do not influence the conceptual view of the business by definition. Generalization/specialization as a natural tool for working with hierarchies of concepts allows focusing just on those aspects which are important in the given context. This makes the potential necessary changes in the system contents minimal.

The presented four process abstraction levels are designed with the main respect to the mutual independence of main factors that influence the potential changes in the system as well as to their hierarchical relationships. Structure of the *Business Function* is derived from the business goals and it is independent of the organization of processes inside particular functions and other subordinated aspects. Only the change in business goals may cause the need for a change of the business function. *Process Map* reflects the collaborative organization of processes and it is independent of the internal structure of particular processes. Besides the superordinated changes in the business goals, just the changes in business conception that require the different collaboration of processes may cause the need for the change in their mutual arrangement. *Process Steps level* of the detailed process model is derived from (and thus dependent on) the way of the collaboration of processes reflecting its internal process consequences. The structure of process steps is independent of the internal structure of particular steps. *Process Tasks level* of the detailed process model is derived from the causality related to the relevant Real World objects expressed as their life cycles. Any possible change of the causality related to the given object should not influence the structure of process steps, process map nor the business function unless it influences also the needed way of the collaboration of processes, their membership in the business functions or the business goals. Any Real World change then can be located to just relevant level(s) of process abstraction and implemented with minimal changes in the system, leaving its irrelevant aspects untouched.

The practical benefits of the resilience and adaptivity of the minimal business architecture can be illustrated also in the example case presented in this chapter. For instance, there may be introduced an idea to extend customers' evaluation in the *Sale of Books* process with a blacklist, which would warn the operators when dealing with a particular customer. As the rules for levels of a process abstraction for minimal business architecture are set up, this change would have no effect on the current *Sale of Books* process model. It is because this extension would not add another important state of an object and this new evaluation would be represented only by a new operation, which would be part of the customer evaluation task. The consistency rules between the models specified in the meta-model then guide one to think the extension throughout. Although no modification of the process model is necessary, the blacklist would have to be added into the class diagram and all relevant operations defined so that it is clear what the blacklist looks like and what one can do with it.

Another case may be adding gift packing of the books. In this case, if the *Book Order* object state "gift packed" was considered as significant by the business, there would be added one gift packing task into the detail of the *Book Order Expedition* process step, the rest of the *Sale of Books* process model would stay intact. Again the consistency rules between the models then guide one to thing the extension throughout and to modify the class diagram and object lifecycle of the *Book Order* accordingly.

Finally, there may be introduced a new service, for instance obtaining autographs. Thanks to the focus of the minimal architecture on key processes, which are "end-to-end", this change does not require any significant change of the process model either. There would be necessary only to add a new autograph obtaining business process into the abstract process map and in the detail of the *Sale of Books* process to add synchronization with the new autograph obtaining process (so-called process state). Nothing else in the actual process model would have to be changed. The autograph obtaining process would be then specified in detail in its detailed diagram. The consistency rules between the models, again, guide one to define the business rules completely—how the new objects or attributes fit the existing class diagram and how their life cycles, defining the business rules of the new objects, look like.

In all cases, the minimal architecture keeps the individual perspectives neatly separate yet interconnected, reduces the necessary changes of the business process model and guides one to specify the necessary minimum for each proposed change, so that the necessary analytical minimum for its implementation is properly specified.

Minimal architecture is also a certain way of exploiting the "digital" opportunities. MMABP methodology, presented in this chapter, is aimed on the clear understanding of the essence of business. As the business system, we understand the complex of two basic characteristics: *causality*, represented by the conceptual model, and *intentionality*, which we model as a system of business processes (see [12] for more detail). Unlike the causality, which is relatively stable and independent of human intentions, the intentionality is closely tied with the changes in the business system towards the use of new technology possibilities. In accordance with the essential work in the field of business process management [19], we believe that the technology evolution allows changing the current practices in terms of the simplification of business processes towards their natural essence. Therefore, in the process model, we pursue to uncover the natural essence of business processes, represented on various levels of abstraction, by abstracting all non-essential aspects like technological, organizational, qualification, etc. limitations. This way, we create the model of the simplest possible process that cannot be more simplified without the change of its essential meaning. Such model is an ideal that can be used for the navigation of the technology innovation activities towards the changes that are consistent with the given business strategy as well as with the given business ontology (objective rules). We believe that such approach also helps non-IT, business people understand the essence of so-called "digital change" and the practical role of the Enterprise Architecture (EA) in this process.

Although the discussed models of the minimal business architecture are relatively abstract in order to keep their resilience and adaptivity, they are still a good basis for the specification of requirements on IS(s). The method presented in this chapter shows how to proceed from business architecture to aligned application architecture with its own fitting level of detail.

The presented approach has been applied in various evolutionary stages in several transformation projects in the fields of small **business** as well as public administration. Experience from each project has been always used for the improvement of some of its features as the evaluation in projects is an essential part of the never-ending methodology development process.

General limitations of our approach are mainly related to the current state of common knowledge and they are also closely connected with common mistakes when applying it. One of the most significant aspects is the relatively low maturity of existing modeling standards which also naturally influence the quality of supporting modeling tools. Except for UML [9], which can be regarded as relatively matured, all other related standards have many insufficiencies from our methodology point of view. For instance, BPMN [10] as a common business process modeling standard does not respect the essential difference between the system of processes (process map) and the single process (process flow). Therefore, we have to complete BPMN with the additional standard diagram from TOGAF [11]. BPMN is also inconsistent in basic meta-elements like operation, task, and event which in various BPMN elements typically overlap. The problem in dominant standards for Enterprise Modeling TOGAF and ArchiMate [11, 14] is their low preciseness and level of detail which are also discussed above. We believe all those insufficiencies can disappear in the future. But currently, these features complicate the use of standards in accordance with the methodology which can be often achieved only at the price of very complicated ways of use of these languages and the need of a high level of abstraction.

In the future development of the methodology, we would like to focus on the deeper integration of our approach with other related fields in informatics, particularly with the ontology modeling where we plan to follow the work of G. Guizzardi (UntoUML [20]). Another relevant future task is the improvement of the support of modeling tools where we currently work with Modelio.

References

1. Andersson, H., Tuddenham, P.: Reinventing IT to Support Digitization, McKinsey & Company (2014). Retrieved from http://www.mckinsey.com/insights/business_technology/reinventing_it_to_support_digitization
2. Kohnke, O.: It's not just about technology: the people side of digitization. In: Oswald, G., Kleinemeier, M. (eds.) Shaping the Digital Enterprise. Trends and Use Cases in Digital Innovation and Transformation, pp. 69–91. Springer International Publishing (2017)
3. Markovitch, S., Willmott, P.: Accelerating the Digitization of Business Processes, pp. 1–4. McKinsey-Corporate Finance Business Practise (2014)
4. Ashworth, C.M.: Structured systems analysis and design method (SSADM). Inf. Softw. Technol. **30**, 153–163 (1988). https://doi.org/10.1016/0950-5849(88)90062-6
5. Yourdon, E.: Modern Structured Analysis. Yourdon Press Computing Series. Yourdon Press, Englewood Cliffs (1989)
6. Jacobson, I., Booch, G., Rumbaugh, J.: The Unified Software Development Process. The Addison-Wesley Object Technology Series. Addison-Wesley, Reading (1999)
7. Al-Zewairi, M., Biltawi, M., Etaiwi, W., Shaout, A.: Agile software development methodologies: survey of surveys. J. Comput. Commun. **05**, 74–97 (2017). https://doi.org/10.4236/jcc.2017.55007
8. Philippe, D., Raymond, G.: Modeling Enterprise Architecture with TOGAF: A Practical Guide Using UML and BPMN, 1st edn. Morgan Kaufmann Publishers Inc., San Francisco (2014)
9. Object Management Group: Unified Modeling Language (UML) specification v2.5.1 (2017). Retrieved from https://www.omg.org/spec/UML/
10. Object Management Group.: Business Process Model and Notation (BPMN) Specification Version 2.0.2 (2014). http://www.omg.org/spec/BPMN/
11. The Open Group: TOGAF Version 9.2. Van Haren. (2018). ISBN 978-9401802833
12. Řepa, V.:. Business system as an equilibrium of intention and causality. E + M Ekonomie a Manag. **20**, 219–235 (2017). https://doi.org/10.15240/tul/001/2017-4-015
13. Svatoš, O.: A4 Method v1.0 (2019). Retrieved from http://www.e-bpm.org/documents/A4eng.pdf
14. The Open Group: ArchiMate 3.0 Specification. Van Haren (2016). ISBN 94-01-80047-2
15. Svatoš, O.: Business process modeling method for Archimate. In: IDIMT-2017 Digitalization in Management, Society and Economy. pp. 325–332. Linz: Trauner Verlag Universität (2017). ISBN 978-3-99062-119-6
16. Řepa, V.: Business System Modeling Specification: Retrieved from http://opensoul.panrepa.org/metamodel.html
17. Řepa, V.: Modeling objects dynamics in conceptual models. In: Wojtkowski, W., Wojtkowski, W.G., Zupancic, J., Magyar, G., Knapp, G. (eds.) Advances in Information Systems Development, pp. 140–152. Springer US, Boston (2007). https://doi.org/10.1007/978-0-387-70802-7_12

18. Svatoš, O., Řepa, V.: Working with process abstraction levels. In: Perspectives in Business Informatics Research—15th International Conference, BIR 2016, Prague, Czech Republic, 15–16 Sept Proceedings. pp. 65–79 (2016). https://doi.org/10.1007/978-3-319-45321-7_5
19. Hammer, M., Champy, J.: Reengineering the Corporation: A Manifesto for Business Revolution. Brealey (1993)
20. Guizzardi, G.: Ontological Foundations for Structural Conceptual Models. Enschede; Telematics instituut: Enschede: University of Twente. Centre for Telematics and Information Technology (2005)

Chapter 12
Adaptive Integrated Digital Architecture Framework: Risk Management Case

Yoshimasa Masuda

Abstract There are a lot of problems leading to the loss of profits because of less strategic alignments and non-standardization in application, technology and data in Digital Transformation in global corporations. The above problems are crucial ones for Enterprise Architecture to minimize risks in Digital Transformation in global corporations. Therefore, the author hypothesizes that the Resiliency Architecture model—the Risk Management with Architecture Board for Digital Transformation can be effective for mitigating risks for Digital Transformation in terms of full lifecycle in software development lifecycle (SDLC), and this hypothesis needs to be investigated. The purpose of this chapter is also to propose and brief the unique and new solution of the Resiliency Architecture model—risk management model as well as the architecture assessment model, knowledge sharing model and social collaboration model, which are related to the "Adaptive Integrated Digital Architecture Framework—AIDAF," for the above crucial problems in Digital Transformation, in this chapter. As the results of Architecture Board reviews, the above solution of Resiliency Architecture model can show a practical guidance for companies that consider Resilience for Digital Transformation. Finally, the author verifies the above Resiliency Architecture model in the case study and clarifies the effective strategy elements to mitigate risks in Digital Transformation, to assist EA practitioners with realizing Digital IT strategies.

Keywords Digital IT strategy · Enterprise architecture · Case study · TOGAF · AIDAF · Risk management · Digital transformation · Architecture Board

Y. Masuda (✉)
Carnegie Mellon University, Pittsburgh, USA
e-mail: ymasuda@andrew.cmu.edu; yoshi_masuda@keio.jp

Keio University, Tokyo, Japan

A. Zimmermann et al. (eds.), *Architecting the Digital Transformation*,
Intelligent Systems Reference Library 188,
https://doi.org/10.1007/978-3-030-49640-1_12

12.1 Introduction

The author proposed and briefed the "Adaptive Integrated Digital Architecture Framework—AIDAF" as the framework of Digital Enterprise Architecture in the previous chapter. In this chapter, the author will propose and brief the models related the AIDAF—global digital transformation communication (GDTC) model, social collaboration model (SCM) and STrategic Risk Mitigation Model (STRMM), etc. to meet the requirements of the digital transformation in relation to the agility-related aspects described in the previous chapter, in the following sections. The author found existing EA frameworks to be inappropriate to achieve digital transformation, while the necessity of implementing EA in parallel in the mid-/long term (roadmaps and target architectures, etc.) in the era of cloud/mobile IT/digital IT should be emphasized in terms of promoting the alignment of IS/IT projects with management strategy/IT strategy. On the other hand, crucial problems exist, which should lose profits as results of less strategic alignments and non-standardization in application, technology and data in Digital Transformation in global corporations. In this chapter, the author hypothesized the process for Architecture Review and Evaluation with mitigating risks and verified them in the case study in global organizations in the global enterprise. As a whole, the research question employed in this chapter is:

RQ1 How can the Resiliency Architecture model—the "STRMM model" be effective for mitigating risks for Digital Transformation in terms of full lifecycle in SDLC?

The author proposed Resiliency Architecture model—risk management model as well as the architecture assessment model, knowledge sharing model and social collaboration model should be the unique and innovative approach as the solution. This is how there is only the description regarding Architecture Board in the TOGAF9 documentation, while processes and criteria for Architecture Board have not existed in previous researches [1]. On the other hand, COBIT5 also guided the way of implementing architecture governance, involving Architecture Board, aligned with IT governance [2], but this guide didn't refer to processes for Architecture Board, either. Furthermore, from standpoints of ISO standards on Architecture, the draft of overview for Architecture Process—ISO/IEC42020 [3] and Architecture Evaluation—ISO/IEC42030 [4] were shown, but the above ISO standards on Architecture have not been published at the current time. Besides, from standpoints of Business Architecture (BA), Data Architecture (DA) and Application architecture (AA), there should be the requirement of coping with delays in standardization projects of Business Process and Master Data Management platforms flexibly, and one of coping with the difficulties of demand managements for Digital IT systems and projects. In the following sections, the author proposes the above "AIDAF" and the Resiliency Architecture model to minimize the risks in Digital Transformation and verifies them in the case study in the global enterprise.

This chapter is organized as follows: the next section presents the overview of the Architecture Board in the AIDAF and global company case, followed by the proposals and overview of Assessment Meta-model, GDTC model, SCM model and Resiliency Architecture model—STRMM model for Digital Transformation and description of research methodology. The results and data analysis in the case study are presented, and the Solutions proposing the Resiliency Architecture model—the Risk Mitigation model are verified. The Strategy elements to mitigate Risks in Digital Transformation are then formulated. Finally, the limitations of the current study and directions for future research are outlined.

12.2 Assessment Meta-Model in Architecture Board

12.2.1 Architecture Board and Global Healthcare Company Case

Our preliminary research [5] for this chapter proposes an adaptive integrated EA framework to align with IT strategy, promoting cloud, mobile IT, big data and digital IT and is verified by our case study [6]. The author of this chapter has named the EA framework suitable for the era of Digital IT as "Adaptive Integrated Digital Architecture Framework—AIDAF" [7, 8]. Figure 12.1 illustrates the AIDAF proposed model with Architecture Board.

This AIDAF begins with the Context Phase, while referencing the Defining Phase (i.e., architecture design guidelines related to Digital IT aligned with IT strategy). During the assessment and architecture review, the Architecture Board reviews the initiation documents and related architectures for the IT project.

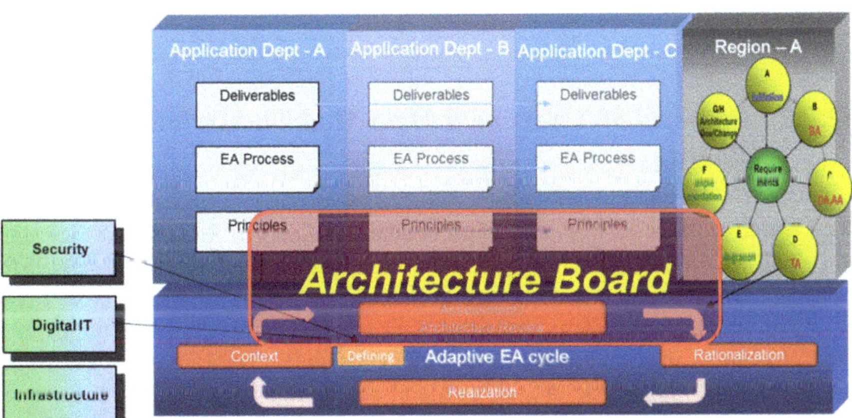

Fig. 12.1 AIDAF proposed model with Architecture Board (ex: TOGAF and Adaptive EA framework, etc.)

Table 12.1 Responsibilities and tasks of the Architecture Board

Responsibilities of Architecture Board	Tasks of Architecture Board
– Operational/Functional Aspect and Viability of solution architectures with Enterprise level Standard conformance are assessed properly, and the projects related architecture risks should be coped with – Architecture integration issues across IS/IT organization are assessed, reviebd and resolved – Enterprise architecture risks are assessed and mitigated timely and consistently	– Review solution architecture/standard/common platform of new IS/IT projects, involving new technologies – Endorse signed architecture of new IS/IT projects for the month – Share Architecture strategy in each departments, initiatives and their status

In particular the Architecture Board, the example of the EA framework structure in a certain global healthcare company examined in this chapter is explained. This global pharmaceutical company is the largest Japan-headquartered global company in the industry in Asia based on a sales basis. In a global EA rollout, they are handling cloud, mobile IT and big data strategic projects and systems that took priority in Europe and US Group companies well by structuring and implementing EA with the above Adaptive Integrated Digital Architecture Framework—AIDAF to be consistent with global IT strategy focusing on cloud, mobile IT, big data and digital IT [6, 7].

Actually, one of the authors works in this global healthcare company and conducted all phases of structuring and implementing in this EA framework and was the facilitator and managed the coordination of the global Architecture Board. In this global pharmaceutical company, one of the authors had the above responsibilities in the Assessment/Architecture Review phase on the global Architecture Board as shown in Table 12.1, which we were focusing on, and performed the above tasks [6, 7, 9, 10].

12.2.2 Proposal of Assessment Meta-Model in Architecture Board

As a result of investigating tasks of architecture reviews in Architecture Board, in this case, we confirmed that project planning documents of almost all new IS/IT projects have been submitted into the Architecture Board. This Architecture Board performs architecture reviews regarding their solution architectures on the basis of the following evaluation criteria of four elements—(1), (2), (3), and (4) and issued the action items there. One of authors found out the risks connected to each action item and defined equivalent solution for each risk in or after this Architecture Board [7, 9].

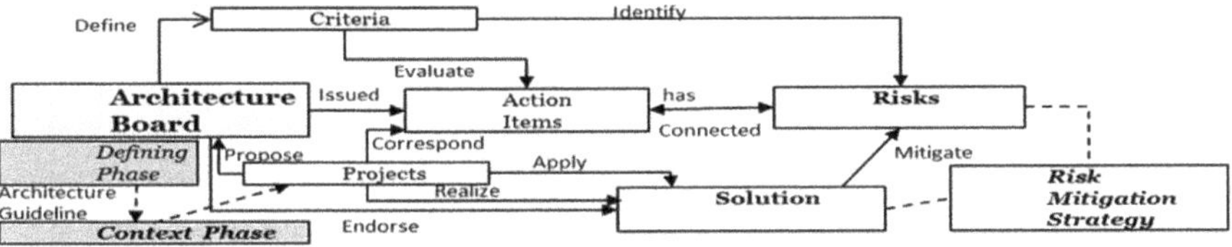

Fig. 12.2 Assessment meta-model in Architecture Board

(1) Enterprise level conformance—align with architecture roadmap, standard, each architecture principle.
(2) Functional aspect—clarify solutions/architecture specs, integration points, with standard master data.
(3) Operational aspect—alignment with the service level, security design and privacy, system availability.
(4) Viability—application rationalization, risks/countermeasures, data migration/testing strategy.

We propose the "Assessment meta-model in Architecture Board" in Fig. 12.2, based on the above process and descriptions in the case study [9, 10].

12.3 GDTC Model for Knowledge Management on Digital Platforms

12.3.1 Intermediary Knowledge Model

In organizational computer-mediated communication (CMC), such as portals and social networks, it has been established that experiential knowledge is visualized and shared through text that is generalized as explicit knowledge [11–13].

Fragmented knowledge that is disseminated horizontally across the organization by its employees, through corporate digital media, is referred to as intermediary knowledge [11, 12, 14]. The knowledge-sharing process based on intermediary knowledge is visualized through publication, resonant formation, collaboration, sophistication, and fragmentation. The intermediary knowledge model is shown in Fig. 12.3.

In this study, we show that knowledge sharing through intermediary knowledge can effectively and efficiently promote global communication as a result of digital IT transformation, by means of corporate bodies conducting global communication accompanying digital IT transformation through enterprise portals [15]

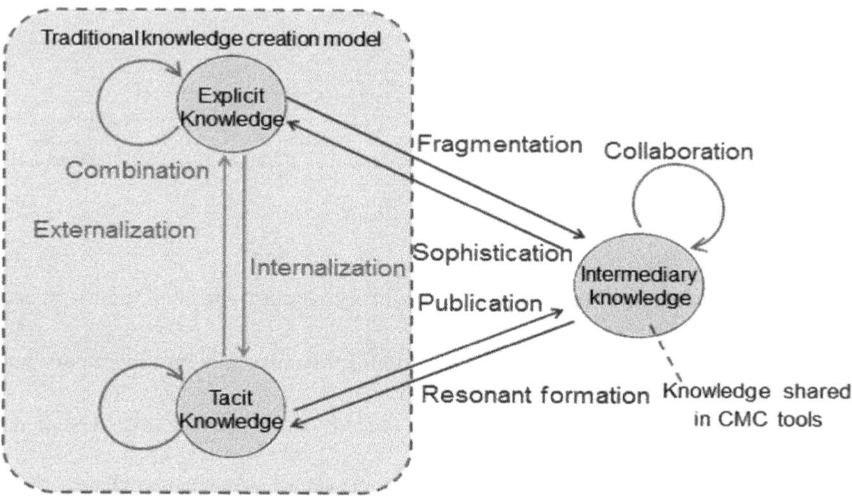

Fig. 12.3 Intermediary knowledge model

12.3.2 Communication Model

There are two types of communication models: the linear information model, which transfers activities from the speaker to the receiver, and the convergence model, in which an iterative process occurs until a mutual understanding of the information has been reached [16]. In addition, Yamamoto et al. proposed the cooperative model, focusing on the process of organizational activities for the purpose of achieving goals [13].

In this book, we assume that enterprise portals introduced as a method of global communication accompanying digital IT transformation follow the convergence model of global organizations and communities. Furthermore, we anticipate that they will follow the cooperative model, with the combined use of a corporate social network, as a result of global communication [15].

12.3.3 GDTC Model Overview

The author proposes the GDTC model for global communication on an enterprise portal, which will be able to promote the digital transformation in the global community. The GDTC model consists of three layers: the tacit knowledge group (TKG), intermediary knowledge group (IKG), and explicit knowledge group (EKG), as shown in Table 12.2 [15].

TKG contains roles for exchanging tacit knowledge, and its group node is human. TKG is related to organizational structures, member roles, and decision-making

Table 12.2 Elements of layers in GDTC model

Knowledge group	Group node	Media	Documentation	Examples of deliverables
Tacit knowledge group	Human	Telephone, web conference	No documentation	Discussions, web meeting
Intermediary knowledge group	Portal contents	Portal, SNS, e-mail	Just in time documentation	Architecture Board logs, SNS logs, e-mail logs
Explicit knowledge group	Document	Document management services	Full documentation	Architecture Board results, architecture guidelines

process, among others, and is generated in telephone or web conference communications. It appears that TKG does not generate any documentation for global communication, and we assume that it does not create any formal documents. The deliverables of TKG are discussions and meetings; however, it does not always produce tangible deliverables that can be observed.

IKG performs the role of exchanging intermediary knowledge, and its group node is portal contents, which is related to the portal, SNS and e-mail within the global community. IKG provides just-in-time documentation to global community members, and CMC tools can be used if one member needs to coordinate with others. Just-in-time documentation refers to necessary knowledge becoming documents when global community members communicate with each other using CMC tools. The deliverables of IKG are web conference, SNS, and email logs [15].

The role of EKG is the exchange of explicit knowledge. The group node of EKG is document, and the document group results from document management services, whose functions are historical management, enterprise document searching, and document file sharing. EKG provides full documentation to global community members, and its deliverables are documents such as conference results and guidelines [15].

As a whole, the TKG group undergoes human communication during discussions and meetings. Communication between TKG and IKG involves the processes of intermediary knowledge provisions and acquisitions in the portal, SNS, and e-mail, while communication between IKG and EKG consists of the processes of quotations and documentations of explicit knowledge in the portal, SNS, and e-mail [15].

12.3.4 *Proposal of GDTC Model for Global Communication on Enterprise Portal*

The Traditional global communications employ TKG and EKG in the knowledge process; however, a great deal of communication in meetings is necessary among the

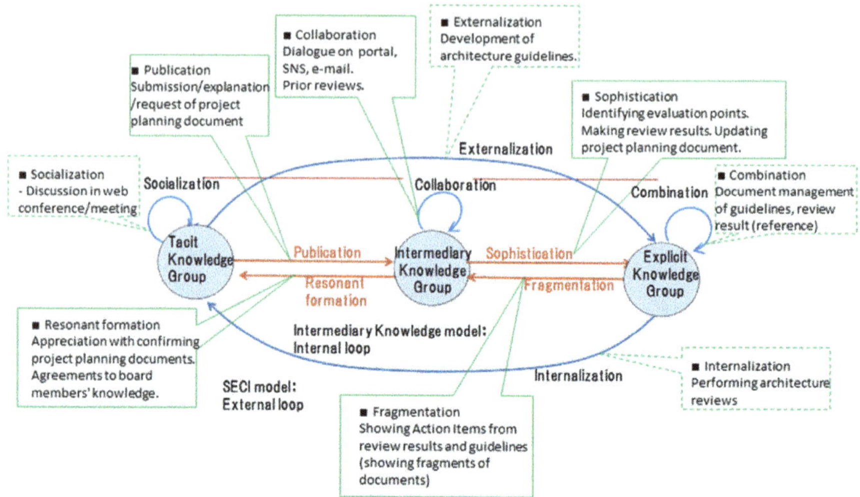

Fig. 12.4 GDTC model with knowledge sharing

stakeholders in the case of covering many regions, such as Europe, Asia, the United States, South America, the Middle East, and Africa.

Figure 12.4 shows our proposed GDTC model, which adds IKG to the traditional knowledge process in global communications. CMC tools, such as the portal and SNS, are used to support IKG. The square balloons illustrate the representative knowledge process of the GDTC model, while the dotted square balloons represent the knowledge processes of traditional global communication styles [15].

The knowledge processes in IKG consist of open and agile communication in the portal, SNS, and e-mail, and these CMC tools facilitate the communication within global communities and organizations. IKG records meeting, SNS, and e-mail logs, which are not formal documents, but still very effective knowledge for global communication. Global community members can read each other's knowledge processes in these logs. The knowledge processes in the GDTC model correspond to the knowledge transfer modes in the intermediary knowledge model in Fig. 12.4. Tacit, intermediary, and explicit knowledge correspond to the TKG, IKG, and EKG, respectively [15].

12.4 Social Collaboration Model for Architecture Review in Architecture Board

12.4.1 Proposal and Overview of Social Collaboration Model for Architecture Review in Architecture Board

Our latest research of global communication case study suggested that the global communication steps on the enterprise portal should be divided into the types such as request, reference, consolidation, confirmation, evaluation, and understanding and that each new project planning document was made/submitted and reviewed/evaluated in the Architecture Board, and in another the review results were published and the project was endorsed after dealing with the necessary action items [9, 10]. The above global communication process of Architecture Board review on the enterprise portal between the leader, Architecture Board members and new project manager, for the purpose of review in the Architecture Board, can be equivalent to Business Architecture (BA) for Architecture Review in Architecture Board. The author proposes the "Social Collaboration Model for Architecture Review in Architecture Board" in Fig. 12.5 in this book, while the above global communication process is shown as BA of this SCM model in the upper part of Fig. 12.5.

The back-colored scope in Fig. 12.5 suggests that this scope of BA should conform to Assessment meta-model shown in Fig. 12.2 for Architecture reviews in Architecture Board.

Furthermore, the above global communication process as BA of Architecture Review in Architecture Board should be activated on the enterprise portal application and social tool application, as Application Architecture (AA), built on cloud based technology platforms like private cloud, PaaS and SaaS as Technology Architecture (TA) as depicted in Fig. 12.5.

12.5 Resiliency Architecture Model—STRMM Model for Digital Transformation

12.5.1 Proposal of Resiliency Architecture Model—STRMM Model for Digital Transformation

As a result of investigating tasks of architecture reviews in the Architecture Board, in this case, we confirmed that project planning documents of almost all new IS/IT projects have been submitted to the Architecture Board. This board performs an Architecture Review on the basis of the following evaluation criteria of 4 elements, namely, (1), (2), (3), and (4) and issued the action items there. One of the authors found out the risks connected to each action item and defined an equivalent solution for each risk, and started the Risk Management process with the Architecture Board

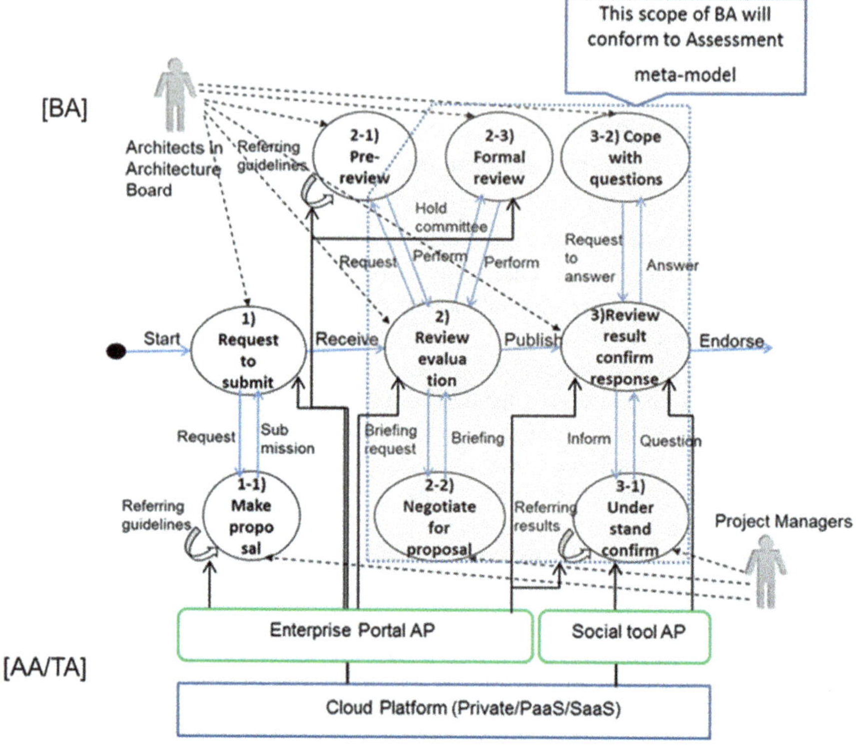

Fig. 12.5 SCM model for architecture review in Architecture Board

based on proper Mitigation Strategy [7, 8, 10]. The mitigation of these risks can lead to resiliency in Digital Transformation.

(1) Enterprise level conformance—align with architecture roadmap, standard, each architecture principle.
(2) Functional aspect—clarify solutions/architecture specifications, integration points with standard master data.
(3) Operational aspect—alignment with the service level, security design and privacy, system availability.
(4) Viability—application rationalization, risks/countermeasures, data migration/testing strategy.

We propose the Resiliency Architecture model—the "STrategic Risk Mitigation Model for Digital Transformation" in Fig. 12.6 in this case study [7, 8, 10].

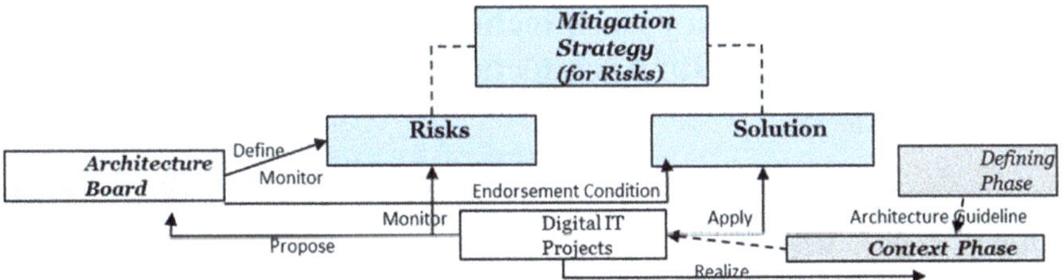

Fig. 12.6 Resiliency Architecture model—STRMM model for digital transformation

12.5.2 Research Methodology

To investigate Architecture Risks and solutions in enterprise systems (ES) and EA, we investigate the case study within a large scaled global enterprise, where we built and implemented the "Adaptive Integrated Digital Architecture Framework—AIDAF" and started up the Architecture Board. In the global Architecture Board, all new IS/IT project architecture designs were reviewed and action items for the next steps were raised by architects, top managements, and PMO members. After the Architecture Board was held, we defined the risks connected to each action item raised and considered solutions to mitigate them.

As the next step, we propose the resiliency architecture model—the strategic risk mitigation model for digital transformation in the Architecture Board as a hypothesis. Furthermore, we analyse the data of activities for digital IT projects in the Architecture Board in this global enterprise, and we evaluate our proposed model in this empirical research. In our analysis, we mapped solution categories into the previously defined risk categories in the Architecture Board. We analysed the interrelationships between these risks and solutions that were defined in the architecture risk analysis for digital IT projects involving Big Data related digital IT ones in the Architecture Board.

Based on the above research, we defined the effective strategy elements to mitigate risks with the Architecture Board for EA practitioners as a hypothesis. Moreover, we conducted further research regarding these solutions spreading across the Digital IT Systems, such as cloud (SaaS, PaaS, and IaaS), mobile IT applications, and specific application layer on cloud and mobile IT (big data and analytics) in this case of global enterprise. We evaluated our defining elements in the empirical research [7, 8, 10]. Specifically, the research questions employed in this study are:

RQ1-1 How can the Architecture Board control the Solutions with Resiliency Architecture model—Risk Mitigation model in Digital Transformation involving big data?

RQ1-2 What Strategy elements can be clarified to mitigate Risks for Digital Transformation?

12.6 Case of Resiliency Architecture Model—"Risk Management with Architecture Board" in Global Enterprise

This chapter proposes the "Adaptive Integrated Digital Architecture Framework—AIDAF" to align with IT strategy, promoting cloud, mobile IT, big data and digital IT, and the "STRMM model with AIDAF" is verified by our case study [7]. The first author has named the EA framework suitable for the era of Digital IT as "Adaptive Integrated Digital Architecture Framework—AIDAF" at this time. Figure 12.7 illustrates this AIDAF proposed model. This AIDAF begins with the Context Phase, while referencing the Defining Phase (i.e., architecture design guidelines related to Digital IT aligned with IT strategy). During the assessment and architecture review, the Architecture Board reviews the initiation documents and related architectures for the IT project.

In a particular Architecture Board, the example of the EA framework structure in a certain global company examined in this chapter is explained. This global pharmaceutical company is the largest Japan-headquartered global company in that industry in Asia based on a sales basis. In a global EA rollout, they are handling cloud/mobile IT/big data strategic projects and systems that took priority in Europe and US Group companies well by structuring and implementing EA with the above Adaptive Integrated Digital Architecture Framework—AIDAF to be consistent with global IT strategy focusing on cloud, mobile IT, big data and digital IT [7].

Actually, one of the authors works in this global company and conducted all phases of structuring and implementing in this EA framework and was the facilitator and managed the coordination of the global Architecture Board. In this global pharmaceutical company, one of the authors had the above responsibilities in the

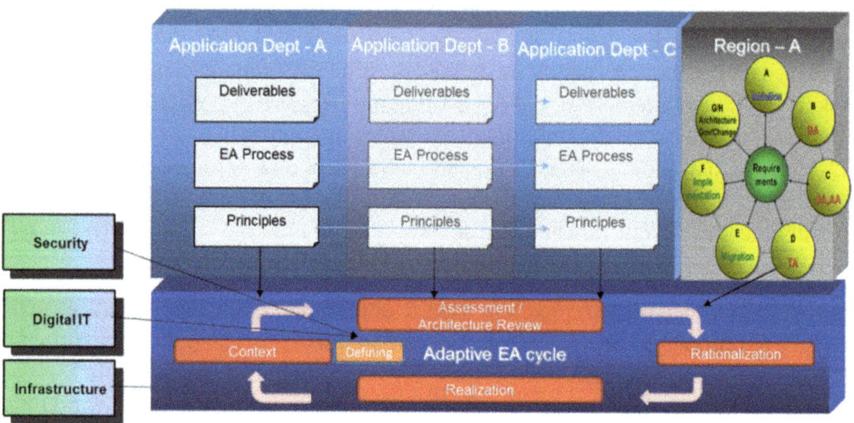

Fig. 12.7 AIDAF proposed model (ex: TOGAF and adaptive EA framework, etc.)

Table 12.3 Responsibilities and tasks of the Architecture Board

Responsibilities of Architecture Board	Tasks of Architecture Board
– Operational/Functional aspect and viability of solution architectures with enterprise level standard conformance are assessed property, and the projects related architecture risks should be coped with – Architecture integration issues across IS/IT organization are assessed, reviewed and resolved – Enterprise architecture risks are assessed and mitigated timely and consistent	– Review solution architecture/standard/common platform of sew IS/IT projects, involving new technologies – Endorse signed architecture of new IS/IT projects for the mouth – Share Architecture strategy to each departments, initiatives and their status

Assessment/Architecture Review phase on the global Architecture Board as shown in Table 12.3, which we were focusing on, and perform the above tasks [7].

As a result of investigating tasks of architecture reviews in the Architecture Board, in this case, we confirmed that project planning documents of almost all new IS/IT projects have been submitted to the Architecture Board. This board performs an Architecture Review on the basis of the following evaluation criteria of 4 elements, namely (1), (2), (3) and (4) and issued the action items there. One of the authors found out the risks connected to each action item and defined an equivalent solution for each risk, and started the Risk Management process with the Architecture Board based on proper Mitigation Strategy, as the Resiliency Architecture model. The mitigation of these risks can lead to resiliency in Digital Transformation.

(1) Enterprise level conformance—align with architecture roadmap, standard, each architecture principle.
(2) Functional aspect—clarify solutions/architecture specifications, integration points with standard master data.
(3) Operational aspect—alignment with the service level, security design and privacy, system availability.
(4) Viability—application rationalization, risks/countermeasures, data migration/testing strategy.

We propose the Resiliency Architecture model—the "STrategic Risk Mitigation Model for Digital Transformation" in Fig. 12.6 in this case study [7, 8, 10].

12.7 Case of Evaluation and Analysis of Case Study

12.7.1 Data Analysis—Risk Categories for Digital IT Areas

As a result of data analysis on the results of reviews for Digital IT projects in the Architecture Board in this global enterprise for 1 year and 8 months, 118 items/risks were raised and categorized into 10 risk categories. The revealed risk categories for Digital IT are presented, with percentages of each category indicated in parentheses [7, 8, 10].

(1) Security (23%)—risks mainly related to the information security and privacy, cyber security, security architecture, security solutions, such as user authentication, and access control in digital IT systems.
(2) Architecture Conformance (17%)—risks mainly connected with solution architecture conformance for digital IT systems in terms of business, process, technical, and sourcing strategy. 11% of risks are related to architecture standard conformance in digital IT area.
(3) Technology Architecture (12%)—risks mainly connected with technology architecture and target architecture in digital related platforms and mobile applications, digital IT systems, and others.
(4) Project Management (11%)—risks mainly related to the project management, project scope and cost structure, project scheme and project planning, and project definition.
(5) Compliance and Validation (8.5%)—risks mainly connected with compliance and validation, such as regulatory compliance and System Development Life Cycle (SDLC) related compliance, cloud data centers, cloud vendors, digital IT related vendors, GxP validation, and testing ones.
(6) Application Architecture (7.5%)—risks mainly related with application systems architecture and target architecture, roadmap of digital IT systems, in terms of functional aspects and viability.
(7) Data Architecture (7.5%)—risks mainly connected with data architecture and target architecture, a roadmap of digital IT systems with Master Data Management platforms and data migration strategy.
(8) Application Rationalization (7.5%)—risks mainly related with application system's rationalization in digital IT areas to prevent from increasing the number of applications in vain.
(9) Strategic Alignment (5%)—risks mainly connected with the alignment with IT strategy and architecture strategy in digital IT areas and high level of target architecture and roadmap.
(10) System Development (1%)—risks mainly connected with system development and enterprise system implementations in digital IT areas, and smooth transitioning to the new ES.

12.7.2 Data Analysis—Solution Categories for Digital IT Areas

In the Architecture Board in this global enterprise, 118 action items with architecture risks have been raised as a result of the architecture reviews for Digital IT projects. Moreover, each solution to mitigate these 118 risks has been defined in the risk assessment process in this global enterprise. During the analysis, we distinguished

10 solution categories for Digital IT areas. These solutions are presented in the following description, which contains short definitions of categories and percentages of adopted solutions as each category. The solution categories for Digital IT areas have been ordered in decreasing order of percentages of the adopted categories. The largest number of categories encompasses solutions connected with various approaches for assigning appropriate employees. It is followed by a category that includes solutions related to the involvement of Top Management, such as Global IT Executives, CISO, and regional CIO. Then, with similar frequency of appearance, the Application Architecture and Project Scope Definition categories occur. They are followed by System, Implementation Process, Technology Architecture, and Security Architecture categories. The Architecture Board adopted the least frequent solutions that belong to Vendor Contract Definition and System Vendor categories [7, 8, 10].

(1) Employees (42%)—solutions that assign proper employees.
(2) Top Management (18%)—solutions connected with top management involvement.
(3) Application Architecture (9.5%)—verification of appropriate application architecture and integration standard.
(4) Project Scope Definition (8%)—solutions connected with the definition of project scope.
(5) System (7%)—system customization, modification, and optimization, decommission.
(6) Implementation Process (6%)—solutions connected with the management of implementation process.
(7) Technology Architecture (4.5%)—verification of appropriate technology architecture and interface standard.
(8) Security Architecture (4.5%)—verification of appropriate security architecture and security solution adoption.
(9) Vendor Contract Definition (0.2%)—solutions connected with vendor contract definition.
(10) System Vendor (0.1%)—solutions connected with vendor support and consultants

12.7.3 Interrelation Between Solutions and Risks for Digital IT

Figure 12.8 presents the mapping of solution categories onto risk categories. In this figure, some solution categories are coped with many risk categories. Specifically, the Employees solution category, which was coped with for risks that belonged to Security, Architecture Conformance, Technology Architecture, Data Architecture and Project Management, Compliance and Validation risk categories. Similarly, solutions from Top Management solution category have been coped with for risks from Architecture Conformance and Data Architecture, Project Management, and

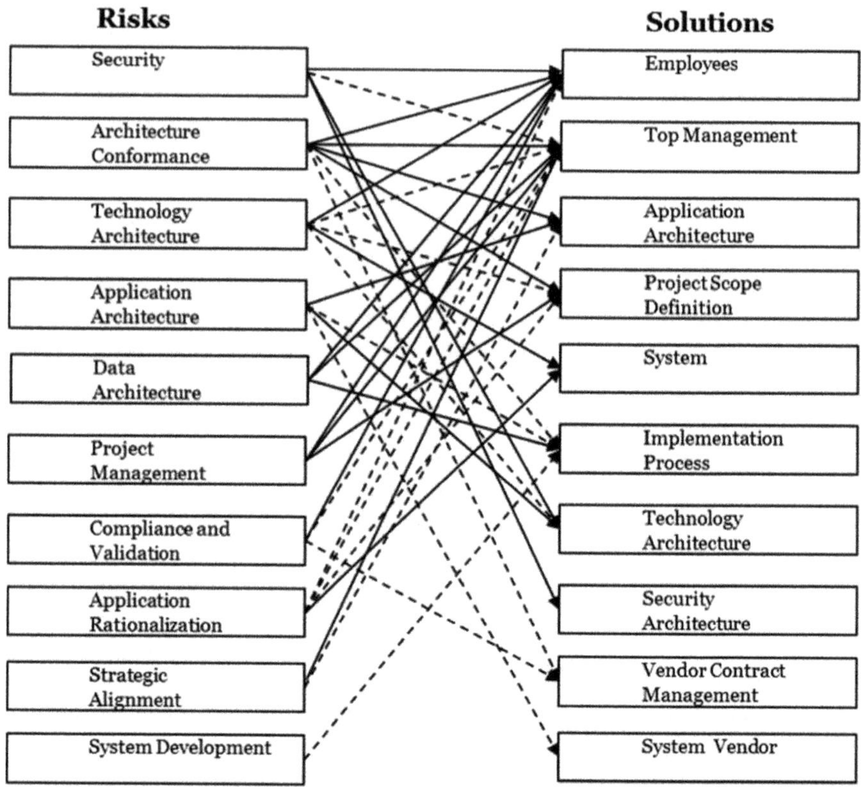

Fig. 12.8 Solution categories mapped onto architecture risk categories in digital IT projects. *Note* Solid line denotes direct correspondence between solutions and risk categories more than 3 times. Dotted line shows a direct correspondence between solutions and risk categories more than once

Strategic Alignment risk categories. Therefore, the element of "Role Assignments of experts" will be suitable for mitigating risks [7, 8, 10].

On the other hand, the Security risk category had been addressed by solutions that belonged to Technology Architecture and Security Architecture categories. For this reason, the element of "Solution Architecting for Security" will contribute to these situations to mitigate risks. Similarly, risks from Application Architecture risk category obtained their solutions from the Application Architecture and Technology Architecture solution categories. Therefore, the element of "Solution Architecting for Application" will contribute to these situations to mitigate risks. Moreover, the Data Architecture risk category had been addressed by solutions that belonged to the Implementation Process solution categories. The element of "Data Architecting" may be suitable for mitigating risks. Furthermore, solutions from the System solution category addressed risks from the Application Rationalization risk category. The element of "System Decommission" should be suitable for mitigating risks. Finally, solutions from Project Scope Definition solution category addressed risks from Project Management risk category. The element of "Project Scoping" should contribute to these situations to mitigate risks [7, 8, 10].

12.8 Verifications and Summary

12.8.1 Cloud, Mobile IT with Resiliency Architecture Model—STRMM Model

We proposed the Resiliency Architecture model—the "STrategic Risk Mitigation Model for Digital Transformation" in Fig. 12.6 for RQ1-1. As a result of reviews for Digital IT projects in the Architecture Board in this global enterprise for 1 year and 8 months, more than 110 kinds of Action Items for new IS/IT projects involving implementations on cloud, such as SaaS, PaaS, and IaaS, were issued, while 50 kinds of Action Items for new projects involving mobile IT applications were issued in the Architecture Board.

These new IS/IT projects' documents are developed with reference to architecture guidelines and abovementioned action items connected to risks summarized in the aforementioned "Risk Categories." To obtain endorsements for each new IS/IT project involving implementations on cloud from the Architecture Board, the Solutions summarized as "Solution Categories" described in Sect. 12.7.2 were discussed and defined to mitigate the Risks, and each new IS/IT project has been endorsed/realized by applying the above-mentioned Solutions. The mitigation of these risks can lead to resiliency in Digital Transformation.

The status of Solutions for each area of Cloud is presented on the left side of Table 12.4.

First, almost all Solution Categories are defined and controlled for "SaaS Cloud" related projects in the Architecture Board. Extremely high level of solution occurrence is found in the "Security of Employees Solution Category", and high level of solution occurrences are found in the "EA/Data and Compliance/Validation of Employees Solution Category", "Strategy of Top Management Solution Category", "EA/Conformance of Application Architecture Solution Category", and "Project Management of Project Scope Definition Category." Moreover, the medium levels of solution occurrence are found in the "Project Management of Employees Solution Category", "EA/Data and Project Manager of Top Management Solution Category", "Application Solution of Application Architecture Solution Category", "System Solution Category", "Implementation Solution Category", "Technology Architecture Solution Category", and "Security Architecture Solution Category."

Second, approximately half of the Solution Categories are defined and controlled for "PaaS Cloud" related projects in the Architecture Board. High level of solution occurrence is found in the "Security of Employees Solution Category." On the other hand, low level of solution occurrences is found in the Solution Categories

Table 12.4 Solutions for digital IT areas with mitigation strategy elements

Solution / Digital IT Areas	<Cloud & Mobile IT>				<Specific Application Layer on Cloud & Mobile IT>					<Mitigation Strategy >						
	SaaS Cloud	PaaS Cloud	IaaS Cloud	Mobile IT Application	Social/SNS	Big Data Solution	Business Analytics	Analytics for R&D	Internet Of Things	Role Assignments	Security Architecting	Application Architecting	Data Architecting	System Decommission	Project Scoping	Vendor Relationship
Employees																
Security	●●	●	●	●	○	●	◑	○	○	●	●●	◑	○		○	
EA/Data	●	○		○	○	○	◑			●	○	○	◑	○		
Compliance/Validation	●	○	○	●	○	◑	○	○	○	●						
Project Management	◑		○	○			○			◑						
Top Management																
Strategy	●			○		○	○	○		◑		○		○		
EA/Data	◑		●	○		○	◑			◑		●	◑			
Security/Compliance	○			○		○				○	○					
Project Manager	◑	○	○	○		○				○					◑	
Application Architecture																
Strategy	○	○		○		○						○			○	
Application Solution	◑			○		○	○					◑				
EA/Conformance	●	○	○	○	○		○					●				
Project Scope Definition																
Project Management	●	○	◑	◑	○	◑		○	○						●	
Application Rationalization	○	○		○		○	○								○	
Technology Architecture	○	○	○	○	○	○			○						○	
System																
Application Rationalization	◑			○			○							◑		
Technology Architecture	◑			◑		○	○					○				
Implementation Process																
Application	○			○		○	○					○				
Data	◑			○		○	○						◑			
EA/Conformance			○				○					○				
Technology Architecture																
Application Technology	◑		○	○			○					◑				
Integration Architecture	◑		○	◑		○	○					◑				
EA/Conformance	○			○								○				
Security Architecture																
Security/Privacy overall	○		○			○					○					
User Authentication	◑	○	○	○		○					◑					
API controls		○		○							○					
Vendor related																
Vendor Contract	○															○
Application Architecture	○		○									○				

of “Employees”, “Top Managements”, “Application Architecture”, “Project Scope Definition”, and “Security Architecture.”

Third, most of the Solution Categories are defined and controlled for “IaaS Cloud” related projects in the Architecture Board, except for the “System Solution Category.” High level of solution occurrences is found in the “Security of Employees Solution Category” and “EA/Data of Top Management Solution Category.” Medium level of solution occurrence is found in the “Project Management of Project Scope Definition Solution Category.”

The status of Solutions for the area of Mobile IT Application is also presented on the left side of Table 12.4.

Almost all Solution Categories are defined and controlled for "Mobile IT Application" related projects in the Architecture Board. High level of solution occurrences is found in the "Security" and "Compliance/Validation" of "Employees Solution Category." Moreover, medium levels of solution occurrence are found in the "Project Management of Project Scope Definition Solution Category", "Technology Architecture of System Solution Category", and "Integration Architecture of Technology Architecture Solution Category" [7, 8, 10].

This empirical case study provides evidence how it is possible to manage risks with solutions and strategy elements using the Resiliency Architecture model—the "STRMM model for Digital Transformation" in Fig. 12.6 in the "Cloud/Mobile IT" areas to cope with RQ1-1.

12.8.2 *Specific Application Layer on Cloud/Mobile IT—Big Data with Resiliency Architecture Model—STRMM Model*

In the current study, we investigate the specific Application Layer on Cloud and Mobile IT, such as Big Data, social network services, business analytics, analytics for R&D, and IoT.

The status of solutions for each area of "Specific Application Layer" under "Cloud and Mobile IT" is also presented in the middle (light grey) of Table 12.4. First, most of the solution categories are defined and controlled for the Big Data solution-related projects, except for "vendor-related solution." In this Global Enterprise, eight kinds of Big Data projects, such as Key Opinion Leaders (KOL), occur. Four kinds of KOL management-related Big Data projects involving mobile IT applications for business intelligence (BI)/customer relationship management (CRM) are raised, all implemented on SaaS cloud, except for one developed on PaaS. Two other kinds of Big Data mobile IT projects were also managed, both implemented on SaaS cloud; one project was implemented on IaaS cloud as well. Another Big Data project was partially-implemented on SaaS and IaaS. The new Big Data Digital IT project documents were proposed and reviewed; 29 action items were raised. High levels of solution occurrence are found in the "Security" under "Employees". Moreover, medium levels of solution occurrences are found in "Compliance/Validation" under "Employees" and "Project Management" under "Project Scope Definition." Moreover, low levels of solution occurrences are found in solution categories under "Employees," "Top Managements," "Application Architecture," "Project Scope Definition," "Implementation Process," and "Security Architecture."

Most of the solution categories are defined and controlled for business analytics-related projects, except for "Security Architecture" and "Vendor-related solution." Medium levels of solution occurrences are found in "Security and EA/Data" under "Employees" and "EA/Data" under "Top Management." Less than three solution categories involving "Employees" and "Project Scope Definition" are defined and controlled for "social/SNS," "analytics for R&D," and "IoT" projects. The number of the projects in these areas is less than three and limited in this global enterprise.

Our current research verifies that the Architecture Board can control solutions with the Resiliency Architecture model—the STRMM model for digital transformation. Table 12.4 shows the "Specific Application Layer on Cloud and Mobile IT," including "Big Data solution" and "business analytics," to cope with RQ1-1 [7, 8, 10].

12.8.3 Strategy Elements to Mitigate Risks in Digital Transformation

Based on the research described in the previous section, 6 kinds of main strategy risk mitigating elements can be formulated to answer RQ1-2 that are "Role Assignments", "Security Architecting", "Application Architecting", "Data Architecting", "System Decommission" and "Project Scoping." The mitigation of risks can lead to resiliency in Digital Transformation. The status of each mitigation strategy element is presented in the right side of Table 12.4 [7, 8, 10]. Furthermore, the details of the above risk mitigation elements will be explained in the following section.

This chapter described our proposed the Resiliency Architecture model—STRMM model for Digital Transformation and verified this model in the above case study in the global company. Furthermore, as a result of this case study, Strategy elements to mitigate risks in Digital Transformation were also formulated appropriately. Section 12.9 will brief the valuation of Strategy elements to mitigate Risks in digital transformation, and Sect. 12.10 will describe the conclusion on the Resiliency Architecture model—Risk Management for Digital Transformation.

12.9 Valuation of Strategy Elements to Mitigate Risks in Digital Transformation

Based on the research described in Sect. 12.8, the following strategy risk mitigating elements can be formulated to answer RQ1-2 defined in Sect. 12.5. The mitigation of these risks can lead to resiliency in Digital Transformation. The status of each mitigation strategy element is presented in the right side of Table 12.4 [7, 8, 10]. The detail of valuation results for these Strategy risk mitigating elements in Digital Transformation are as follows.

(1) Role Assignments
Two types of solution categories, "Employees" and "Top Management," are adopted to mitigate risks with this Strategy element. High levels of solution occurrences are found in "Security," "EA/Data," and "Compliance/Validation" of the "Employees solution category." Moreover, medium levels of solution occurrences are found in "Strategy" and "EA/Data" under "Top Management

solution category" and "Project Management" under "Employees solution category." In terms of Big Data projects, high levels of solution occurrence are found in "Security" under "Employees solution category," and medium levels of solution occurrence are found in "Compliance/Validation" under "Employees solution category" [7, 8, 10].

(2) Security Architecting
Several kinds of solution categories, including "Employees," "Security Architecture" and "Top Management," are adopted to mitigate risks with this Strategy element. High levels of solution occurrence are found in "Security" under "Employees." Medium levels of solution occurrence are found in "User Authentication" under "Security Architecture." In terms of Big Data projects, high levels of solution occurrence are found in "Security" under "Employees" [7, 8, 10].

(3) Application Architecting
Seven kinds of solution categories, including "Employees," "Top Management," "Application Architecture," and "Technology Architecture" are adopted to mitigate risks with this Strategy element. High levels of solution occurrence are found in "EA/Data" under "Top Management" and "EA/Conformance" under "Application Architecture solution." Medium levels of solution occurrence are found in "Security" under "Employees," "Application Solution" under "Application Architecture," and "Application Technology" and "Integration Architecture" under "Technology Architecture" [7, 8, 10].

(4) Data Architecting
Three kinds of solution categories, including "Employees," "Top Management,"and "Implementation Process"were adopted to mitigate risks with this Strategy element. Medium levels of solution occurrence are found in "EA/Data" under "Employees solution category," "EA/Data" under "Top Management solution," and "Data" under "Implementation Process solution" [7, 8, 10].

(5) System Decommission
Three kinds of solution categories, including "Employees," "Top Management," and "System" were adopted to mitigate risks with this Strategy element. Medium levels of solution occurrence are found in "Application Rationalization" under "System solution" [7, 8, 10].

(6) Project Scoping
Four kinds of solution categories, including "Top Management," "Employees" and "Project Scope Definition" are adopted to mitigate risks with this Strategy element. High levels of solution occurrence are found in the "Project Management" under "Project Scope Definition," and medium levels of solution occurrence are found in the "Project Manager" under "Top Management solution." For Big Data projects, medium levels of solution occurrences are found in the "Project Management" under "Project Scope Definition" [7, 8, 10].

(*) Vendor Relationship
The Vendor-related solution category was adopted to mitigate risks with this Strategy element. However, only low-level solutions are found in the "Vendor Contract" [7].

12.10 Conclusion—Resiliency Architecture Model, Risk Management for Digital Transformation

This chapter proposed the Resiliency Architecture model—the "STRMM model for Digital Transformation" as the Risk Mitigation model in the Architecture Board based on the current research with the global enterprise case study, where we built and implemented the "Adaptive Integrated Digital Architecture Framework—AIDAF." Based on the empirical data gathered from the result of the Architecture Board reviews in the global enterprise, this research verified that the Architecture Board can control the Solutions with the Resiliency Architecture model—the "STRMM model for Digital Transformation" in areas of "Cloud/Mobile IT" and "Specific Application Layer on Cloud/Mobile IT", such as "Big Data Solution" to cope with RQ1-1 in the research methodology defined in Sect. 12.5.2 [7]. Furthermore, we clarified the Strategy elements to mitigate the Risk for Digital Transformation, such as "Role Assignments", "Security Architecting", "Application Architecting", "Data Architecting", "System Decommission", and "Project Scoping", which can lead to the answer for RQ1-2 [7]. Besides, in the case study of global enterprise, one of the authors monitored the status of each risk defined in the risk assessment process with Architecture Board in the full-lifecycle of SDLC on the basis of the Resiliency Architecture model—the "STRMM model for digital transformation" with "Strategy elements to mitigate risks," while the above risks were mitigated then,that can answer for the Research Question of "How can the Resiliency Architecture model—the "STRMM model" be effective for mitigating risks for Digital Transformation in terms of full lifecycle in SDLC?" in this chapter.

References

1. TOGAF® Version 9.2: an Open Group Standard (2018); refer to: www.opengroup.org/togaf
2. Information Systems Audit and Control Association—ISACA: COBIT5: A Business Framework for the Governance and Management of Enterprise IT (2012)
3. Martin, J.N.: Overview of an emerging standard on architecture processes—ISO/IEC 42020. In: The 26th Annual INCOSE International Symposium (IS 2016)
4. Martin, J.N.: Overview of an emerging standard on architecture evaluation—ISO/IEC 42030. In: The 27th Annual INCOSE International Symposium (IS 2017). Adelaide, Australia, July 2017
5. Masuda, Y., Shirasaka, S., Yamamoto, S.: Integrating mobile IT/cloud into enterprise architecture: a comparative analysis. In: Proceedings of the 21th Pacific Asia Conference on Information Systems (PACIS 2016), p. 4
6. Masuda, Y., Shirasaka, S., Yamamoto, S., Hardjono, T.: 7 1. An adaptive enterprise architecture framework and implementation: towards global enterprises in the era of cloud/mobile IT/digital IT. Int. J. Enterp. Inf. Syst. **13**(3), 1–22 22 p. https://doi.org/10.4018/ijeis.2017070101
7. Masuda, Y., Shirasaka, S., Yamamoto, S., Hardjono, T.: Risk management for digital transformation in Architecture Board: a case study on global enterprise. In: EAIS International Conference 2017, 6th IIAI International Congress on Advanced Applied Informatics (IIAI-AAI), IEEE Conference Proceedings, pp. 255–262, July in 2017

8. Masuda, Y., Shirasaka, S., Yamamoto, S., Thomas Hardjono, T.: Risk Management for digital transformation and big data in architecture board: a case study on global enterprise. In: Information Engineering Express (IEE) Vol 4, No.1, in March in 2018
9. Masuda, Y., Shirasaka, S., Yamamoto, S., Hardjono, T.: Architecture board practices in adaptive enterprise architecture with digital platform: a case of global healthcare enterprise. Int. J. Enterp. Inf. Syst. IGI Global. February 2018
10. Masuda, Y., Murlikrishna, V.: Enterprise Architecture for Global Companies in a Digital IT Era: Adaptive Integrated Digital Architecture Framework (AIDAF), Springer Nature, February 2019
11. Kanbe, M., Yamamoto, S., Ohta, T.: A proposal of TIE Model for Communication in Software Development Process. In: Nakakoji, K., Murakami, Y., McCready, E. (eds.): JSAI-is AI, LNAI 6284, pp. 104–115. Springer, Berlin, Heidelberg (2010)
12. Yamamoto, S., Kanbe, M.: Discussions on models for digital knowledge sharing in enterprises. In: 3rd Research Meeting On Knowledge Sharing Networks, Japanese Society of Artificial Intelligence, 2008
13. Yamamoto, S.: Organizational Communication Change Through CMC: Learning From Practice Through Enterprise SNS. NTT Publishing (2010)
14. Yamamoto, S., Kanbe, M.: Knowledge creation by enterprise SNS. Int. J. Knowl., Cult. Change Manag. (2008)
15. Masuda, Y., Shirasaka, S., Yamamoto, S., Hardjono, T.: Proposal of GDTC model for global communication on enterprise portal. In: KES 2017, 21st International Conference on Knowledge-Based and Intelligent Information & Engineering Systems, France, September in 2017. Journal of Procedia Computer Science, Elsevier, pp. 1178–1188 (2017)
16. Everett. M. Rogers, translated by Toshiaki Yasuda, "Multimedia Technology: The New Media in Society", Kyoritsu Publishing, 1992.

Chapter 13
Automated Enterprise Architecture Model Maintenance via Runtime IT Discovery

Martin Kleehaus and Florian Matthes

Abstract Enterprise architects struggle to cope with rapid architectural changes and to document them accordingly in their architecture models due to missing adequate tool support. For that reason, the documentation of Enterprise Architecture (EA) models is still mostly achieved by manual effort and often contains outdated information. In this work, we present a novel approach for EA model creation that leverages runtime service instrumentation of the existing IT architecture to automatically create, update, and enhance static EA models with runtime information. We introduce a new integration layer that synchronizes static and runtime data from different data sources. The hereby implemented prototype allows different stakeholders to explore information from both perspectives (static and runtime) based on a linked enterprise knowledge graph, which supports new use cases and analysis capabilities. We evaluate our prototype by implementing it in a big German retailer. The introduction of service naming conventions and validation workflows enables fully automated data integration, which minimizes the effort for manual tasks. Based on interviews we conducted with 17 experts from two different companies, we could prove that the tool is capable to automate EA model maintenance.

Keywords Enterprise architecture · EA model maintenance · IT discovery · GraphQL · Microservice · Runtime · Monitoring

13.1 Introduction

The Enterprise Architecture (EA) describes the interaction between Information Technology (IT) and the business processes in a company. The hereby performed management discipline—EA management (EAM)—is an instrument to improve the

---

M. Kleehaus (✉) · F. Matthes
Chair of Software Engineering for Business Information Systems, Technical University of Munich, Garching Bei München, Germany
e-mail: martin.kleehaus@tum.de

F. Matthes
e-mail: matthes@tum.de

A. Zimmermann et al. (eds.), *Architecting the Digital Transformation*,
Intelligent Systems Reference Library 188,
https://doi.org/10.1007/978-3-030-49640-1_13

alignment of business and IT, to realize cost-saving potentials, and to increase availability and fault tolerance [1, 2]. While classical software engineering approaches focus on details, EAM conveys a holistic view of the entire organization including information about the business, IT, and relationship between both layers.

IT-landscape modelling, as a sub-area of EAM, tries to discover and document the IT-landscape of an organization. It aims to generate and maintain a virtual representation—a digital twin—of the whole organization [3]. With this digital twin, it becomes much more convenient for the enterprise architects to assess the *as-is* EA. This is important for decision support especially when it comes to IT transformation and modernization.

EAM tools like Iteraplan,[1] Alfabet[2] or LeanIX[3] support the modelling of the EA digital twin, however, the creation of these models is a time-consuming task [4]. Enterprise architects struggle to collect EA-relevant information to unveil the current business- and IT-landscape [5]. This leads to out-of-date EA models and, in turn, to decisions made on wrong or obsolete data basis [6].

The reasons for this problem are manifold: (1) Organizations try significantly to shorten their software release cycles nowadays. This is achieved by introducing agile practices, continuous deployment, and delegating architectural decisions to development teams [7]. As a result, architecture designs suffer more than ever from the problem of capturing the full essence of an ever-changing landscape. (2) Due to new software development methodologies like microservice architectures (Lewis, James; [8], the number of to be managed IT services increase significantly. Hence, enterprise architects need to spend more time to maintain an increasing number of models. (3) Last, the most important reason, most of the EA-relevant data is still collected manually, which does not represent a reliable approach considering the first two points [4].

Runtime data of the IT landscape represents the true *as-is* IT architecture at a specific point in time. Many monitoring solutions [9–12], that create those runtime data have been developed to account for a layered architecture and provide metrics for every EA layer [13–15] including infrastructure, application and business-related aspects. Especially Application Performance Monitoring (APM) had been adapted significantly to meet the new monitoring requirements that emerged with microservice architectures [16]. New APM vendors, like Dynatrace,[4] AppDynamics,[5] Instana[6] just to name a few provide powerful instrumentations to gain insights from an end-to-end perspective. Based on this development it becomes obvious, why runtime data is defined as a good information source for delivering EA-relevant data as identified by studies conducted by [17, 18]. In particular, the actuality and correctness of the data are perceived as being high among the survey

[1] https://www.iteraplan.de/.

[2] http://alfabet.softwareag.com/.

[3] https://www.leanix.net.

[4] https://www.dynatrace.com.

[5] https://www.appdynamics.com.

[6] https://www.instana.com.

participants. However, only a few research endeavours [19–21] leverage runtime data for this purpose as the low maturity level of monitoring tools in this time did not allow to capture the complete IT landscape.

In this work, we present a novel approach for automating the maintenance of EA models (in the following *architecture models*) via the extraction of those models from runtime data and federated information systems. Both data sources are mapped on the level of the runtime artefact as the most granular entity in the EA. For this purpose, we introduce in a prototypical implementation, called *Microlyze* a new integration layer that is incorporated into the continuous delivery pipeline and allows the synchronization of low-level monitoring information to high-level architecture models maintained in EAM tools. Hereby, we aim to answer three research questions:

RQ1 Due to an inherent disconnect between the design of architectures and the development of the concrete runnable software artefacts, the relationship between both entities is not apparent immediately. This raises the following question: *How to map runtime artefacts provided by monitoring tools with design-time entities reside in federated information sources*?

RQ2 Monitoring data is by nature non-permanent, as it is an expression of many changes in the runtime architecture and therefore in contrast to EAM information, which is rather static and expected to be stable. For this reason, it becomes important to identify *how to synchronize architecture changes discovered from runtime data with maintained architecture models?*

RQ3 As EA-relevant information is dispersed in the IT landscape, it must be properly collected for EA documentation. Hereby, it is important to validate *where and how to integrate the tool into the IT environment?*

The remainder of the paper is organized as follows: In Chap. 2, we list the related work. Chapters 3 and 4 shows the overall architecture of our solution. In Chap. 5, we dive deeper into the aspects of how we synchronize runtime data with architecture models. Afterwards, we continue in Chap. 6 to discuss our evaluation results and finish the paper with a conclusion and a description of our future work in Chap. 7.

13.2 Related Work

There exist a few concepts on how to integrate runtime data from existing data sources for EA model maintenance [21], as well as [22] make use of network analysis tools to infer information on the IT infrastructure [19], on the other hand, interpret the configuration of an Enterprise Service Bus (ESB) to include knowledge on communicating information systems in the EA model. These approaches have in common that they are limited to a specific layer of the EA and the authors describe their concepts about how to connect the data to the EA model rather superficial.

Farwick et al. [20] presented a similar approach by tackling the challenge of EA model synchronization with different technologies and applications. In their approach, all infrastructure elements are tagged manually with a Universally Unique

Identifier, as soon as they are planned in the project management tool. After recognizing changes on these elements, they will be propagated to the EAM tool. The main difference to this research is the manual efforts which have to be applied in initially capturing tags and keeping them up to date. On microservice level, this approach is not feasible, as the number of services would exceed the required manual effort.

A further approach for synchronization between runtime and design time architecture is proposed by Breu et al. [23] and similarly by Akkiraju et al. [24]. In comparison to the traditional agile way of software development where changes are implemented iteratively, they consider the design to be done upfront and manifest itself in architecture models. All running artefacts are the result of generation from these models and changes are directly reflected on these respective models. The nature of generating code from models is framed by the term "living models". While this idealistic approach is sensible in theory, it requires a lot of design work upfront, which is not compatible with agile practices. It remains to be verified whether this works in practice, as no larger scale evaluations are presented.

A class of publications recognizes the importance of receiving events from other information systems that indicate architecture-relevant changes [25–28]. Some of the publications mention the importance of recognizing external events. Others, such as the thesis of Buckl [25] consider triggers for manual action in the provided meta-model. Only one paper discusses implementation issues of utilizing external events to drive automated EA documentation processes [26].

Most of the papers reviewed do not concern themselves with documenting the architecture in an existing tool, but rather build their type of documentation in contrast to using existing EAM software [29, 30]. For the part of identifying the IT components either custom integration has to be built into the software or it requires manual efforts [20]. Overall this research focuses completely on the extraction of IT components from runtime data in order to automate EA model maintenance.

13.3 System Design

For our objective to discover the EA, we created a software called *Microlyze* [30, 31] that recovers the detailed IT landscape architecture by continuously analyzing runtime data, exposed by APM tools. The backend of *Microlyze* consists of four main components: (1) a runtime data collection part referred to as Microlyze.Collect, (2) an analysis part, defined as Microlyze.Analyze, that provides the corresponding infrastructure to analyze runtime data and to extract architecture-related components and their relationships, (3) a component for storing the architecture models in a database named as Microlyze.Store and (4) finally a component Microlyze.Expose that provides an interface for exposing the recovered architecture for the end-users. *Microlyze* unveils all running IT components from the infrastructure and application layer, the interrelationship between those layers, like which application runs on which hardware and the intrarelationship that means the communication behaviour and dependencies between applications.

An important prerequisite that the APM solution must fulfill is *distributed tracing* [32]. This technique became popular with the rising interest in microservice architectures [33]. Distributed tracing monitors the consumed time for request processing between microservice calls by injecting tracing information into the request HTTP headers. The main purpose is to analyze application performance. *Microlyze* leverages this runtime information in order to unveil the communication behavior within the architecture.

During the analysis of runtime data, *Microlyze* must manage the following four challenges: (1) In order to reduce network overhead, most tracing techniques are based on sampling, i.e. only a percentage of requests is traced and forwarded to the monitoring server. In the worst case, specific communication paths are rarely seen. (2) Due to resource limitations, most monitoring tools can only provide a small timeframe of runtime data, e.g. last 6 h, depending on the frequently incoming data volume. A request for runtime data for a longer duration would be too resource intensive and cannot be served. (3) Most monitoring tools store runtime data for only a specific period of time and archive or even delete older data in order to ensure free storage capacity is always available. (4) The history does also contain old components and communication paths that were already removed several sprints ago. This legacy data must be filtered in order to unveil the real architecture.

Considering the listed challenges, we developed a concept that discovers the architecture of the current IT landscape based on a two-step process, which is detailed in Kleehaus et al. [34]. First, we developed the algorithm *backwardDiscovery* that reconstruct an incomplete architecture by analyzing historical data. In the second step, the algorithm *forwardDiscovery* gets triggered on a fixed time interval and consumes new incoming runtime data for continuously adapting the final architecture.

13.4 Federated Data Management

The described architecture discovery concept that uses runtime data to automate the IT landscape documentation focuses primarily on the technical aspect of an enterprise. However, an overall documentation of an IT landscape from an EA perspective also encompass the allocation of the architecture models within the business layer. This involves the status of ongoing projects, software development progress, performance of products and business domains [35] as well as team definitions, which are not available in runtime data. This information resides in federated information systems as Hauder et al. [18] elaborated. Consequently, it is necessary to extract this information and map it with runtime data to achieve a complete and holistic view on the enterprise architecture. With this federated setup, we follow the approach proposed by Fischer et al. [36] who state that each stakeholder in an enterprise (e.g. systems operation, project management, etc.) needs their own set of tools which they are familiar with.

13.4.1 Configuration File

We realize the mapping with the introduction of a configuration file that is stored in each application folder and exposed during deployment. This configuration file contains meta-information about the particular application and maintains all references to the related federated information systems. The mapping is established on the application or where applicable on the microservice level as the smallest unit. The configuration file is located in the root directory of the project. We hereby leverage the technology pivio.io[7] and modified it according to our needs. This file contains the following information.

- name: An understandable unique name of the artefact (mandatory).
- description: More information about the purpose of the artefact and which task it is supposed to solve (mandatory).
- application: The application name as the parent IT artefact, in case the application was split into microservices (mandatory).
- product: The product assignment is important to understand in which business-specific relation the application was built (mandatory).
- boundedcontext: The bounded-context represents the domain in which the artefact "lives" (mandatory).
- references: The abbreviation of the information source, like "pm", "eam", "cr", "cmdb", "wiki" (optional).

The reference contains the following attributes:

- tool: The name of the federated information system (mandatory).
- domainURL: The URL address pointing to the information source (mandatory).
- apiURL: The URL for the API interface (mandatory).
- id: The ID of the artefact representation (mandatory).
- apitoken: Access credential API token (optional).

All attributes indicated by the tag "mandatory" are required for the functionality of the system. Based on the application name the mapping between the documented, and the runtime artefacts is established. Hereby, we do not maintain a copy of the data in a central database but store only the references to the original source in order to realize the mapping. This concept ensures referential integrity, i.e. changes on applications like IP address, port, name, etc. do not affect the recovered architecture model.

The reference list can be extended by any other information source as long as an adapter for accessing the tool is available. In the current state, we support the tools Iteraplan, PlanningIT, Jira, Github and Jenkins. Further important information sources that might contain EA-relevant data and could be referenced by the configuration file was elaborated by Farwick et al. [17].

[7] http://pivio.io/.

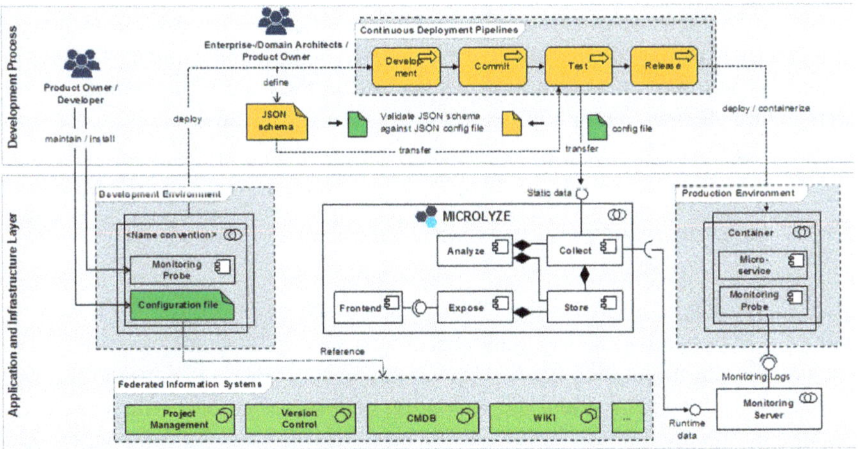

Fig. 13.1 Overall architecture composition described in Archimate 3.0.1

13.4.2 *Integration Pipeline*

In order to ensure that (1) the configuration file was created and attached to the software project, as well as (2) the content follows a predefined schema, we suggest integrating our tool in a continuous integration (CI) pipeline. The workflow is illustrated in Fig. 13.1. First, a consensus between Enterprise-, and Domain Architects must be found on which references are at least included in the configuration file. The agreement or contract between the Architects is persisted as JSON schema.[8] With JSON Schema it is simple to define a meta-model specifying mandatory and optional fields, attribute types and their properties. An additional test case in the CI pipeline ensures that the configuration file exists and meets the predefined schema. If not, the file is rejected which leads to a failed test case. If the developers do not adhere to regulations, the service cannot be deployed to production. A successful test case sends the configuration file to *Microlyze* which registers the artefact for the first time. Afterwards, *Microlyze* establishes a connection to the monitoring server and maps the static and the runtime representation of the same artefact based on the provided naming convention.

13.5 Architecture Model Synchronization

The primary goal of the synchronization process is to identify changes in the IT landscape and synchronize them with the maintained architecture model. The process results in either the creation of new entities, the update of existing and referenced

[8]https://json-schema.org/.

entities, or the deletion of entities. The synchronization process itself is designed to be idempotent as long as no changes have occurred in the architecture itself, therefore running it multiple times has no further impact on the result. The identification of architecture changes can be triggered at different points in time.

13.5.1 Creation of IT Landscape Entities

First, the prototype discovers the IT landscape based on the consumed monitoring data [30]. Afterwards, the tool validates whether an architecture model is already known, and a configuration file has been maintained. If the model has not yet been inserted into the database and the validation of the configuration file successfully results in the extraction of all provided business-related meta-information, it will be inserted as a new entity. In case the product, application and domain name, as well as their relationships do not exist yet, they will be created accordingly. If the application maintains predecessors and successors this information will also be inserted as communication relationships. Furthermore, all provided references to the related entities in the federated information systems are mapped to the application and stored in the database. In fact, *Microlyze* does not export the entities but keeps only a technical reference which ensures independence from any kind of changes done inside the federated information system, making also renaming possible.

13.5.2 Update of IT Landscape Entities

New releases of applications or performed updates in the referenced federated information systems do not affect the architecture models maintained in *Microlyze*. However, if new communication paths (new predecessors or successors) are recognized by the monitoring tool, or the deployment environment has changed, *Microlyze* updates the relationship information automatically. However, an update does not mean overwriting information. In fact, we create revisions that basically represents the history of performed modifications, i.e. the architecture evolution. A revision is reflected by the validity of each maintained architecture model, determined by two timestamps "validFrom" and "validTo". A performed change set the "validTo" to the current time and a new entity for the affected architecture model is created with an empty "validTo" attribute. Hence, we never delete entities, but set them as invalid. Based on this "only insert" approach, we are capable to move through time and unveil the architecture evolvement or compare different architecture states.

In case of performed changes to the main name of a runtime artefact, it will inevitably be considered as a new artefact, which will trigger the creation flow. At some point in time, the deletion flow (see Sect. 13.5.3) will remove the references to the old artefact which was renamed.

13.5.3 Deletion of IT Landscape Entities

In addition to the validity attributes, we also introduce two attributes that determine a potential deletion state of an architecture model. *Microlyze* regularly polls the monitoring system and compares the discovered IT components with the maintained components. In case, an architecture model is encountered whose "lastSeen" timestamp is beyond a threshold, *Microlyze* will mark the artefact as to be retired by setting the "retiredAt" timestamp. After a defined while the artefact and all associated relationships will be removed. Both time thresholds can be defined individually to the needs of different environments [34].

13.6 Enterprise Architecture Knowledge Graph

In order to access the discovered architecture models including the relationship structure, we create the *Enterprise Architecture Knowledge Graph.* It represents an interface for querying the data that enables to retrieve an holistic picture on the IT landscape resources. Hereby, we follow the idea of an "Enterprise topology graph" (ETG) that was introduced by Binz et al. [37] The purpose of an ETG is to capture all entities of an IT landscape including logical, functional, and physical relationships and to model this structure in a graph-based representation. The nodes in the graph define IT landscape entities and the edges their semantic relationships.

With the support of GraphQL, we can provide stakeholders with a query language that enables them to traverse through the knowledge graph. The graph leverages the references stored in *Microlyze* to all the respective federated information systems in order to expose an interface for traversing through the architecture models that live and are maintained in those systems. As we do not copy the architecture models, all data exposed is always "live" and retrieved via the interfaces.

The metamodel which was built for the Enterprise Architecture Knowledge Graph is independent of any of the connected federated information systems as long as the necessary adapter for calling the interface is developed.

As depicted in Fig. 13.2, the "ArchitectureModel" is the recovered IT landscape artifact, which is assigned to one of the three-layer structure (business, application, technology). Those models are either recovered from runtime data or from the content of the configuration file. The "RuntimeService" is the smallest unit in the IT-landscape and represents the runtime process of an application. IT contains multiple "ModelReferences" to all provided data sources. Each model reference contains important information about this runtime artefact. This allows connecting an arbitrary amount of references to other systems. The detailed implementation of a model reference is dependent of the referenced data source and can be extended as requested.

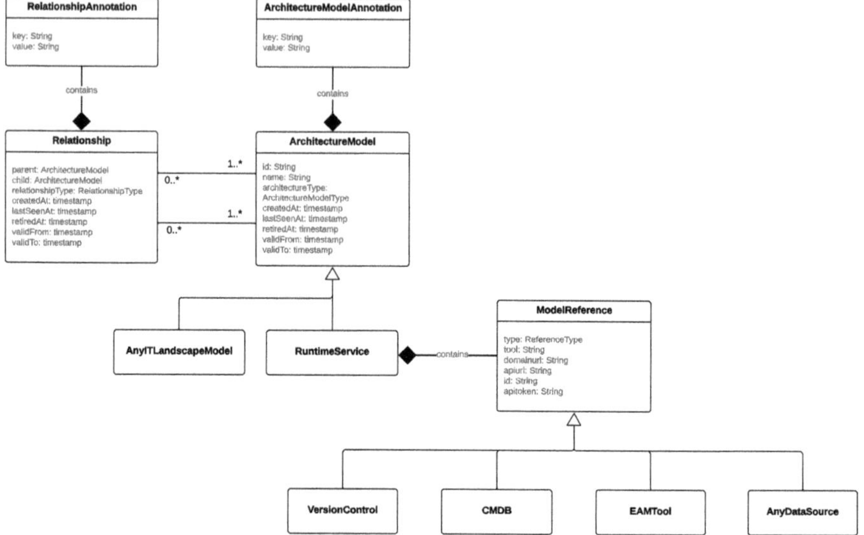

Fig. 13.2 Meta-model of the enterprise graph

13.7 Evaluation

The evaluation of our approach is two-folded: First, we developed a prototype as a result of our design science research and evaluated in a large European retailer. Second, we interviewed 17 experts from two different companies that rated the concept based on the criteria collected by Prat et al. [38]. However, we were not able to incorporate all the criteria as some of them demand a long-term evaluation.

13.7.1 Prototype Integration Result

The company we chose for evaluating the prototype is a large European retailer for electronics with an omnichannel presence. The company, with the headquarter in Germany, offers its customers the ability to buy products online or in brick and mortar stores.

The IT of the company is organized as a "product-based" organization, where IT services are no longer delivered for business units but rather as products in different domain contexts. The domain boundaries served as the basis for splitting the core applications into smaller units (products).

The company uses Dynatrace as the APM solution for its core applications covering about 40% of the whole application landscape, but especially the part of self-developed software that changes frequently and is challenging to manage. The responsibility for keeping the documentation up-to-date is delegated to the product owner and the respective team. In conclusion, many lower level changes to application architecture are delegated to people outside of the EAM division, which results in outdated, and wrong documentation.

As the ground truth for validating the business-related entities, we leverage the in use EAM tool PlanningIT. In total, 185 running microservices were identified that are supposed to be recovered by *Microlyze*. After starting the synchronization process, *Microlyze* discovered the IT landscape including all running services and their communication behaviour and established the references to the monitoring system. 96 out of 185 services could be automatically assigned to product and domain entities. 89 services did not find any product relationship. Six products out of the 89 services were identified that had no corresponding entity in the EAM tool. Hence, the products were created and assigned to two existing domains automatically. Further, four products were created in PlanningIT as *Microlyze* recognizes them as missing entities. However, those products did exist in PlanningIT but with different names. They were updated manually. In conclusion, after little manual adaptions to the EA models, all 185 services were successfully created and could be assigned to the particular product and domain levels. Missing domains were not discovered.

13.7.2 Interview Result

We presented the prototype to 17 experts from two different companies. One represents the described integration partner and the other one is a big German insurance company. Especially, insurance companies are obliged to comply with several regulatory requirements (e.g. ISO 22301 or VAIT regulation) that enforce the continuous documentation of the IT landscape. The interviewees are depicted in Table 13.1.

The interview experts rated the evaluation criteria within the scope of 1 = strongly agree, 2 = agree, 3 = disagree and 4 = strongly disagree. To each criteria group, we asked one or more questions in order to obtain more insights about the rationale of the selected answers. The result of the interview is depicted in Table 13.2 and described in the following in more detail.

All interviewed experts confirmed that our prototype is a valid approach for documenting the EA in an automated manner. However, the prototype requires a set of tools to become fully operational, at least a CI pipeline and a fully monitored environment. In case both tools are not present in the company, they must be integrated beforehand which lead to an inevitable additional expenditure.

According to the "Feasibility" of the prototype, we asked the question *to what extent does the proposed configuration file present a valid approach?* The general idea was well appreciated by the practitioners. In particular, the possibility to link all business-related architecture models to the application artifact was well accepted. A slight majority fully supported this idea as software artifacts become directly associated with business layer relationships instead of having this information maintained decoupled in an EAM repository. Several interview partners came up with ideas how

Table 13.1 Interview population, N = 17

No	Company ID	Role of the expert	Experience (in years)
1	C1	Backend Developer	3
2	C1	Full-Stack Developer	5
3	C1	DevOps/SRE	4
4	C1	Product Owner	4
5	C1	Enterprise Architect	6
6	C2	Enterprise Architect	20
7	C2	Enterprise Architect	2
8	C2	Enterprise Architect	17
9	C2	Enterprise Architect and Product Owner	10
10	C2	Enterprise Architect	5
11	C2	Enterprise Architect and IT Management Expert	20
12	C2	Enterprise Architect	18
13	C2	Enterprise Architect	16
14	C2	Enterprise Architect	30+
15	C2	Product Owner and Head of Product Architecture	11
16	C2	Product Owner	1
17	C2	Product Owner	3

to expand the configuration file to realize further use cases, like the analysis of cloud-ready applications. One enterprise architect from C2 stated, that the configuration file adds a high value for EAM as it allows to interpret, filter and report on applications recovered at runtime from a business layer perspective.

All experts agree that the tool could positively impact their performance on maintaining EA models and the purpose of the software is aligned with the organization's goals. However, some of them initially had difficulties to understand the JSON schema for validating the configuration files, hence some experts disagree that the software is easy to use. In addition, some doubts were expressed whether the software can be easily integrated into the IT landscape as it depends heavily on the availability of monitoring tools.

In addition to the criteria group "Environment", we wanted to know *whether the instrumentation of deployment pipelines to trigger the data collection process is a practicable and reasonable approach?* Most experts endorsed the idea to instrument CD pipelines for triggering the documentation process for two reasons: 1) It brings the process close to the actual knowledge carriers, i.e. the agile development teams, and 2) it ensures architecture model updates are performed at the most important points in time, which is the deployment of application releases. However, the enterprise architects from C2 also raised the concern that the solution could lead to a

Table 13.2 Tool evaluation result

Criteria	Description	1	2	3	4	ø
Goal						
Efficacy	The software fulfills its purpose	71%	29%	0%	0%	1.3
Efficiency	Good ratio between cost and benefit	6%	71%	24%	0%	2.2
Feasibility						
Technical feasibility	The concept is from a technical point of view a feasible approach	47%	43%	0%	0%	1.5
Operational feasibility	Would you support the proposed artefact and operate it	24%	59%	6%	0%	1.6
Economic feasibility	The benefits of the proposed artefact outweigh the costs of operating it	18%	82%	0%	0%	1.8
Environment						
Usefulness	The artifact positively impacts your task performance	29%	71%	0%	0%	1.7
Ease of use	The software is easy to learn and to use	0%	59%	35%	6%	2.5
Business alignment	The artefact goal is aligned with the organization and its strategy	71%	29%	0%	0%	1.3
Technology fit	It is easy to integrate the artefact into the organization's IT infrastructure	18%	65%	18%	0%	2.0
Structure						
Completeness	The artifact can maintain all necessary EA models and relationships	0%	76%	24%	0%	2.2
Simplicity	The artifact maintains the most important EA models	41%	59%	0%	0%	1.6
Activity						
Accuracy	The degree of the accuracy of the expected output is satisfying	71%	29%	0%	0%	1.3
Performance	The artifact accomplishes its functions within an acceptable time	76%	24%	0%	0%	1.2

1 = strongly agree, 2 = agree, 3 = disagree and 4 = strongly disagree, N = 17

complex fragmentation of the overall documentation process, due to the decentralized approach. In addition, they claim that the solution does not consider important preceding EA phases like planning stages.

In addition, related to the e "Environment" section, we wanted to know whether *the shift in responsibility for IT landscape documentation towards agile development teams is a practicable and reasonable approach?* All experts agreed that this approach is the correct way to implement the concept. The maintenance of architecture models must be performed by the knowledge carriers, i.e. the agile teams. In the current state, the required information is collected from agile teams in labor-intensive and error prone work as one enterprise architect from C2 stated. By shifting the documentation responsibility to the development teams, awareness is raised for the need of a proper understanding of business-related models and their relationships. Based on the feedback received from the development teams, the solutions preconditions are easy to integrate as part of regular sprint planning and sprint execution processes, which was inspired in exchange with the agile teams that adopted the solution. A product owner and a developer from C1 appreciated that only a onetime effort is needed.

According to the "Structure" of *Microlyze*, most of the experts agree that the tool can discover and maintain the most problematic EA models that are applications, their relationships and their allocation to the business domains. In addition, the link to the federated information systems is perceived as useful as otherwise, the references must be searched manually.

Next, regarding the criteria group "Activity", we wanted to know *which additional use cases can be derived from the concept?* The developers in C1 both stated that the concept could help in understanding how the exposed APIs rely on each other in the context of domain-driven design. This was summarized by one of the interview participants "even though there are good API collaboration tools available, most of the tools we used did not allow to explore how APIs are in context to each other. This could be achieved with the presented prototype." The DevOps/SRE role stated that the concept "bears a lot of potentials when it comes to automated linking incidents to the relevant teams/products after the root cause is identified by the APM system." This potential also unveils the impact of incidents on other domains as soon as there are cross-domain dependencies. Furthermore, the concept could support Product Owners to understand which changes were done recently without bothering the particular team, as one of the Product Owner in C2 stated. Moreover, another Product Owner in C2 regards "the linkage of architectural changes to the respective user stories maintained in Jira as relevant. As user stories are also captured in the context of enabling some part of a business process this could possibly enhance the business process documentation as well." In addition, one Enterprise Architect in C2 regards "API management to become a very important topic and we see potential extension points for *Microlyze* here."

13.8 Conclusion and Future Work

In this research, we proposed a solution that is built on the foundations of existing tools in companies to support the automation of EA model maintenance. By discovering the IT landscape from runtime data and linking the runtime artefacts with the documented models in federated information sources we are able to capture most of the EA-relevant information. Our prototype does not duplicate the information in a separate database but stores only references to the particular information systems via a configuration file that is incorporated in the root path of every runnable IT artefact. To enable fully automated data integration, validation workflows were introduced so that manual efforts could be kept to a minimum. With the integration layer between all federated information sources, this research further introduced the notion of an enterprise graph, that represents the idea of providing a query language for traversing the whole enterprise architecture for different responsibilities.

The prototype was implemented in one environment of a big German retailer and evaluated by 17 experts from two different companies. In general, the feedback for the proposed solution has been positive. Some of the necessary assumptions, such as the configuration file introduction and the existence of an APM tool were considered feasible.

Based on the evaluation some potential future work could be identified: 1) The prototype can only query the information or create and update EA models in EAM tools without visualizing them. In our future work, we want to provide different charts that use the GraphQL interface for visualizing any information and their dependencies. 2) Furthermore, on a more granular level, services and their interfaces represent different activities of an overall business process. We want to incorporate business process mining in our EA discovery approach and link the identified business activities with service executions. This could further help in analyzing the business impact from any type of operational problem.

References

1. Langenberg, K., Wegmann, A.: Enterprise Architecture: What Aspects is Current Research Targeting (2004)
2. Ross, J.W., Weill, P., Robertson, D.: Enterprise Architecture As Strategy: Creating a Foundation for Business Execution. Harvard Business School Press (2006)
3. Tao, F., Sui, F., Liu, A., Qi, Q., Zhang, M., Song, B., Guo, Z., Lu, S., Nee, A.: Digital twin-driven product design framework. Int. J. Prod. Res. **57**(12), 3935–3953 (2019)
4. Buckl, S., Matthes, F., Schweda, C.: Investigating the state-of-the-art in enterprise architecture management methods in literature and practice. In: The 5th Mediterranean Conference on Information Systems, MCIS (2010)
5. Roth, S., Hauder, M., Farwick, M., Breu, R., Matthes, F.: Enterprise architecture documentation: current practices and future directions. In: 11th International conference on Wirtschaftsinformatik Proceedings (2013)
6. Lucke, C., Krell, S., Lechner, U.: Critical issues in enterprise architecting—a literature review. In: Americas Conference on Information Systems (AMCIS) (2010)
7. Dingsøyr, T., Nerur, S., Balijepally, V., Moe, N.: A decade of agile methodologies. J. Syst. Softw. **85**(6), 1213–1221 (2012)
8. Fowler, J. (2014). Microservices—a definition of this new architectural term. www.martinfowler.com
9. Agrawal, R., Gunopulos, D., Leymann, F.: Mining process models from workflow logs. In: Advances in Database Technology. EDBT 1998. Lecture Notes in Computer Science, vol. 1377. Springer, Berlin (1998)
10. Fallis, A.: Building a Monitoring Infrastructure with Nagios. Prentice Hall (2013)
11. Hoorn, A. Van, Waller, J., Hasselbring, W.: Kieker: A framework for application performance monitoring and dynamic software analysis. In: Proceedings of the 3rd Joint ACM/SPEC International Conference on Performance Engineering, ICPE (2012)
12. Ly, L.T., Maggi, F., Montali, M., Rinderle-Ma, S., van der Aalst, W.: Compliance monitoring in business processes: functionalities, application, and tool-support. Inf. Syst. **54**, 209–234 (2015)

13. Ranshous, S., Shen, S., Koutra, D., Harenberg, S., Faloutsos, Ch., Nagiza, F.: Anomaly detection in dynamic networks: a survey. J. WIREs Computat. Stat. **7**(3), 223–247 (2015)
14. Standard, O.G., Group, T.O.: Open Group Standard The Open Group. The TOGAF® Standard, Version 9.2 (2013)
15. Zachman, J.A.: A framework for information systems architecture. IBM Syst. J. (2010)
16. Heinrich, R., van Hoorn, A., Knoche, H., Li, F., Lwakatare, L., Pahl, C., Schulte, S. Wettinger, J.: Performance Engineering for microservices: research challenges and directions. In: WOSP-C@8th ACM/SPEC International Conference on Performance Engineering (ICPE'17) (2017)
17. Farwick, M., Breu, R., Hauder, M., Roth, S., Matthes, F.: Enterprise architecture documentation: empirical analysis of information sources for automation. In: 46th Hawaii International Conference on System Sciences. pp. 3868–3877 (2013)
18. Hauder, M., Matthes, F., Roth, S.: Challenges for automated enterprise architecture documentation. In: Lecture Notes in Business Information Processing (2012)
19. Buschle, M., Ekstedt, M., Grunow, S., Hauder, M., Matthes, F., Roth, S.: Automated enterprise architecture documentation using an enterprise service bus. In: 18th Americas Conference on Information Systems (AMCIS), 2012
20. Farwick, M., Agreiter, M., Breu, R., Häring, M., Voges, K., Hanschke, I.: Towards living landscape models: automated integration of infrastructure cloud in enterprise architecture management. In: Proceedings—2010 IEEE 3rd International Conference on Cloud Computing. pp. 35–42 (2010)
21. Holm, H., Buschle, M., Lagerström, R., Ekstedt, M.: Automatic data collection for enterprise architecture models. J. Softw. Syst. Model. **13**(2), 825–841 (2014)
22. Alegria, A., Vasconcelos, A.: IT Architecture automatic verification: a network evidence-based approach. In: Fourth International Conference on Research Challenges in Information Science (RCIS). pp. 1–12 (2010)
23. Breu, R., Agreiter, B., Farwick, M., Felderer, M., Hafner, M., Innerhofer-Oberperfler, F.: Living models—ten principles for change-driven software engineering. Int. J. Softw. Inf. **5**, 267–290 (2011)
24. Akkiraju, R., Mitra, T., Thulasiram, U.: Reverse Engineering Platform Independent Models from Business Software Applications (2012)
25. Buckl, S.: Developing organization-specific enterprise architecture management functions using a method base. Technische Universität München (2011)
26. Farwick, M., Pasquazzo, W., Breu, R., Schweda, C. M., Voges, K., Hanschke, I.: A meta-model for automated enterprise architecture model maintenance. In: Proceedings of the 2012 IEEE 16th International Enterprise Distributed Object Computing Conference (2012)
27. Hanschke, I.: Strategisches Management der IT-Landschaft: ein praktischer Leitfaden für das Enterprise-architecture-Management, Hanser (2009)
28. Sousa P., Gabriel R., Tadao G., Carvalho R., Sousa P.M., Sampaio A.: Enterprise transformation: the Serasa experian case. Practice-Driven Research on Enterprise Transformation. PRET 2011. Lecture Notes in Business Information Processing, vol. 89 (2011)
29. Granchelli, G., Cardarelli, M., Francesco, P. D., Malavolta, I., Iovino, L., Salle, A. D.: Towards recovering the software architecture of microservice-based systems. In: IEEE International Conference on Software Architecture Workshops (ICSAW). Gothenburg, pp. 46–53 (2017)
30. Kleehaus, M., Schaefer, P., Matthes, F.: MICROLYZE: a framework for recovering the software architecture in microservice-based environments. In Lecture Notes in Business Information Processing (2018)
31. Kleehaus, M., Hauder, M., Uludag, O., Matthes, F., Corpancho, N.: IT landscape discovery via runtime instrumentation for automating enterprise architecture model maintenance. In: Americas Conference on Information Systems (AMCIS) (2019)
32. Sigelman, B., Barraso, L.-A., Burrows, M., Stephenson, P., Plakal, M., Beaver, D., Jaspan, S., Shanbhag, C.: Dapper, a Large-Scale Distributed Systems Tracing Infrastructure. Google Inc. (2010)
33. Silva, F., Rich, Ch., Ganguli, S.: Magic Quadrant for Application Performance Monitoring. Gartner Research (2019)

34. Kleehaus, M., Corpancho, N., Matthes, F., Huth, D.: Discovery of microservice-based IT landscapes at runtime: algorithms and visualizations. In: 52th Hawaii International Conference on System Sciences (HICSS) (2020)
35. Evans, E.: Domain-Driven Design: Tackling Complexity in the Heart of Software. Addison-Wesley Professional (2003)
36. Fischer, R., Aier, S., & Winter, R.: A Federated approach to enterprise architecture model maintenance. In: Enterprise Modelling and Information Systems Architectures (2007)
37. Binz, T., Fehling, T., Leymann, F., Nowak, A., Schumm, D.: Formalizing the cloud through enterprise topology graphs. In: IEEE Fifth International Conference on Cloud Computing, pp. 742–749 (2012)
38. Prat, N., Comyn-Wattiau, I., Akoka, J.: A taxonomy of evaluation methods for information systems artifacts. J. Manag. Inf. Syst. (2015)

Chapter 14
Digital Enterprise Architecture for Global Organizations

Yoshimasa Masuda

Abstract This chapter explains the direction of Digital IT such as Cloud, Mobile IT, and specific application layer on cloud and mobile IT like Big Data and Internet of Things (IoT). This chapter introduces and briefs enterprise architecture (EA) frameworks such as TOGAF, FEAF, DoDAF, MODAF, Adaptive EA, etc. and analyses the selected EA frameworks from standpoints of Cloud/Mobile IT integration, as well. In the latter of this section, the author investigated problems in digital transformation and related enterprise architecture, and we show solutions to cope with these problems in digital transformation and enterprise architecture—"Digital Enterprise Architecture." Furthermore, the author shows the Strategic Architecture framework aligned with Digital IT strategies on the basis of the above solutions, as the framework of Digital Enterprise Architecture.

Keywords Digital IT · Cloud computing · Mobile IT · Enterprise architecture · EA frameworks · Digital agility · Big data · Internet of things

14.1 Introduction

Over the past decade, EA has become an important method for modeling the relationship between the overall image of corporate and individual systems. In ISO/IEC/IEEE42010:2011, architecture framework is defined as "conventions, principles, and practices for the description of architecture established within a specific domain of application and/or community of stakeholders." Furthermore, in the TOGAF (2011) technical literature, it is defined as "a conceptual structure used to develop, implement, and sustain an architecture." In addition, EA visualizes the current corporate IT environment and business landscape to promote a desirable future IT model [6]. EA is required as an essential element of corporate IT planning;

Y. Masuda (✉)
Carnegie Mellon University, Pittsburgh, PA, USA
e-mail: ymasuda@andrew.cmu.edu; yoshi_masuda@keio.jp

Keio University, Tokyo, Japan

A. Zimmermann et al. (eds.), *Architecting the Digital Transformation*, Intelligent Systems Reference Library 188,
https://doi.org/10.1007/978-3-030-49640-1_14

it is not a simple support activity [1], and it offers many benefits to companies, such as coordination between business and IT, improvement in organizational communication, information provision, and reduction in the complexity of IT [34]. In order to continue to deliver these benefits, EA frameworks need to embrace change in ways that adequately consider the emerging new paradigms and requirements that affect EA, such as the paradigm of cloud computing or enterprise mobility [2].

Mobile IT computing is an emerging concept in general that uses cloud services provided over mobile devices [26]. In addition, Mobile IT applications are composed of Web services. There is not much literature that discusses EA integration with Mobile IT and the relationship between the two; however, integration with service oriented architecture (SOA) has been discussed greatly. Many organizations have invested in SOA as a crucial approach for achieving agility as an organization that can manage rapid change [8]. In the meantime, there has been a recent focus on Microservices architecture, which allows rapid adoption of new technologies, such as mobile IT applications and cloud computing [28]. This chapter considers both perspectives.

In terms of cloud computing, mobile devices also widely use cloud computing capabilities, and many mobile IT applications also operate with Software as a Service (SaaS) cloud-based software [26]. There is literature that concerns the integration and relationship between EA and Cloud computing, but it is scarce. Although cloud computing formats consist of three general services—SaaS, PaaS, and IaaS—under the current EA framework, there is merely a modeling of only this computing format and the business components managed by this company. Considering recent dynamic moves in business and the characteristics of cloud computing, it is necessary for companies to link the service characteristics (those similar to the above mobile IT characteristics) of EA and cloud computing [19]. It is said that the traditional EA approach requires months to develop an EA that allows cloud technology in order to realize a Cloud adoption strategy, and organizations will demand adaptive enterprise architecture to iteratively develop and manage an EA adaptive to the cloud technology [15].

In addition, the Open Platform 3.0 Standard was developed and approved by The Open Group, and it focuses on emerging technological trends, such as cloud computing and mobile IT, that create new business models. In this environment, many basic architecture models are noted, including mobile IT and cloud computing. Furthermore, the core elements of mobile devices, applications, device management, and application management, as well as those of cloud computing, which are SaaS, Platform as a Service (PaaS), and Infrastructure as a Service (IaaS), have been proposed [4]. On the other hand, the pubic standards group OASIS [20] has introduced the SOA Reference Model, which presents SOA core elements of service and service interface.

The overall goal of this chapter and research is to show the Strategic Architecture framework aligned with Digital IT strategies based on the solutions to cope with problems in digital transformation and enterprise architecture, as the framework of Digital Enterprise Architecture.

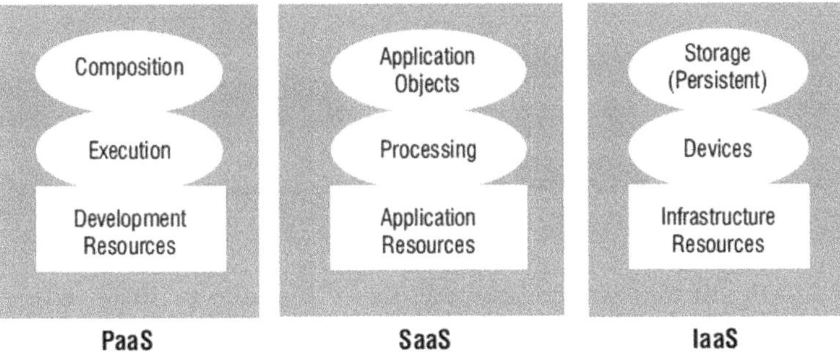

Fig. 14.1 High-level architectural components of cloud computing from EA perspective. *Source* Cutter IT Journal, November 2009 [19]

14.2 Directions of Cloud, Mobile IT and Digital IT

14.2.1 Cloud Architecture

The NIST (National Institute of Standards and Technology) cloud computing definition emphasizes three cloud service models: SaaS, PaaS, and IaaS. PaaS is an IaaS platform that includes both system software and an integrated development environment. SaaS is a software application that is developed, implemented, and operated on a PaaS foundation. The SaaS interface is accessed through the client and API interface. IaaS accommodates PaaS and SaaS by offering infrastructure resources such as computing network storage memory through specific centers (Gill 2015). The Open Group developed Open Platform 3.0 standard that presents basic architecture models by focusing on trends in cutting-edge technology that creates new business models such as cloud, mobile IT.

Figure 14.1 shows the "high-level architectural components of Cloud computing from an enterprise point of view.

14.2.2 SOA and Microservices

SOA and Microservice vary greatly from the viewpoint of service characteristics [32]. SOA is a cooperative design method in which h multiple services provide different functions and has been used in large-scale, monolithic applications [28]. The SOA architecture pattern defines the four basic forms of business service, enterprise service, application service, and infrastructure service [32]. OASIS, a public standards group [20], introduces an SOA reference model that presents elements and interfaces for each service. Microservice is defined as an approach for distributed systems that promote the use of finely grained services with their own lifecycles while integrating newly emerging technologies. Microservice is an approach for

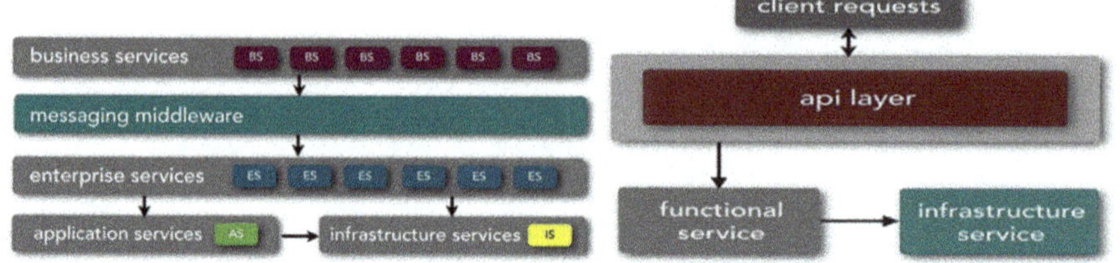

Fig. 14.2 SOA and Microservices taxonomy. *Source* Microservices versus service-oriented architecture, O'Reilly, November 2015 [32]

dispersed systems defined from the two basic forms of functional services through an API layer and infrastructure services. As these services operate cooperatively through integrating newly emerging technologies, they resolve many problems that existed in SOA implementation [28]. Microservice enables early adoption of new technologies such as Mobile IT and Cloud computing [28]. Multiple Microservices cooperating to function together enable implementation as a Mobile IT application [9]. We show the service structures of SOA and Microservices in Fig. 14.2 [32].

14.2.3 Specific Application Layer on Cloud, Mobile IT—Big Datas

In terms of big data, new computing trends require data with far greater volume, velocity, and variety than ever before. Big data is utilized in ingenious methods to predict customer buying behaviors, detect fraud and waste, analyze product opinion, and react quickly to changes in business conditions (a driving force behind new business opportunities) (Chappelle 2013). The term "big data" refers to data that is so large, it is difficult to process using currently-available IT systems. There is a growing opportunity for analysis, visualization, and distributed processing software to enable users to extract useful information from such data [4]. Sources of big data include the following.

- Corporate data in SQL databases, data in cloud-based SQL or NoSQL databases.
- Data provided by social networks, by sensors in the internet-of-things (IoT).

Big data applications include visualization functionality for effective user presentation of analytical results. Furthermore, big data applications leverage web services that make the results of their analyses available to other applications through APIs; objects in the IoT can be data generators [4].

14.2.4 Specific Application Layer on Cloud, Mobile IT—Internet of Things

The term of "Internet of Things (IoT)" is used to mean "the collection of uniquely identifiable objects embedded in or accessible by Internet hosts" [4]. A "uniquely identifiable object" can be described as follows: these objects are connected with real world interaction devices, smart homes and cars, and other SmartLife scenarios. The IoT fundamentally revolutionizes digital strategies with innovative business operating models [33], and holistic governance models for business and IT [31], under fast changing markets [3].

- A sensor, such as a temperature sensor (thermometer)
- A control; for example, to control a valve in a heating system
- A combination of sensor and control (for example, a thermostat)
- An object identifier, such as an RFID tag or a barcode.

The current state of research for the Internet of Things architecture [30] lacks an integral understanding of EA and Management [27, 29, 36, 37, 39, 40], and shows a number of physical standards, methods, tools and a large number of heterogeneous IoT devices [3]. A first reference architecture (RA) for the IoT is proposed by [44] and can be mapped to a set of open source products. This RA covers aspects like "cloud server-side architecture," "monitoring/management of IoT devices, services," "specific lightweight RESTful communications," and "agent, code on small low power devices." Layers can be instantiated by suitable technologies for the IoT [3, 24].

14.3 EA Frameworks with the Integration of Cloud/Mobile IT

To start, the first step is to select an EA framework for this research. The criteria for this selection are: (A) widely used and highly evaluated EA frameworks and (B) an EA framework that supports Mobile IT/Cloud computing and Web service elements. According to a survey in the Journal of Enterprise Architecture (2013), from the perspective of the "widely used" criterion, TOGAF, Zachman [46, 47], Gartner, Federal Enterprise Architecture Framework (FEAF), and DoDAF are the most widely used EA frameworks, and it was decided that TOGAF, FEAF, and DoDAF are "highly evaluated." Furthermore, according to Microsoft (2007), Zachman, TOGAF, FEAF, and Gartner are the most commonly used EA frameworks.

In this chapter, the second criterion for EA framework selection is integration with the elemental framework of Mobile IT/Cloud computing and Services (part of SOA). It is argued that FEAF, TOGAF, Zachman, and the Adaptive Enterprise Architecture framework [14] are suitable from the perspective of integrating the elements of Mobile IT and strongly directly linked Cloud computing [15]. In addition, TOGAF, FEAF, DoDAF, and the British Ministry of Defence Architecture Framework (MODAF) are discussed from the perspective of integration with SOA elements, which have Web services [2, 10, 43].

In addition, because the Gartner framework is limited to commercial use, complete access is not possible and it is therefore outside of our scope [12]. Moreover, because the Zachman framework does not provide an enterprise architecture process for implementing and operating an enterprise architecture capability [17], this is also out of our scope at this time.

What follows is an explanation of the five EA frameworks, selected in the above steps, that were the subject of a comparative survey in this research for the book.

14.3.1 TOGAF (the Open Group Architecture Framework)

As TOGAF is a framework for developing enterprise architecture with a detailed method and supporting tools, developed and maintained by members of The Open Group. Architecture Development Method (ADM) is the core of TOGAF. Figure 14.3 shows this ADM in TOGAF9 as below. It describes a step-by-step approach to developing Enterprise Architecture [13]. TOGAF is not attached to government enterprises. It is a generic and comprehensive framework that can be tailored for the development of effective enterprise architecture capability for technology-enabled enterprise adaptation [17]. With regard to the remaining parts of TOGAF, "the content framework" provides a conceptual meta-model for describing architectural artifacts. "The enterprise continuum" is a virtual repository for storing architectural models and architectural descriptions. "The TOGAF reference models" are divided into the TOGAF Technical Reference Model and the Integrated Information Infrastructure Reference Model [5].

14.3.2 FEAF (Federal Enterprise Architecture Framework)

FEAF [11] is a comprehensive framework for developing and maintaining the enterprise architecture capability of the Federal Government. FEAF provides a common and standardized approach and principles for developing and sharing architecture information between agencies [17]. FEAF was developed by the US Federal Chief Information Officers Council [10]. The core of FEAF is a Collaborative Planning Methodology (CPM) and Consolidated Reference Model (CRM). CRM specifies six interrelated reference models: Performance Reference Model (PRM), Business Reference Model (BRM), Data Reference Model (DRM), Application Reference Model (ARM), Infrastructure Reference Model (IRM), and Security Reference Model (SRM) [17].

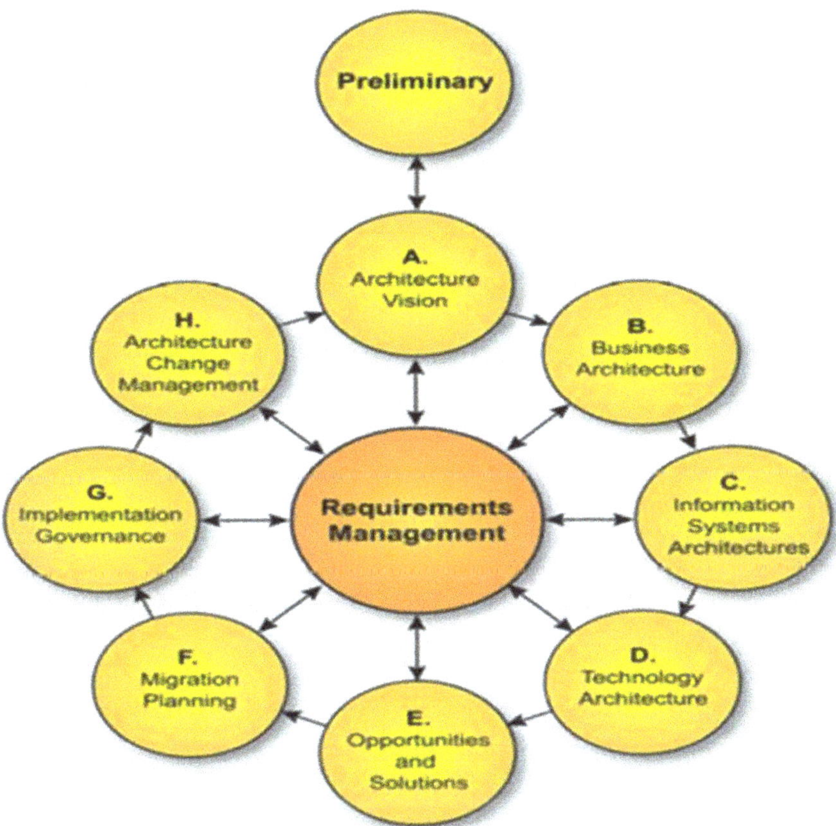

Fig. 14.3 The architecture development method (ADM) in TOGAF9 [13, 40]

14.3.3 DoDAF (Department of Defense Architecture Framework)

As TOGAF is a framework for developing enterprise architecture with a detailed method and supporting tools, developed and maintained by members of The Open Group. Architecture Development Method (ADM) is the core of TOGAF. Figure 14.3 shows this ADM in TOGAF9 as below. It describes a step-by-step approach to developing Enterprise Architecture [13]. TOGAF is not attached to government enterprises. It is a generic and comprehensive framework that can be tailored for the development of effective enterprise architecture capability for technology-enabled enterprise adaptation [17]. With regard to the remaining parts of TOGAF, "the content framework" provides a conceptual meta-model for describing architectural artifacts. "The enterprise continuum" is a virtual repository for storing architectural models and architectural descriptions. "The TOGAF reference models" are divided into the TOGAF Technical Reference Model and the Integrated Information Infrastructure Reference Model [5].

14.3.4 MODAF (British Ministry of Defence Architecture Framework)

MODAF defines a normalized way of conducting Enterprise Architecture, and it was originally developed by the UK Ministry of Defence (MOD). The UK's MOD defined an architecture framework named MODAF in 2006–2008 with adapting DoDAF. MODAF is an internationally normalized EA framework developed by MOD to support Defence planning and change management activities [13]. MODAF provides a consistent set of rules and templates, known as views, that present a textual and graphical visualization of an area of the business. Each view provides a special perspective of the business in order to meet various stakeholder concerns. The views are divided into seven categories: strategic, operational, service-oriented, systems, acquisition, technical, and all views. MODAF includes a meta-model that defines the relationship between all data in all the views [41].

14.3.5 Adaptive Enterprise Architecture Framework

The adaptive enterprise architecture framework (also known as the Gill Framework) is a meta-framework that can be used to support the tailoring of a situation-specific adaptive enterprise architecture capability or framework [14]. Figure 14.4 shows the "Adaptive Enterprise Architecture Framework" as below. This framework provides support for developing and managing adaptive or agile enterprise architecture in a modern context, including adaptive Cloud technology-enabled enterprise architecture. This framework has its foundation on the new "adaptive enterprise service system" theory, which extends the system of systems (SoS), agility, and service science approaches for defining, operating, managing, supporting, and adapting a modern enterprise as an "adaptive enterprise service system" [14]. This framework has two main layers: outer and inner. The outer layer presents five adapting capabilities (i.e., context awareness, assessment, rationalization, realization, and unrealization) to guide the continuous adaptation of the adaptive enterprise architecture as an adaptive enterprise service system in response to internal and external changes. The inner layer assists in defining, operating, managing, and supporting the complex enterprise as an adaptive enterprise service system in response to changes or requirements reported by the outer layer [15].

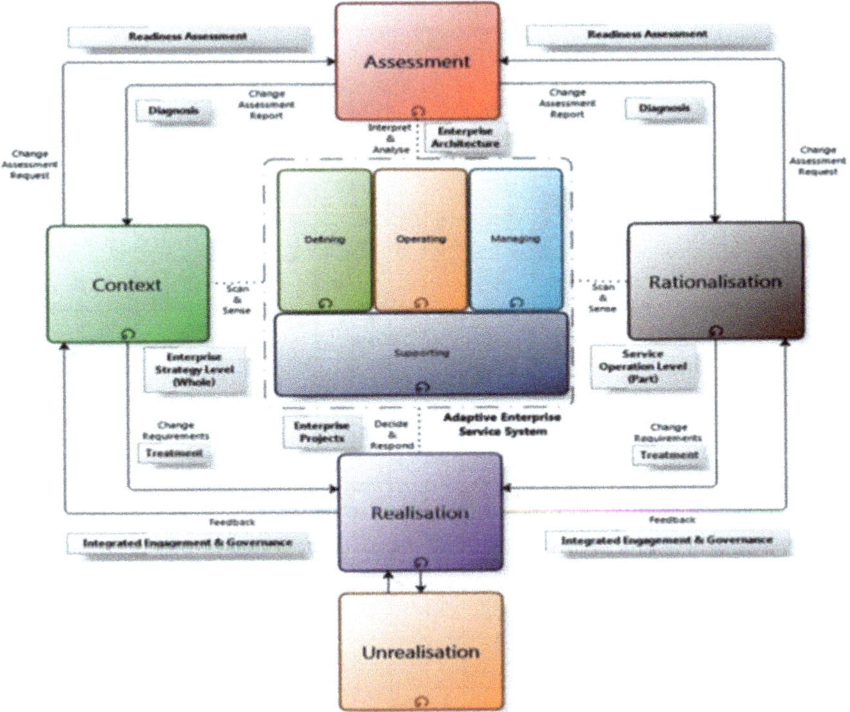

Fig. 14.4 The adaptive enterprise architecture framework. *Source* Agile enterprise architecture: a case of a cloud technology-enabled government enterprise transformation. Proceeding of the 19th Pacific Asia Conference on Information Systems (PACIS 2014), Gill [15]

14.4 EA Framework Analysis

The five selected EA frameworks are compared based on the key elements of Mobile IT/Cloud computing and Services in order to present an overview of the status in terms of the mobile IT/cloud elements and the position of these elements in the layers (viewpoints) of the five frameworks as shown in Table 14.1. Perspectives and conclusions based upon this comparison are presented in the following paragraphs [21].

First, concerning the integration of mobile IT elements in technology architecture (TA), the preliminary research [21] of this study identified mobile IT-related elements in all subject EA frameworks, as shown in Table 14.1. Mobile devices are identified in TOGAF, DoDAF, MODAF, and adaptive EA frameworks and are also seen in FEAF meta-models. A description of the API is seen in TOGAF and applied EA framework documentation as well as in FEAF meta-models. However, no description of mobile device managers or mobile application controllers is seen in any subject EA frameworks.

Table 14.1 Mobile IT/cloud elements in EA frameworks

EA Frameworks		**TOGAF 9**			**FEAF**					**DoDAF v2.0**		**MODAF**		**Adaptive EA**	
Layers (views) Mobile IT /Cloud elements		**Business**	**Information Systems**	**Technology**	**BRM**	**DRM**	**ARM**	**IRM**	**SRM**	**Capability View**	**Services View**	**Strategic /System Viewpoint**	**Service Oriented Viewpoint**	**ES System**	**Cloud EA Capability**
Mobile IT Category (TA)	Mobile Device			*	*		*	* *	* *	*		*			*
	API		*	*		*	* *								*
	Mobile Device Manager														
	Mobile Application Controller														
Mobile IT-related Cloud Category (TA)	SaaS							*		*	*				**
	PaaS							*		*					**
	IaaS							*		*					**
	Cloud Interface														*
	Other						*	* *	*	*					**
Services Category (Application Architecture - AA)	Service									*	**		**	* *	**
	Business Service (micro service)	* *			* *	*	*				*			* *	**
	Application Service (functional service)		* *				*			*	*		*	* *	**
	IS Service (functional service)		* *											* *	**
	Enterprise Service (functional service)				*									*	
	Infrastructure Service (infrastructure service)			*				*						* *	**
	Platform Service (functional service)			* *										* *	**
	Mobile Application+														*
Actors Category	Actor	*			*				*					* *	**
	Service Consumer	*		*	*								**	*	
	Service Provider	*	*	*	*								*		
	Performer									*	**				
Interfaces Category	Business Interface														**
	Application Interface						* *								*
	Infrastructure Interface														*
	Service Port										**				
	Service Interface	*	*	*	*		*						**	*	

Foot Notes Gray areas: the results of further prior research [2]; yellow areas: resurveying results in the preliminary research accompanying the FEAF 2013 model definition change; *Denotes written content, and **Denotes a meta-model viewpoint architecture process existing in a layer in the framework. +This mobile application is service based

Next, concerning the integration of cloud-computing elements, according to the preliminary research on which this study is based, many cloud-computing elements including SaaS, PaaS, and IaaS are identified in the adaptive EA framework as shown in Table 14.1. In addition, all of SaaS, PaaS, and IaaS are seen in the FEAF document IRM and DoDAF documents. However, the description of a cloud interface is only seen in the adaptive EA framework.

According to the preliminary research of this study, mobile IT and cloud architecture models and guide process elements that are effective in pro-moting EA are rarely found in the subject EA framework. On the other hand, the cloud interface indispensable in a hybrid cloud-based implementation is important to companies and is only identified in the adaptive EA framework. Based on prior research and an observation we made, an analysis of this EA framework suggests a method that integrates an adaptive EA framework that provides further support to cloud elements in corporate entities already implementing EA. This is suggested to be possible by applying frameworks such as TOGAF and FEAF along with a TOGAF mobile IT/cloud guide and adaptive EA framework mobile IT/micro-service architecture meta-model for additional effectiveness [22, 25].

14.5 Problems and Solutions

14.5.1 Problems' Structure and Their Factors in Digital Transformation and Enterprise Architecture

As results of investigations regarding problems in Digital Transformation and related Enterprise Architecture in global corporations and previous research in Australia [18] and worldwide, we consider the following 7 kinds of factor's categories in their problems' structure [25].

1) Factor of Architecture Strategy and Governance
(2) Factor of Business Architecture
(3) Factor of spanning Business Architecture and Application Architecture, their dependencies
(4) Factor of spanning Application Architecture and Technology Architecture, their dependencies
(5) Factor of Data Architecture
(6) Factor of spanning Data Architecture and Technology Architecture, their dependencies
(7) Factor of Technology Architecture.

Therefore, there are a lot of cross-functional problems in Digital Transformation and EA , that will lead to the loss of profits because of less strategic alignments and

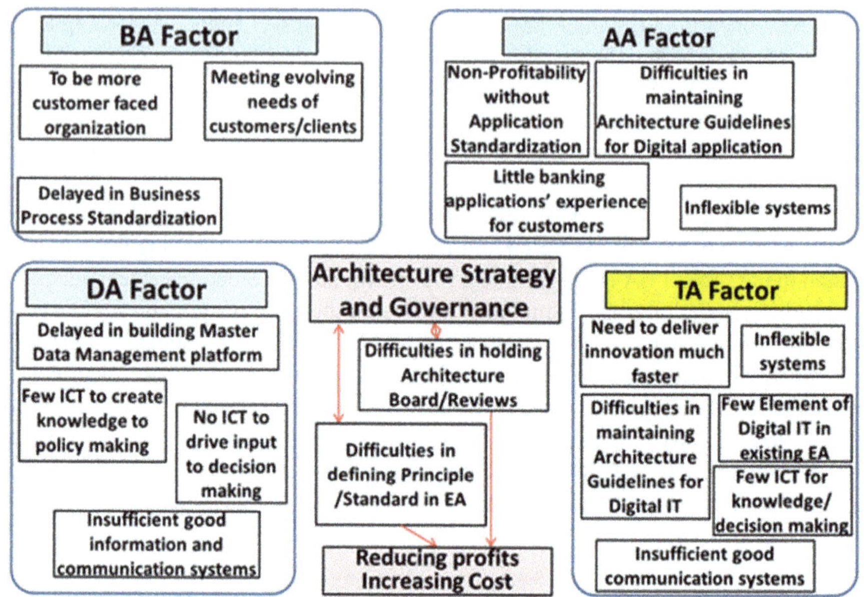

Fig. 14.5 Problems' structure in digital transformation and EA. *Source* The Australian Business Review, July 2012/Springer Nature [18, 25]

non-standardization in application, technology and data [18, 25]. Figure 14.5 shows the results of our investigation regarding problems' factors and grouping them as below.

14.5.2 Solutions to Cope with Problems in Digital Transformation and Enterprise Architecture

Though the author of this book mentions prior notice basis, the solutions proposed and verified in this book, which should consist of the Adaptive Integrated Digital Architecture Framework (AIDAF) with related 5 kinds of agility elements and four kinds of models, can correspond to the above problems in Digital Transformation and EA. Figure 14.6 shows the relationship between the above solutions and problems in Digital Transformation and EA, while the left side of problem items show architecture domains related to each problem and the right side of problem items show problem categories applicable to each one as below.

The solutions in Fig. 14.6 will be explained and proposed in the following sections and next chapter in this book. The relationship between problems and solutions for digital transformation is described and briefed in Sect. 14.7.2. Furthermore, each solution in Fig. 14.6 will be evaluated and verified in the next chapter in this

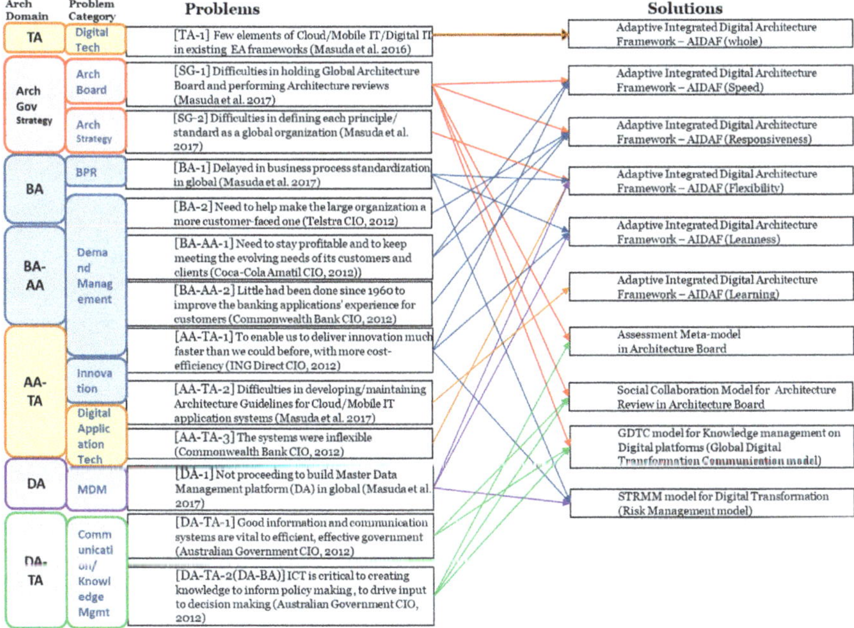

Fig. 14.6 Problems' structure in digital transformation and EA. *Source* The Australian Business Review, July 2012/Springer Nature [18, 25]

book, especially in terms of the STRMM model for digital transformation (Risk Management model) with the AIDAF.

14.6 Overview of Strategic Architecture Framework in the Era of Digital IT (AIDAF Covering Related Models)

14.6.1 Overview and Positioning of AIDAF and Related Models

For the purpose of proceeding with IT strategy promoting cloud, mobile IT, digital IT and big data, the author introduces strategic architecture approach and framework in the era of Digital IT in this research. This approach and framework suit the requirements of digital IT related application systems, that needs the agility elements, while coping with each life cycle defined in System Development Lifecycle (SDLC). According to the definitions of agility elements published by Gill [15, 16] (2014), agility elements consist of "Speed," "Responsiveness," "Feasibility," "Leanness" and

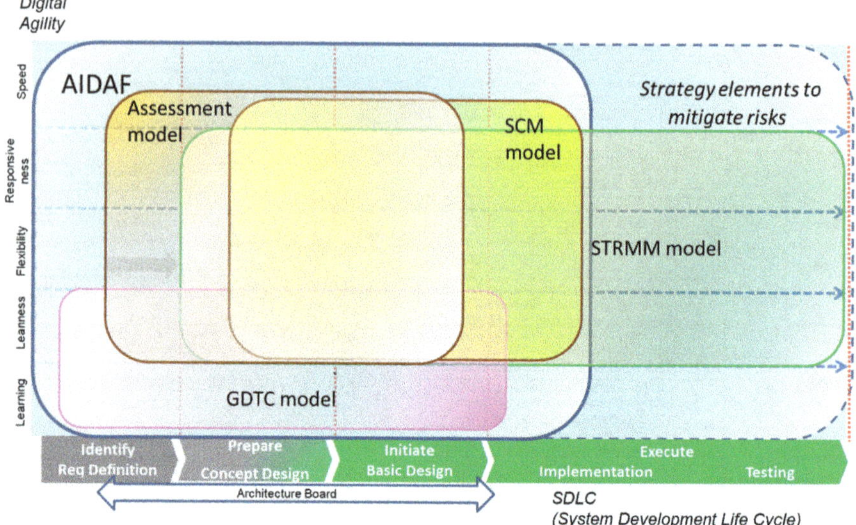

Fig. 14.7 Overview of strategic architecture framework in the era of digital IT (AIDAF and related models). *Source* The Australian Business Review, July 2012/Springer Nature [25]

"Learning." Figure 14.7 illustrates the overview of Strategic Architecture Framework in the era of Digital IT (AIDAF covering related models) while the vertical axis shows "Digital Agility" and the horizontal axis shows SDLC in Fig. 14.7.

The author proposed an "Adaptive Integrated Digital Architecture Framework—AIDAF" supporting all elements of Digital Agility, and the author proposed and divided related models, involved in AIDAF, into several ones: (1) Adaptive Integrated Digital Architecture Framework—AIDAF, (2) Assessment meta-model in Architecture Board, (3) GDTC model for global communication on enterprise portal, (4) Social Collaboration Model for Architecture Review in Architecture Board, (5) STrategic Risk Mitigation Model for Digital Transformation, with Strategic elements to mitigate risks.

First, the "Adaptive Integrated Digital Architecture Framework—AIDAF" is the overall Architecture Framework that will support and promote an IT strategy toward the Digital IT. This Architecture framework can be applied in most of phases from "Requirement Definition" to the mid of "Implementation" in SDLC. In this Architecture framework, architecture board (AB) can be held on short term basis to respond to business user's requirements flexibly with lean structure of architectural processes and deliverables, while developing and utilizing architecture guidelines for Digital IT architectures with learning basis.

The scope of dotted line in Fig. 14.7 shows the AIDAF in the phases from the latter of "Implementation" to "Testing," where mechanisms of architecture change management need to be fully defined and processed in AIDAF.

Second, the "Assessment meta-model in Architecture Board" is covered in AIDAF and the one that can perform architecture reviews regarding solution architectures of new IS/IT projects on the basis of a defined evaluation criteria. This Assessment model will support the AB reviews in the early phases from the mid of "Requirement Definition" to "Basic Design" in SDLC, that can be held on short term basis to respond to business user's requirements flexibly with lean structure of architectural deliverables such as target architectures, current architectures and roadmaps.

Third, "GDTC model for global communication on enterprise portal" is the effective knowledge management process and model on digital platforms for AB reviews and involved in AIDAF. This model can be applied in the early phases from the "Requirement Definition" to "Basic Design" in SDLC with the lean structured processes on digital platforms involving architecture guidelines for Digital IT architectures on learning basis.

Fourth, "Social Collaboration Model for Architecture Review in Architecture Board" is the Architecture model on digital platforms for AB reviews, covering Business Architecture, Application Architecture and Technology Architecture, and this model is covered in AIDAF as well. This model can be applied in the phase from the mid of "Conceptual Design" to the early "Implementation" in SDLC with the lean structured processes on digital platforms to respond to business user's requirements flexibly on short term basis.

Fifth, "STRMM model for Digital Transformation" is the risk mitigation model with Architecture Board toward the digital transformation while involved in AIDAF [23]. This model can be applied in the full life cycle and phases from the mid of "Requirement Definition" to the "Testing" in SDLC with the lean structured processes for Architecture reviews, while responding to the questions and requests from the risk's stakeholders flexibly. Furthermore, in Digital Transformation, Strategy elements to mitigate risks, as well as Scaling Agile Frameworks can be effective mainly in the execution phase (implementation and testing) of SDLC and will sometimes support the all elements of Digital Agility, while covered in AIDAF.

Finally, the AIDAF will support the all elements of Digital Agility, and STRMM model will be applied in all phases from "Requirement Definition" to "Implementation" and "Testing" as shown in Fig. 14.7. Therefore, in the next chapter, the Resiliency Architecture model—the STRMM model covered in AIDAF will be evaluated in terms of full lifecycle in SDLC.

14.7 Necessary Elements and Requirements in EA Frameworks for the Era of Cloud/Mobile IT/Digital IT

14.7.1 Necessary Elements in Enterprise Architecture Framework for the Era of Cloud/Mobile IT/Digital IT

When considering the necessary elements of the EA framework for the era of cloud, mobile IT and digital IT, the EA should have the ability to accommodate agility-related elements. However, the TOGAF is criticized for its size, lack of agility and complexity [15]. Table 14.2 contains the results of efforts to identify the elements defined in each of the architecture domain categories and agility-related elements in all subject EA frameworks below. In Table 14.2, TOGAF9, FEAF, MIT EA, and our proposed "Adaptive Integrated EA framework" are included as all-subject EA frameworks [22]. Moreover, the author of this book has named this EA framework suitable for the era of Digital IT as "Adaptive Integrated Digital Architecture Framework—AIDAF" [22]. DODAF and MODAF were excluded from this Table 14.2 because these frameworks do not contain a specific description of agility-related elements [41–43]. In addition, because the Gartner framework is limited to commercial use, complete access is not possible and it is therefore outside of our scope [12]. Moreover, because the Zachman framework does not provide an enterprise architecture process for implementing and operating an enterprise architecture capability [17],

Table 14.2 Elements of each architecture domain and agility in EA frameworks [22]

EA Frameworks			TOGAF 9	FEAF	MIT EA	AIDAF (Adaptive Integrated Digital Architecture Framework)
Necessary elements of EA for Cloud/Mobile IT/Digital IT		Review Criteria				
Elements of each Architecture Domain Category	BA – High level	Business process policy	○ (definable)	○ (definable)	○ (definable)	○ (definable)
	BA - Detailed level	Business function chart Business Process flow	○ (definable)	○ (definable)	× (none)	○ (definable)
	AA – High level	Application optimization policy	○ (definable)	○ (definable)	○ (definable)	○ (definable)
	AA – Detailed level	Application function chart, Application user location & Communication diagram	○ (definable)	○ (definable)	× (none)	○ (definable)
	DA – High level	Data Integration Policy	○ (definable)	○ (definable)	○ (definable)	○ (definable)
	DA – Detailed level	Standard Logical Data model, Standard Interfaces, BI/DWH Specifications	○ (definable)	○ (definable)	× (none)	○ (definable)
	TA – High level	Technology Platform Integration Policy	○ (definable)	○ (definable)	○ (definable)	○ (definable)
	TA -Detailed level	Technology Standard, Technology Reference Model, Logical diagram	○ (definable)	○ (definable)	× (none)	○ (definable)
Agility related Elements	Speed	Rapid flexible response of enterprise in timely manner	Extensible with IT4IT RA (Requirement to Deploy)	× (none) No specific description	-Faster develop in stage 4 (Business Modularity)	○ (definable) -Accommodates expected or unexpected changes rapidly by short-term cycle
	Responsiveness	Appropriate responses to deal with changes while sensing situations	Extensible with IT4IT RA (Detect to Correct)	- Service Responsiveness in Performance Reference Model	-IT responsiveness improve in stage 2(Standardize) & 4	○ (definable) - scans, senses and reacts properly to expected and unexpected changes by short-term cycle
	Flexibility	Adapt to changing complex business demands in defining Principles in EA.	Extensible with IT4IT RA (Requirement to Deploy)	Extensible with SOA (Service Oriented Architecture	-"Foundation for execution" is this base.	○ (definable) -Able to define Principles flexibly
	Leanness	EA operation with optimal /minimal resources without compromising quality	× (none) No description	× (none) No description	-Optimal core business process and data in stage 3	○ (definable) -With optimal EA deliverables, focus on reducing waste and cost without compromising on quality
	Learning	EA using up-to-date knowledge/experience with continuous growth/adaptation	× (none) No description	× (none) No specific description except for education systems	-Management Practice in stage 4 accelerate learning	○ (definable) -With knowledge sharing, focuses on enterprise fitness, improvement, transformation and innovation.

this is also out of our scope at this time. Moreover, when describing the review criteria of "elements in each Architecture Domain Category" in Table 14.2, we referred to the definitions of each element in the EA framework development project (in this case the global pharmaceutical company) because there were no specific definitions for these elements in existing EA frameworks. On the other hand, regarding the review criteria of "agility-related elements" we referred to the definitions of agility elements published by Gill (2014).

First, all the elements of each architecture domain such as Business Architecture (BA), Application Architecture (AA), Data Architecture (DA), and Technology Architecture (TA) are identified in TOGAF9 [35, 39, 40]. On the other hand, agile-related elements can be realized by extension by IT4IT, although there is no specific description regarding agile-related elements such as "leanness" and "learning" in TOGAF9 itself.

IT4IT can be used to cover the agile-related elements that would extend the capabilities of TOGAF, whereas the logical service model defined in IT4IT should be equivalent to parts of the adaptive EA framework [38].

Second, all the elements of each architecture domain (BA, AA, DA, and TA) are also identified in FEAF [11]. However, in terms of agile-related elements, there is no specific description regarding "speed," "leanness," and "learning" defined in Table 14.2, while "flexibility," which may be realized by extending SOA and "responsiveness," is identified in the PRM (Performance Reference Model) in FEAF itself [11].

Third, all high-level elements of each architecture domain, such as the "business process policy" in BA, "application optimization policy" in AA, "data integration policy" in DA and "technology platform integration policy" in TA, are identified in the MIT EA [33]. However, almost none of the detailed elements in each architecture domain (BA, AA, DA and TA) are found in the MIT EA [45]. On the other hand, descriptions regarding the agility-related elements of "speed," "learning," and "responsiveness" are found in stage 4 of the "business modularity" section in the MIT EA, and a description of "responsiveness" is also found in stage 2 of "technology standardization." A description concerning the agility-related element of "leanness" is found in stage 3 of "optimized core" and the description of the agility-related element of "flexibility" is found in the "foundation for execution" in the MIT EA [33].

Forth, all the elements of each architecture domain (BA, AA, DA, and TA) should be identified in the Adaptive Integrated Digital Architecture Framework (AIDAF) proposed in this study, because this EA framework is designed to include long-term principles and target architectures in addition to an adaptive EA framework. Moreover, descriptions regarding all the agility-related elements of "speed," "responsiveness," "flexibility," "leanness," and "learning" are identified in both the adaptive EA framework [17] and the proposed AIDAF.

Based on the above comparison, the "Adaptive Integrated Digital Architecture Framework —AIDAF" we propose in this study should have capabilities for all the elements of each of the architecture domain categories and all of the agility-related elements defined in Table 14.2, to address the limitations of TOGAF9, FEAF, and MIT EA [22].

14.7.2 Requirements in Enterprise Architecture Framework for the Era of Cloud/Mobile IT/Digital IT)

According to Fig. 14.6 of Problems and Solutions in this chapter, several kinds of requirements in Enterprise Architecture Framework in the era of Digital IT are identified.

First, in terms of Architecture Strategy and Governance, there should be the requirement of holding architecture board (AB) and reviews in global organization, etc., while defining each principle/standard in global ones for EA framework and models in the era of Digital IT. For the purpose of coping with the above requirements, the author can propose "Adaptive Integrated Digital Architecture Framework—AIDAF" from standpoints of speed, responsiveness and flexibility of agility-related elements described in the previous section, and "Assessment model," "Social Collaboration Model" and "Global Digital Transformation Communication model" in the following section and next chapter.

Second, according to Fig. 14.6, from standpoints of Business Architecture (BA) and Data Architecture (DA), there should be the requirement of coping with delays in standardization projects of Business Processes and Master Data Management platforms flexibly. In the following sections, the author can propose the AIDAF from standpoints of flexibility, leanness of agility-related elements described in the previous section, and "Strategic Risk Mitigation Model" to minimize the related risks.

Third, according to Fig. 14.6, in terms of BA and AA, there should be the requirement of coping with the difficulties of demand managements for Digital IT systems and projects. In the following sections, the author can propose the AIDAF from standpoints of speed, responsiveness and leanness of agility-related elements described in the previous section, and "Strategic Risk Mitigation Model" to minimize the related risks.

Forth, according to Fig. 14.6, from standpoints of AA and Technology Architecture (TA), there should be the requirement of keeping up with the rapid progress of Digital application technologies for cloud, mobile IT, big data and digital IT. In the following sections, the author can propose the AIDAF from standpoints of learning, speed and flexibility of agility-related elements described in the previous section.

Finally, according to Fig. 14.6, from standpoints of DA and TA, there should be the requirement of good communication systems with knowledge management for the decision making. The author can propose "Assessment model," "Social Collaboration Model" and "Global Digital Transformation Communication model" for this requirement in the next chapter.

14.8 Proposal of Adaptive Integrated Digital Architecture Framework—AIDAF

The preliminary research of this study promoted the strategic use of cloud/mobile IT. This suggests that corporate entities that implement EA by having applied frameworks such as TOGAF and FEAF, could adopt a framework that enables the integration of an adaptive EA framework to provide further support for cloud elements as one possible solution. Accordingly, this study proposes an "Adaptive Integrated Digital Architecture Framework—AIDAF" based on this suggestion for an EA framework that can even be used by corporate entities to promote a cloud/mobile IT strategy. Figure 14.8 illustrates the proposed model of the AIDAF. The proposed model is an EA framework integrating an adaptive EA cycle in the lower part of the diagram with TOGAF or a simple EA (framework) 1 for different business division units in the upper part of the diagram.

The adaptive EA cycle in the proposed model makes provision for initiation documents (including conceptual architecture designs) for new cloud/mobile IT related projects that are continuously drawn up on a short-term basis (monthly, etc.). This begins with the Context Phase, which is prepared for referencing the Defining Phase (architecture design guidelines related to all types of security/cloud/mobile IT consistent with the IT strategy) in line with the needs of business divisions. In the next phase of the assessment/architecture review, the architecture committee/organization reviews the architecture by focusing on the conceptual design portion of the initiation documents for this IT project. In the Rationalization Phase, the stakeholders and AB differentiate/decide upon information systems that will be replaced by the proposed new information system structure or that are no longer necessary and can be abandoned. In the Realization Phase, this project team begins to implement the new IT project agreed upon as a result of deliberating these issues/action items. This

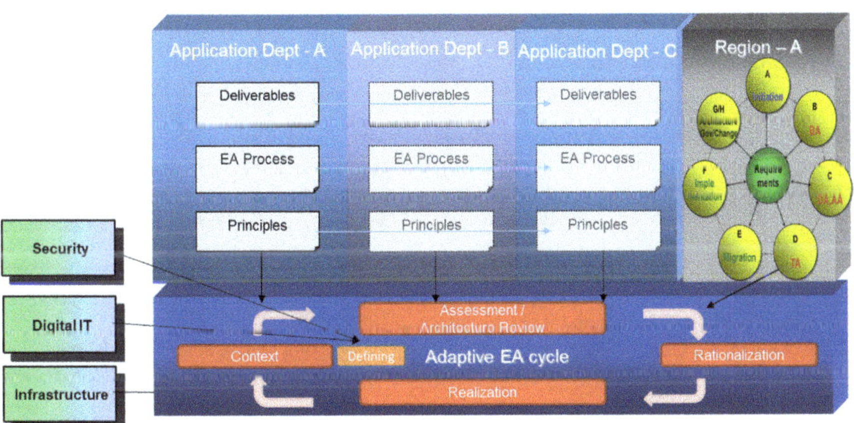

Fig. 14.8 AIDAF proposed model (ex: TOGAF and adaptive EA framework, etc.)

enables the corporate entity to adopt an EA framework capable of flexibly adapting to new cloud/mobile IT projects that continuously occur, and which are composed of these four phases.

Moreover, the "TOGAF" and "simple EA (framework)" based on an operational division unit in the top part of Fig. 14.8 is able to respond to differing policies and strategies in business divisions from a mid-long-term perspective. This part of the framework has a structure that can select the above EA framework in line with the characteristics of business division unit operational processes and future architecture. This part also enables application.

Further, the framework should align EA guiding principles with the definitions of these principles for business divisions to ensure consistency between the adaptive EA cycle in the lower portion of this Fig. 14.8 and the "TOGAF" and "simple EA (framework)1" in the upper portion.

14.9 Conclusion

The author briefed that the "Adaptive Integrated Digital Architecture Framework—AIDAF" is developed to meet the requirements of the era of cloud/mobile IT/digital IT, which involves the five agility-related elements of "Speed/Responsiveness/Flexibility/Leanness/Learning." Furthermore, the author explained that the "Adaptive Integrated Digital Architecture Framework—AIDAF" can solve problems occurring in the era of cloud/mobile IT/digital IT.

The STRMM model with AIDAF will be verified with the case study in global organizations in next chapter.

Notes

(1) A simple EA (framework) is a simple mid- to long-term perspective EA structure composed of EA principles, EA processes, and EA deliverables (target architecture, roadmaps, etc.).

References

1. Alwadain, A., Korthaus, A., Fielt, E., Rosemann, M.: Integrating SOA into an enterprise architecture: a comparative analysis of alternative approaches. In: The Fifth IFIP International Conference on Research and Practical Issues of EIS (2010)
2. Alwadain, A., Fielt, E., Korthaus, A., Rosemann, M.: A comparative analysis of the integration of SOA elements in widely-used enterprise architecture frameworks. Int. J. Intell. Inf. Technol. **9**(2), 54–70 (2014)
3. Zimmermann, A., Schmidt, R., Sandkuhl, K., Jugel, D.: Digital enterprise architecture—transformation for the internet of things. In: Enterprise Distributed Object Computing Workshop (EDOCW), IEEE 19th International (2015)
4. Boardman, S., KPN: Snapshot—Open Platform 3.0™. The Open Group (2015)

5. Buckl, S., Ernst, A.M., Matthes, F., Ramacher, R., Schweda, C.M.:. Using enterprise architecture management patterns to complement TOGAF. In Proceedings of the IEEE International Enterprise Distributed Object Computing Conference, pp. 34–41 (2009). https://doi.org/10.1109/edoc.2009.30
6. Buckl S, Matthes F, Schulz C, and Schweda CM (2010). Exemplifying a framework for interrelating enterprise architecture concerns. In: Sicilia, M.-A., Kop, C., Sartori, F. (eds.) Ontology, Conceptualization and Epistemology for Information Systems, Software Engineering and Service Science, vol. 62, pp. 33–46. Springer, Berlin/Heidelberg
7. Chappelle, D.: Big Data & Analytics Reference Architecture, Oracle Corporation, September 2013
8. Chen, H., Kazman, R., Perry, O.: From software architecture analysis to service engineering: an empirical study of methodology development for enterprise SOA implementation. IEEE Trans. Serv. Comput. **3**(2), 145–160 (2010). https://doi.org/10.1109/TSC.2010.21
9. Familiar, B.: Microservices, IoT, and Azure: Leveraging DevOps and Microservice Architecture to Deliver SaaS Solutions. Apress Media, LLC (2015)
10. Federal CIO Council: A Practical Guide to Federal SOA.V1.1 (2008)
11. Federal Enterprise Architecture Framework: Version 2 (2013)
12. Franke, U., Hook, D., Konig, J., Lagerstrom, R., Narman, P., Ullberg, J., Gustafsson, P., Ekstedt, M.: EAF2- A framework for categorizing enterprise architecture frameworks. In: Proceedings of the 10th ACIS International Conference on Software Engineering, Artificial Intelligences, Networking and Parallel/Distributed Computing, pp. 327–332. IEEE Computer Society (2009)
13. Garnier, J-.L., Bérubé, J., Hilliard, R.: Architecture Guidance Study Report 140430. ISO/IEC JTC 1/SC 7 Software and Systems Engineering (2014)
14. Gill, A.Q.: Towards the development of an adaptive enterprise service system model. In Proceedings of the 19th Americas Conference on Information Systems (AMCIS 2013), pp. 15–17
15. Gill, A.Q., Smith, S., Beydoun, G., Sugumaran, V.: Agile enterprise architecture: a case of a cloud technology-enabled government enterprise transformation. In Proceedings of the 19th Pacific Asia Conference on Information Systems (PACIS 2014), pp. 1–11, United States
16. Gill, A.Q.: Applying agility and living service systems thinking to enterprise architecture. Int. J. Intell. Inf. Technol. **10**(1), 1–15 (2014)
17. Gill, A.Q.: Adaptive Cloud Enterprise Architecture. Intelligent Information Systems. Vol. 4. University of Technology, Australia (2015)
18. Walsh, K.-A.: The deal—from geeks to gurus. Aust. Bus. J. July 2012
19. Khan, K.M., Gangavarapu, N.M.: Addressing cloud computing in enterprise architecture: Issues and challenges. Cut. IT J. **22**(11), 27–33 (2009)
20. MacKenzie, C.M., Laskey, K., McCabe, F., Brown, P.F., Metz, R.: Reference model for service oriented architecture 1.0. (Technical Report), Advancing Open Standards for the Information Society (2006)
21. Masuda, Y., Shirasaka, S., Yamamoto, S.: Integrating mobile IT/cloud into enterprise architecture: a comparative analysis. In: Proceedings of the 21th Pacific Asia Conference on Information Systems (PACIS 2016), p. 4 (2016)
22. Masuda, Y., Shirasaka, S., Yamamoto, S., Hardjono, T.: 2017 7 1. An adaptive enterprise architecture framework and implementation: towards global enterprises in the era of cloud/mobile IT/digital IT: Int. J. Enterp. Inf. Syst. IGI Global. **13**(3), 1 22 22 p. https://doi.org/10.4018/ijeis.2017070101
23. Masuda, Y., Shirasaka, S., Yamamoto, S., Hardjono, T.: Risk management for digital transformation in architecture board: a case study on global enterprise. In: EAIS International Conference 2017, 6th IIAI International Congress on Advanced Applied Informatics (IIAI-AAI), IEEE Conference Proceedings, pp. 255–262, July in 2017
24. Masuda, Y., Yamamoto, S.: A Vision for Open Healthcare Platform 2030: KES 2018, the 6th International KES Conference on Innovation in Medicine and Healthcare (KES-InMed-18), June 2018

25. Masuda, Y., Viswanathan, M.: Enterprise Architecture for Global Companies in a Digital IT Era: Adaptive Integrated Digital Architecture Framework (AIDAF), Springer Nature, February 2019
26. Muhammad, K., Khan, M.N.A.: Augmenting mobile cloud computing through enterprise architecture: survey paper. Int. J. Grid Distrib. Comput. **8**(3), 323–336 (2015)
27. Iacob, M.E., et al.: Delivering Business Outcome with TOGAF® and ArchiMate®, eBook BiZZde-sign (2015)
28. Newman, S.: Building Microservices. O'Reilly (2015)
29. Johnson, P., et al.: IT Management with Enterprise Architecture. KTH, Stockholm (2014)
30. Patel, P., Cassou, D.: Enabling high-level application development for the internet of things, in CoRR abs/1501.05080, submitted to J. Syst. Softw. (2015)
31. Weill, P., Ross, J.W.: IT Governance: How Top Performers Manage It Decision Rights for Superior Results," Harvard Business School Press (2004)
32. Richards, M.: Microservices versus SOA, 1st edn. O' Reilly Media (2015)
33. Ross J.W., Weill, P., Robertson, D.C.: Enterprise Architecture as Strategy—Creating a Foundation for Business Execution. Harvard Business Review Press (2006)
34. Tamm, T., Seddon, P.B., Shanks, G., Reynolds, P.: How does enterprise architecture add value to organisations? Commun. Assoc. Inf. Syst. **28**(1), 10 (2011)
35. The Open Group: Content Metamodel (2009c). Retrieved 15 Oct 2010, from http://www.opengroup.org/architecture/togaf9-doc/arch/chap34.html
36. The Open Group: Archimate 2.0 Specification. Van Haren Publishing (2012)
37. The Open Group: Architecture capability framework (2011). Retrieved from http://pubs.opengroup.org/architecture/togaf9-doc/arch/index.html
38. The Open Group: Open Group Standard—The Open Group IT4IT™ Reference Architecture, Version 2.1 (2017)
39. The Open Group: TOGAF Version 9.2. Van Haren Publishing (2018)
40. TOGAF® Version 9.2: An Open Group Standard; refer to (2018)
41. www.opengroup.org/togaf
42. UK Ministry of Defence: MOD Architecture Framework (MODAF) (2010a). Retrieved 22 Oct 2010, from http://www.mod.uk/DefenceInternet/AboutDefence/WhatWeDo/InformationManage-ment/MODAF/ModafDetailedGuidance.htm
43. UK Ministry of Defence: The MODAF Service Oriented Viewpoint (2010b). Retrieved from http://www.mod.uk/DefenceInternet/AboutDefence/CorporatePublications/InformationManagement/MODAF/TheServiceOrientedViewpointsov.htm
44. US Department of Defense: DoD Architecture Framework Version 2.0 (2009). Retrieved 20 Oct 2010, from http://cio-nii.defense.gov/sites/dodaf20/
45. WSO_2: White Paper: A Reference Architecture for the Internet of Things. Version 0.8.0 (2015). http://wso2.com
46. Yamamoto, S.: Operating Model: Architecture Series, WC Report, Feb 2017 (2017)
47. Zachman, J.A.: A framework for information systems architecture. IBM Syst. J. **26**(3), 454–470 (1987). https://doi.org/10.1147/sj.263.0276
48. Zachman, J.A.: Cloud Computing and Enterprise Architecture. Zachman International (2011)

Part IV
Decision Support

Chapter 15
An Integrative Method for Decision-Making in EA Management

Dierk Jugel

Abstract Due to digitalization, constant technological progress and ever shorter product life cycles, enterprises are currently facing major challenges. In order to succeed in the market, business models have to be adapted more often and more quickly to changing market conditions than they used to be. Fast adaptability, also called agility, is a decisive competitive factor in today's world. Because of the ever-growing IT part of products and the fact that they are manufactured using IT, changing the business model has a major impact on the enterprise architecture (EA). However, developing EAs is a very complex task, because many stakeholders with conflicting interests are involved in the decision-making process. Therefore, a lot of collaboration is required. To support organizations in developing their EA, this article introduces a novel integrative method that systematically integrates stakeholder interests into decision-making activities. By using the method, collaboration between stakeholders involved is improved by identifying points of contact between them. Furthermore, standardized activities make decision-making more transparent and comparable without limiting creativity.

Keywords Enterprise architecture management · Decision support · Viewpoint · EA analytics

15.1 Introduction

Nowadays, organizations are confronted with major challenges [1]. Digitalization, constant technological progress, and ever shorter product life cycles are forcing organizations to adapt their business models more often and faster than in the past [1–3]. Agility is a decisive competitive factor to compete in the markets [4]. This is understood to mean the ability to adapt quickly business models and the organizational structures required to implement them [4]. This includes e.g., business processes, applications and technologies.

D. Jugel (✉)
Herman Hollerith Zentrum, Reutlingen University, Danziger Str. 6, 71034 Böblingen, Germany
e-mail: dierk.jugel@hhz.de

A. Zimmermann et al. (eds.), *Architecting the Digital Transformation*, Intelligent Systems Reference Library 188,
https://doi.org/10.1007/978-3-030-49640-1_15

Products and services offered by the enterprises are not only produced IT-supported, but they themselves have an increasing IT part [5]. As a result, changes in business models have a major impact on the enterprise architecture (EA) [2]. Traditional companies in particular face major challenges in terms of agility in comparison with startups that are increasingly entering the markets.

Historically developed EAs with a multitude of related elements make decision-making a complex task. Startups have the great advantage of being able to implement their business models on a greenfield site. Traditional companies, on the other hand, have to integrate the changed business models into their existing complex business architectures.

Enterprise Architecture Management (EAM) is a systematic approach to develop EAs [6]. The alignment of business and IT structures is a core task. A core characteristic of an EA is providing a holistic view of the enterprise [7]. As a result, an EA encompasses large parts of the enterprise. Due to the multitude of interrelated elements, many different stakeholders from different areas of the organization are involved in their further development [8]. In this context, a stakeholder is defined as a person or group of persons interested in the system of the enterprise [9].

The stakeholders involved in further development look at the EA from a special point of view (so-called viewpoints) due to their interests and responsibilities [9].

Due to many relationships between elements of an EA there are multiple dependencies between viewpoints that are relevant for stakeholders. Therefore, individual stakeholders are not able to change the architecture for their own and without involving other stakeholders that are affected by potential changes [7]. Instead, an integrative decision-making process is required, in which architectural changes are to be planned holistically and collaboratively under consideration of all dependent viewpoints of the stakeholders. In this context, the literature mentions collaboration, especially the conflicting interests of stakeholders, and inadequate tool support as major challenges in the further development of an EA [10].

In order to explore the problem described above, we use Design Science Research [11] and formulate the following research question, which are dealt with in this article:

> How can decision-making in EAM be supported methodically by taking into account the interests of involved stakeholders?

The remainder of this article is structured as follows: In Sect. 15.2, we go into the identification of deficiencies in practice and appropriate requirements. Afterwards, in Sect. 15.3 we revisit the state-of-the-art in decisions and decision processes in general as well as decision-making in EAM and decision modeling. In Sect. 15.4, we introduce the integrative method for decision-making. In Sect. 15.5, we focus on one method component in detail and demonstrate it using a case study in Sect. 15.6. Finally, we conclude with a summary in Sect. 15.7.

15.2 Deficiencies in Practice and Requirements

In order to support stakeholders in their decision-making, we have identified three key aspects through expert surveys and literature analyses:

Decision Process: According to the literature, a high degree of collaboration is required in decision-making due to the large number of stakeholders involved, some of whom have conflicting interests, and the complex corporate architecture [6, 12]. However, this collaboration is disturbed due to undefined decision processes and incomprehensible decision-making [13]. For this reason, we have taken a closer look at the decision-making processes through expert interviews. In addition to the question of what the decision-making processes look like and whether they are defined within the framework of governance, the methodological approach and the use of tools, e.g. automated techniques or visualizations, are of particular interest.

Visualizations: The basis for the further development of the EA is the analysis of the current state (actual architecture) [14]. Visualizations play an important role here as information providers [6]. In this context, our expert surveys provide information about the use of visualizations in decision-making.

Documenting decisions: According to Nakakawa et al. [13], an incomprehensible decision-making process is an obstacle to the collaboration required in the decision-making process. Furthermore, an architecture can only be understood if the reasons for decisions of the past are known [15, 16]. In practice, decisions are often not systematically documented [10, 12, 15]. Therefore, we wanted to investigate through the survey how architectural decisions are documented in practice.

As a result of the surveys and an additional case study to investigate tool support in terms of its visualization capabilities [17], we identified at least five deficiencies in practice that we describe in the following:

Decision processes are often not defined and run ad hoc. As a result, decision-making is almost exclusively a manual process. Support, for example through automated techniques, is difficult because it is unclear how the processes run, and which tools are needed. Furthermore, the processes are difficult to understand.

EAs encompass large parts of the enterprise. Therefore, **many stakeholders** are involved in changes that are affected by them. For this reason, decision-making requires a high degree of collaboration and communication between the stakeholders. The **often conflicting interests** of the stakeholders arising from the various responsibilities further complicate the situation.

Visualizations are often used as information providers. In practice, the **visualizations are often static** and offer few interaction possibilities. In addition, the configuration of visualizations requires a good understanding of the underlying metamodel. Furthermore, the **configuration is often limited** or only possible by special users. Adaptation to changing information requirements during decision-making is therefore difficult to implement. For this reason, the visualizations offered by tools are usually only used as an entry point. Instead, partial models are exported and processed in spreadsheet programs.

Table 15.1 Identified deficiencies in practice and how we want to address them

Deficiency	Requirements
D1. Unclear and undefined decision-making processes	RQ 1. Decision-making activities that can be flexibly combined to form decision-making processes RQ 2. Integration of techniques into decision-making activities
D2. Many stakeholders with conflicting interests	RQ 3. Working environment to support collaboration and communication between stakeholders RQ 4. Parallel consideration of different aspects in order to be able to recognize relationships and dependencies
D3. Static and disconnected visualizations	RQ 5. Interactive visualizations for dynamic information demand during the decision-making process
D4. Loss of context due to sequentially considered aspects	RQ 4. Parallel consideration of different aspects in order to be able to recognize relationships and dependencies
D5. No systematic documentation of architectural decisions	RQ 6. Semi-automated documentation of architectural decisions and their rationale already during the decision-making process

The **sequential consideration of viewpoints** during the decision-making process, caused among other things by different interests of the stakeholders involved, presents them with major challenges. In this context, there is a lack of support in identifying dependencies between viewpoints.

In practice, **architectural decisions are often not documented systematically** and without rationales due to time-consuming ex post documentation. It is difficult to understand an architectural decision from the past without the corresponding justifications. To answer the question why an architecture is as it is, documented architectural decisions and their rationales are essential.

After identifying the deficiencies in practice, we derived appropriate requirements to solve them. Table 15.1 shows the result.

15.3 Related Work

The description of related work is divided into three parts. First, we will look at general decision theory with a focus on decisions and decision-making processes. Subsequently, we discuss approaches that also focus on decision-making but specifically on the domain EAM. In the end, we will discuss work that deals with modeling EA decisions.

15.3.1 Decisions and Decision Processes in General

There is general agreement in the literature on the description of decision-making processes. The processes described are very similar, although they differ in nuances. In the following we will therefore only show a small extract of them.

Dewey [18] first described the steps of solving the problem as "What is the problem?", "What are the alternatives?" and "Which alternative is the best? According to this, Simon [19] defines the decision-making and problem-solving process in three phases. In the first phase, the so-called Intelligence Activity, the environment is searched for situations or states that require action in the sense of a decision. In the second phase, called Design Activity, developing different solutions to solve the identified problem is in focus. Finally, the so-called Choice Activity aims to select the most suitable solution.

Mintzberg [20] defines a decision as a "commitment to action". According to him, a decision process is not merely the selection of a solution but starts with a stimulus that represents a recognized necessity to act and ends with the promise to realize a solution for the improvement of the situation. According to Mintzberg [20, 21], the organizational decision-making process consists of seven activities that can be assigned to the three phases according to Simon [19]. The process ends with the authorization of responsible decision-makers.

Due to the complexity of EAs and the stakeholders involved, decision-making today is no longer conceivable without appropriate tool support. Decision Support Systems (DSS) are specialized tools for this purpose. In the classical normative understanding of decision theory, DSS are suitable for recurring and highly formalized problems whose solution is already known in advance [22–24]. According to this understanding, the decision as a result of the decision process is determined by the DSS [24]. In contrast, a decision according to the understanding in computer supported collaborative work (CSCW) research is left to the user of the DSS and not determined by the system itself [24]. In this context, a DSS is seen as a computer-assisted system that assists people in carrying out semi-structured activities in the context of decision-making but does not take the decision [23].

DSS for groups are often used in combination with meeting support systems, which, in addition to supporting the decision-making process, focus on communication between the participants [24, 25]. The meetings take place in meeting rooms equipped with information technology, the so-called Electronic Meeting Rooms (EMS) [26]. Possible configurations are the Management Cockpit War Room [27] and the Management Cockpit [28]. Both approaches are based on equipping the room with several screens to view and analyze different perspectives on a particular situation in parallel.

15.3.2 Decision-Making in EA Management

There are different approaches that address decision-making in EAM. Although the evolution of an EA is based on a multitude of decisions within the architecture development cycle, TOGAF [14] as the standard work for EAM does not contain a decision-making process. The basic steps of the further development of an EA are however covered by the ADM cycle contained in TOGAF. This supports the collaboration between stakeholders, as the individual activities contain indications for the identification of relevant roles.

Best-Practice EAM [6] is an approach that provides concrete, practice-proven suggestions on how EAM can be established in an organization. The required steps for further development of an EA are described in detail and supported by tools in form of patterns and static visualizations. Although a decision process is not explicitly described, the method provides typical activities for further development of an EA and gives information about which roles and responsibilities there are and when these are to be included.

CEADA [29, 30] includes a collaborative and flexible decision-making process. However, the actual design of a target architecture is considered as a black box, which is carried out by experts. The focus of the method is on a collaborative understanding of the actual problem, a jointly developed solution sketch in the form of requirements and scenarios as well as on the subsequent collaborative selection. The authors consider the interests of the stakeholders and the views derived from them on the architecture as an important starting point for the collaborative decision-making process. However, the views and a systematic integration into the decision-making process, e.g., in the form of a conceptualization, will not be discussed in detail.

GEA [31] is an approach to make the interrelationships essential for organizations explicit and to control them. The defined perspectives allow the identification of stakeholders to be included in the further development of the EA. The explication of the interrelationships between the perspectives supports the stakeholders in the collaboration. Furthermore, the authors outline a process to move from a business problem to a holistic solution. This contains essential activities of decision-making in a very abstract way.

15.3.3 Modeling EA Decisions

ISO Standard 42010 [9] describes an approach for the systematic creation of architectural descriptions of systems based on the needs of corresponding stakeholders. In this context, it also provides modeling decisions. However, this is only rudimentary. A decision can be related to elements of the architecture description, e.g. views. But a decision cannot be related to elements of the architecture. Therefore, the effects of a decision are only very indirectly recognizable.

EA Anamnesis is an approach developed by Plataniotis et al. [32] for ex post modeling of decisions in EA management and decision-making itself. However, decision-making is not represented as a process, but as a strategy. This includes both the criteria on the basis of which potential alternatives are evaluated, their weighting and the evaluation itself. Essential parts of the decision-making process, such as the stakeholders involved in a decision and the findings gathered during the decision-making process, are not covered. The approach is designed for ex post documentation. However, a main reason why decisions are not systematically captured in practice is the time-consuming ex post documentation [33].

Multi-perspective Enterprise Modeling (MEMO) is a framework for enterprise modeling described by Frank [34]. The special feature of this approach is the modularization of different aspects of enterprise modeling into different modeling languages. MEMO addresses the perspectives strategy, organization and information systems. Bock [35] extends MEMO with an additional modeling language to describe decision processes in the organizational context. However, the modeling of individual activities within the decision-making process can only be implemented to a limited extent, since this is only possible by chaining processes. The metamodel consists of a multitude of different concepts, which leads to a high complexity. Similar to Plataniotis et al. [32], Bock also relies on a subsequent, manual modeling of the decision processes.

15.4 Integrative Method for EA Management

The idea of this work is to design a method that specifically address the demands of collaborative decision-making in EAM. The basis for the development of the method are the requirements (RQs) described in Sect. 15.2. The overview of related work presented in Sect. 15.3 did not identify an approach that fully meets all requirements. Therefore, a suitable solution has to be developed.

A key component of the method is a working environment supporting collaboration between stakeholders involved in decision-making (RQ 3), particularly by supporting discussion and providing all relevant information. For this purpose, the idea of the Management Cockpit [28] will be applied.

For method engineering we use the method definition of Goldkuhl et al. [36]. According to this, a method consists of method components for the modularization of decision-making activities, a framework as a ruleset for combining them and forms of cooperation. These forms define roles as well as the working environment in which the activities are carried out. The underlying concepts and relationships are defined by a metamodel. In the following the elements of the method are discussed in more detail. First, we introduce the Decisional Metamodel, which defines the basic mechanisms of the method. Subsequently the method components and the framework follow before we finally explain the cooperation forms.

15.4.1 Decisional Metamodel

The Decisional Model describes fundamental concepts and relationships of decision-making and additionally provides all relevant information for decision documentation (RQ 6). The basis of the metamodel is an extended conceptualization of the architecture description described in [37], which combines decision-making concepts based on related work and own considerations. The result illustrated in Fig. 15.1 is a novel approach to the description of decision-making. Orange colored concepts come from [20], yellow from [9] and blue from [32]. White colored concepts originate from own considerations.

The core concept is the *Decision Process*, which represents the decision-making process. Following [19, 20, 38] we define a *Decision Process* as a logical sequence of activities to solve one or more identified issues. The result of a *Decision Process* are one or more decisions to solve the identified problems taken by the responsible stakeholders.

The particular activities of the process are represented by the concept *Activity*. Since these *Activities* take place with human participation, at least one *Stakeholder* is assigned to an activity, who executes the respective activity. The *Stakeholder* concept is taken from ISO Standard 42010 [9].

In order to support *Stakeholders* in carrying out the respective *Activity*, visualizations (*Architecture Viewpoint*) and so-called *Techniques* can be used. Both concepts are part of the extended conceptualization of the architecture description [37]. *Architecture Viewpoints* are used to visually represent parts of the EA in a stakeholder-oriented way, while *Techniques* contain more detailed recommendations for action or algorithms for the automated execution of a specific task.

Based on CEADA [30], new information (*Process Data*) is created during the execution of an *Activity*. Such information represents the result of an *Activity*. These are, for example, newly identified or refined issues, arguments and discussion results

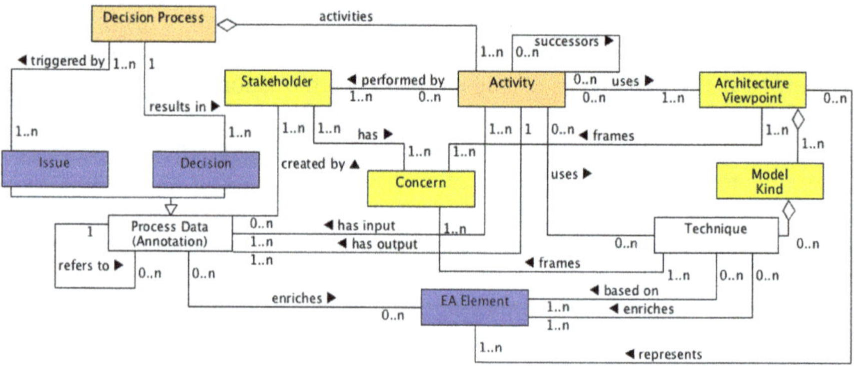

Fig. 15.1 Decisional metamodel

or a decision. In addition, the resulting information is the basis for subsequent *Activities*. Since this information refers frequently to particular elements of the EA, these can be set with one another in relation. These elements are represented by the *EA Element* concept, which also comes from [32].

Specializations of *Process Data* are *Issue* and *Decision*, which both have their origin in [32]. *Issues* represent problems and are the starting point of decision-making. These can be refined by new insights or can emerge during the analysis. The *Decision* in turn represents the result of the *Decision Process*.

The Decisional Metamodel shown is intentionally very abstract. The method components presented in the next section extend the metamodel. For each component there is a specialized *Activity* that defines what possible input and output information (*Process Data*) exists. In addition to *Issue* and *Decision*, there are other forms of the *Process Data* concept.

15.4.2 Method Components and Framework

According to Goldkuhl et al. [36], method components are used to modularize activities to be carried out. In this way, there is the advantage of a flexible sequence depending on the current situation (RQ 1). In order to prevent random combinations, the framework defines rules.

The decision process of Mintzberg [20] is used to identify the method components. This results in the components contained in Table 15.2.

The method components B to F correspond to decision-making activities from Mintzberg [20]. Component A is a special component, because it provides necessary steps to configure the method itself. The steps described there are not part of a decision-making process and are therefore not to be run through during the decision-making process.

In order to break down the complex decision-making into smaller tasks, a cyclical procedure similar to the ADM cycle [14] is possible. For this reason, for example, the successor of component C can be a new execution of the same component. The activities described there are not part of a decision-making process and are therefore not to be run through during the decision-making process.

Table 15.2 Method components and their possibilities of combination

Component	Predecessors	Successors
A: Configuration	–	B, C, D, E, F
B: Define Goals and Requirements	A	C, D
C: Analyze Situation	A, B	B, C, D
D: Develop Solution Candidates	A, B, C	B, C, D, E
E: Evaluate and Choose Solution Candidate	A, D	C, D, E, F
F: Authorize Solution Candidate	A, E	B, C, D

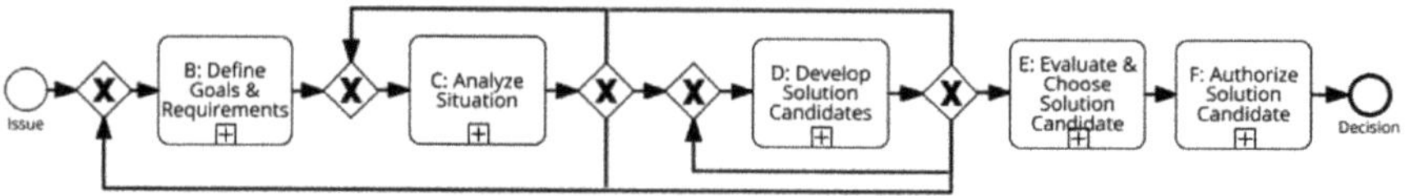

Fig. 15.2 Decision-making process based on the framework

Figure 15.2 illustrates an idealized decision-making process based on the framework's combination rules. We would like to emphasize that this process is not the only one possible. However, we have observed this in the application of the method with experts from practice.

The decision-making process is initiated by one or more identified issues. The process starts with the definition of goals and requirements. The next step is to analyze the current situation with the aim of sufficiently understanding the identified issues in order to be able to design solution candidates. The analysis takes place in an iterative way, which is why the corresponding method component can be applied several times in a row in order to be able to consider different aspects one after the other. By a better understanding of the issues or with the analysis of newly identified issues an adjustment of the goals and requirements could be necessary. This in turn triggers a new analysis. Then the iterative design of solution candidates takes place. For this the analysis and the design of solution candidates can be run through in combination several times and optionally a refinement of the goals and requirements can be inserted before.

After finishing the design of solution candidates, their evaluation and the selection of the candidate best suited to the goals and requirements takes place. Since this is an idealized process, no adjustments are necessary and one of the solution candidates adequately solve the issues. For this reason, no returns to the method components C and D are done at this point. The situation is similar with the authorization of the solution candidate. In case of a positive assessment by the responsible stakeholders, the previously selected solution candidate is authorized. The decision-making process ends at this point because the implementation of the solution is not part of the method.

15.4.3 Cooperation Forms

In order to complete the method on the basis of the method definition of Goldkuhl et al. [36], we finally describe the Cooperation Forms. These describe which roles are required for the application of a method, especially for performing the tasks contained in the method components, and how these should cooperate with each other.

The question which stakeholders are to be included in the particular method component depends very much on the issue at hand. EAM literature describes a multitude of different roles [6, 8]. Based on the role descriptions we identified the following three basic types and assigned the roles to them.

- *Domain Experts* have a profound domain knowledge and are experts in a part of the EA. They are responsible for preparing the decision-making process (components C to E). Typical Domain Experts are Business Architects or Process Owners.
- *Decision Makers* take the main responsibility and make final decisions (component F). Examples are the CIO or committees defined by an organization.
- Unlike domain specialists, *Moderators* work across domains. They guide through the decision-making process and are the mediator between different interest groups in order to reach a common consensus. The typical role of the moderator is the Enterprise Architect.

15.5 Method Component C in Detail

For limited space, we are concentrating on the method component C: Analyze Situation and describe them in detail. This method component is used for focused analysis of previously identified issues. The findings gathered are then used to develop possible solution candidates to solve the issue (method component D). The goals pursued by the analysis are defined by method component B: Determine Goals and Requirements.

According to Goldkuhl et al. [36], a method component consists of (1) a procedure that contains a step-by-step description of actions, (2) concepts for describing the underlying mechanisms and (3) a notation for representing the results. The concepts are used in the method components and specify what is possible. They correspond to the concepts of the Decisional Metamodel or specialize them. Figure 15.3 illustrates the necessary extensions.

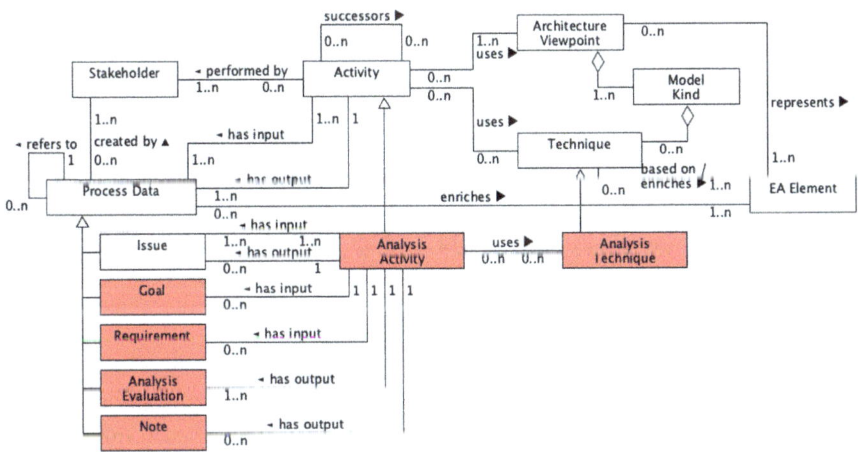

Fig. 15.3 Extended decisional metamodel based on method component C

For representing the analysis, the abstract *Activity* is specialized by the *Analysis Activity*. *Analysis Techniques*, which are a specialization of the abstract *Technique*, can be used to perform the analysis.

Addressed *Issues* are assigned to the *Analysis Activity* as input for the execution. The *Issue* is already included in the original Decisional Metamodel. Further inputs can be *Goals* and *Requirements*, on that basis the analysis is performed. The result of the analysis is so-called *Analysis Evaluations*, which represent the evaluation of results of executed *Analysis Techniques*. Furthermore, discussion results and further information can be documented in the form of *Notes*. New or refined *Issues* can also be the output of the analysis.

In the following, the procedure, which is the second part of a method component, is described. The notation as the third component of the method component is part of the viewpoints. These are based on a modeling language consisting of syntax, semantic, and notation (see [37]).

Step 1: Plan Analysis

This step is about setting up the analysis. For this the moderator is responsible. First of all, the aspect to be analyzed has to be defined. This is derived from goals and requirements of method component B. Stakeholders to be involved must then be identified. Once the stakeholders have been identified, the parts of the EA that are relevant to the analysis of the chosen aspect have to be defined. These parts are motivated by the stakeholder's interests and are represented by viewpoints that they use in their daily work as information providers. The creation of an appropriate viewpoint catalog is part of the configuration of the method (Component A).

Step 2: Analyze and Evaluate

First, the visualizations selected in step 1 have to be analyzed collaboratively. To support this, suitable analysis techniques can be selected and executed, which can run both automatically and manually. Analogous to viewpoints, there is also a catalog for this, which is created in component A. The catalog can be used as a basis for the analysis.

The analysis is based on interactions that can be performed in the selected and parallel visualizations. Some forms of interaction are shown in [39]. These are for example graphic highlighting and filtering as well as the representation of additional information to elements of the EA, which are represented by symbols. The execution of automated analysis techniques is also an interaction. With regard to the identification of dependencies between the viewpoints, dependency analyses can be used.

The results of the analysis have to be evaluated by the stakeholders. Furthermore, conclusions are to be deduced, which are to be documented graphically in the visualizations with the help of interactions. In addition, newly identified or refined issues could be a result of this evaluation. For the documentation of the results, a combination of graphical highlighting and annotation of the associated elements of the EA is used.

The viewpoints selected in step 1 provide the notation as the third component of a method component. In [37] we develop interactive visualizations (RQ 5) and show the interaction with analysis techniques (RQ 2). In [40] we describe which forms of interaction are generally possible. The novel feature of this method is that different visualizations can be viewed in parallel in an appropriately equipped electronic meeting room (RQ 4). Dependencies between the sections of the architecture represented by the visualizations can be made visible by analysis techniques.

15.6 Exemplifying the Approach

To demonstrate the method, we conducted a case study with an insurance company [41]. Insurance companies are exposed to a variety of risks through their core insurance and asset management activities. Including underwriting, operational, strategic but also credit, market, business, liquidity and reputational risks. Internal Governance Risk and Control (GRC) systems as means to actively govern and manage these risks are therefore prevalent in the insurance industry.

The insurance company has a holding structure with over 60 Operating Entities (OEs) represented in more than 70 countries and serving more than 100 million customers. The IT necessary to support the business of the OEs is partly operated by a captive shared service provider, while certain OEs with special situations reserve the right to maintain a local IT. In this context not only efficient and effective but also resilient and above all secure information processing is a key capability for the organization. These demands derived from the company's business model and regulatory requirements are translated into harmonized Global Architecture and Global Security Standards which are mandatory for all OEs and governed centrally in the holding.

In our case study, the CIO instructs the company's Enterprise Architect to improve the eight most important OEs based on predefined architectural indicators. One of these indicators is 'IT debt' that computes the distribution of IT Assets of Standard to non-standard technologies. The IT debt is expressed in the amount of money needed to migrate from non-standard technologies to their standard counterparts is considered the corresponding IT debt. The respective value of the indicator is based on an expert assessment together with the respective OE representatives.

The Enterprise Architect is the moderator and therefore responsible for improving the OEs. The case study takes place in the so-called Architecture Cockpit that is a combination of hardware and software and realizes the collaborative working environment (RQ 3). The hardware is based on the management cockpit concept [28] and includes nine different screens to display several viewpoints in parallel. In addition, in preparation of the case study we implement a software prototype met the support the designed method. In the following we describe the steps given by the method for carrying out the task. For space reasons, we are again focusing on the method component C: Analyze Situation.

Step 1: Plan Analysis

In step one of the method component C: Analyze Situation, the moderator's first task is to determine the aspect to be analyzed. This corresponds to the task assigned to him by the CIO. However, since not all indicators can be considered at the same time and different experts are required in each case, he first wants to analyze the eight most important OEs using the indicator 'IT debt'.

Next the relevant stakeholders have to be identified, their information needs for the present analysis queried and suitable viewpoints selected. Since domain experts are needed for the analysis because they have the necessary background knowledge, the Enterprise Architect involves the appropriate OE representatives who were involved in the expert assessment for the evaluation of the respective IT debt.

Figure 15.4 sketches the configuration of the Architecture Cockpit including the world map and a detail view for each OE.

The world map displayed on the center screen serves as entry point for user interactions. This viewpoint allows to highlight each OE according to a preselected indicator. In our case, all OEs are colored using the IT debt indicator. A legend at the bottom of the viewpoint reflects the category system of 'IT debt' ranging from 'very good' (dark green) to 'very bad' (red). The slider on the right represents an analysis technique that can be used to change the translation into the category system of a values determined by expert assessments. In this way, what-if analyses can be performed against changing global standards.

In contrast to the world map, which aims to provide an overview of the entire company, the detail views on the right and left side provide an overview of a selected OE. The viewpoint shows all existing indicators in the context of the OE. Therefore, there is an individual detail view for each OE. In order to assist users in planning

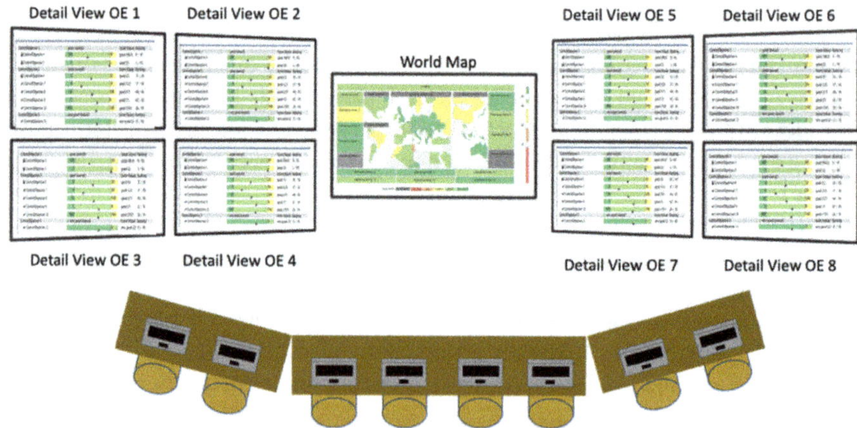

Fig. 15.4 Architecture Cockpit configuration including selected visualizations

standard changes, the exact location of each value within the determined category is shown. Hereby, the required effort to reach the next better category can be estimated.

Step 2: Analyze and Evaluate

Stakeholders first analyze the current situation with regard to 'IT debt'. All relevant OEs are in the good or very good range. In order to continuously improve the OEs, the objectives set for the OEs have to be increased over time. The participants see this potential and document this finding by an interaction on the world map. In this way, an *Analysis Evaluation* object is created in the background as an output of the current activity. This concept is a specialization of *Process Data* and represents findings during the analysis. In this way, semi-automated documentation of decision-making is made possible (RQ 6).

Subsequently, the stakeholders deal with changing the underlying parameters of 'IT debt'. They use a what-if analysis presented in step 1. By using the slider, effects on changed parameters can be analyzed. Since the viewpoints are related, changes made by the slider are displayed in all viewpoints. The indicator is recalculated in the background.

After testing some parameter changes, they choose one. They document this again with an interaction. All previously analyzed what-if scenarios are automatically documented in the background. In this way, you can later see what exactly was tried out and how the decision was made.

Figure 15.5 illustrates the semi-automatically created decision model of the current analysis that is an instantiation of the decisional metamodel (see Fig. 15.1). While activities performed and the *Architecture Viewpoints* and *Techniques* used are automatically documented by the tool. However, stakeholders are responsible for documenting the findings in the form of *Process Data* and the *Stakeholders* involved through interactions. For reasons of complexity, we will limit the decision model illustrated to two of the eight OEs and the first *Analysis Evaluation*, in which the *Stakeholders* have documented that the relevant OEs are in the good to very good range.

The concepts *Goals and Requirements Activity* and *Analysis Activity* are specializations of *Activity* and represent the method components B and C respectively. The *Analysis Technique* is also a specialization of the *Technique* with focus on analysis.

Further Activities

In the example shown, the stakeholders analyzed and planned changes in relation to the 'IT debt' indicator. Since there are further indicators, further iterations with other stakeholders and viewpoints have to be carried out. Each iteration corresponds to an execution of the method component. The individual iterations correspond to individual process steps, which are automatically brought into a sequence in the background.

After all indicators have been analyzed and planned, the aim is to react to the changed indicators (method component D). Actions are necessary so that the respective OEs are still in the good to very good range. For this, the EA have to be optimized. The solution candidates developed for this purpose must then be evaluated

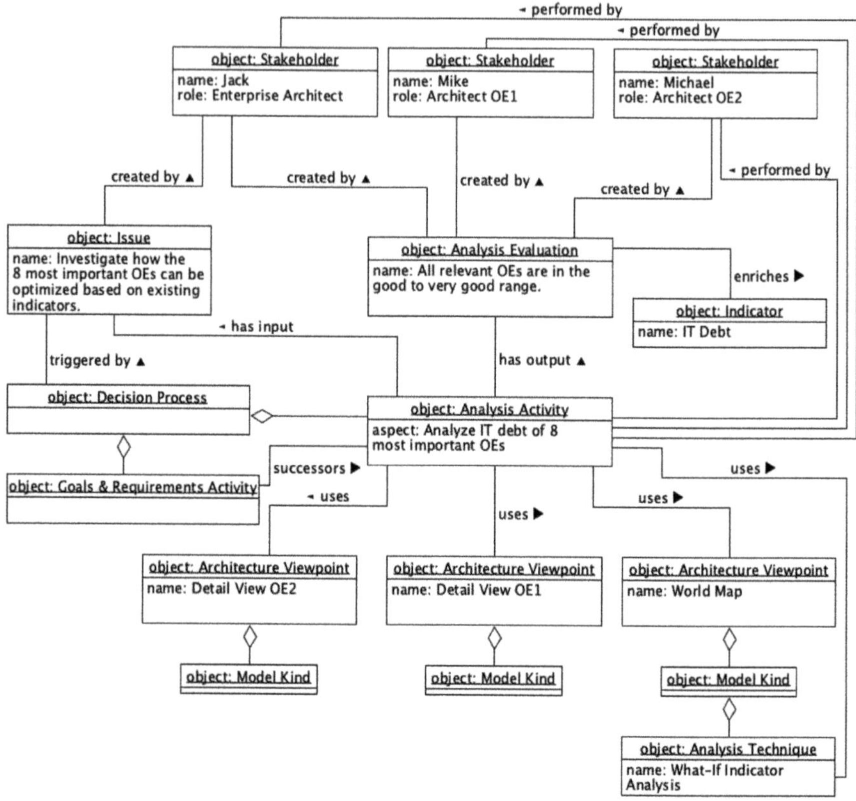

Fig. 15.5 Decision model created during the analysis

by performing method component E. At the end of the process, a decision is made as to which solution will be implemented (method component F).

15.7 Conclusion

The idea of this article is developing an integrative method to support the stakeholders involved in the decision-making process of EAM. The systematic integration of the stakeholders' interests by integrating the corresponding viewpoints on the corporate architecture into the decision-making activities gives the method its novel integrative character.

The interactive viewpoints enable executing manual and automated techniques specialized in the activities. In this way, stakeholders are supported in carrying out the activities.

Stakeholder collaboration is additionally supported by using an electronic meeting approach. This enables considering different viewpoints in parallel and to display dependencies between them. Specialized analysis techniques are used to identify these dependencies.

There are two limitations for the use of this method. On the one hand EAM must already be established in the enterprise, on the other hand an integrated EA model is required. Without an integrated model, automated analyzing techniques based on algorithms would not be possible.

In future work, we want to evaluate the developed method in other companies and further develop it on the basis of the resulting findings. Furthermore, we want to extend the Architecture Cockpit by integrating further techniques and visualization types.

References

1. Dahlström, P., Ericson, L., Khanna, S., Meffert, J.: From disrupted to disruptor: Reinventing your business by transforming the core (2017)
2. Ahlemann, F., Stettiner, E., Messerschmidt, M., Legner, C.: Strategic Enterprise Architecture Management: Challenges, Best Practices, and Future Developments (Management for Professionals). Springer (2012)
3. Land, M.O.L., Proper, E., Waage, M., Cloo, J., Steghuis, C.: Enterprise Architecture: Creating Value by Informed Governance. Springer, Berlin, Heidelberg (2009)
4. Getto, J., Heydrich, U.: Matering digital transformation—how to align your business model and organization (2016)
5. Fleisch, E., Weinberger, M., Wortmann, F.: Business Models and the Internet of Things (2014)
6. Hanschke, I.: Enterprise Architecture Management—einfach und effektiv, 2nd edn. Carl Hanser Verlag München, München (2016)
7. Lankhorst, M.: Enterprise Architecture at Work—Modelling, Communication and Analysis, 3rd edn. Springer, Heidelberg (2012)
8. Wißotzki, M., Köpp, C., Stelzer, P.: Rollenkonzepte im enterprise architecture management. In: Zimmermann, A. and Rossmann, A. (eds.) Digital Enterprise Computing (DEC 2015), Lecture Notes in Informatics (LNI), P-244. pp. 127–138. Gesellschaft für Informatik (GI) (2015)
9. International Organization of Standardization: ISO/IEC/IEEE 42010:2011—Systems and software engineering—Architecture description (2011)
10. Lucke, C., Bürger, M., Diefenbach, T., Freter, J., Lechner, U.: Categories of enterprise architecting issues—an empirical investigation based on expert interviews. In: Multikonferenz Wirtschaftsinformatik. pp. 999–1010 (2012)
11. Hevner, A.R., March, S.T., Park, J., Ram, S.: Design science in information systems research. MIS Q. **28**, 75–105 (2004)
12. Lucke, C., Krell, S., Lechner, U.: Critical issues in enterprise architecting: a literature review. In: 16th Americas Conference on Information Systems (AMCIS) 2010. pp. 1–11 (2010)
13. Nakakawa, A., Van Bommel, P., Proper, H.A.: Challenges of involving stakeholders when creating enterprise architecture. In: CEUR Workshop Proceedings, pp. 43–55 (2010)
14. The Open Group: TOGAF Version 9.1 (2011)
15. Plataniotis, G., Kinderen, S., Linden, D.J.T. van der, Greefhorst, D., Proper, H.A.: An empirical evaluation of design decision concepts in enterprise architecture. In: Proceedings of the 6th IFIP WG 8.1 Working Conference on the Practice of Enterprise Modeling, PoEM 2013, Riga, Latvia, 6–7 Nov 2013. pp. 24–38 (2013)

16. Kruchten, P.: An ontology of architectural design decisions in software-intensive systems. 2nd Groningen Work. Softw. Var. 1–8 (2004)
17. Jugel, D., Schweda, C.M., Zimmermann, A., Läufer, S.: Tool capability in visual EAM analytics. Complex Syst. Informatics Model. Q. (2015)
18. Dewey, J.: Wie wir denken. Eine Untersuchung über die Beziehung des reflektiven Denkens zum Prozess der Erziehung. Morgarten Verlag Conzett & Huber (1951)
19. Simon, H.A.: The New Science of Management Decision. Prentice Hall (1977)
20. Mintzberg, H.: Structuring of Organizations: A Synthesis of Research (Theory of Management Policy). Prentice Hall (1979)
21. Mintzberg, H., Raisinghani, D., Théorêt, A.: The structure of unstructured decision processes. Adm. Sci. Q. **21**, 246–275 (1976)
22. Gluchowski, P., Gabriel, R., Dittmar, C.: Management Support Systeme und Business Intelligence—Computergestützte Informationssysteme für Fach- und Führungskräfte, 2nd edn. Springer, Berlin Heidelberg (2008)
23. Katona, G.: Psychological analyses of business decisions and expectations. Am. Econ. Rev. **36**, 44–62 (1946)
24. Teufel, S., Sauter, C., Mühlherr, T., Bauknecht, K.: Computerunterstützung für die Gruppenarbeit. Addison-Wesley, Bonn (1995)
25. Dennis, A.R., George, J.F., Jessup, L.M., Nunamaker Jr., J.F., Vogel, D.R.: Information technology to support electronic meetings. MISQ. **12**, 591–624 (1988)
26. Petrovic, O.: Workgroup Computing—computergestützte Teamarbeit (1993)
27. Daum, J.H.: Management Cockpit War Room: Objectives, Concept and Function, and Future Prospects of a (Still) Unusual, But Highly Effective Management Tool. Control.—Zeitschrift für die erfolgsorientierte Unternehmensführung. 18, 311–318 (2006)
28. Roth, A.: Management Cockpit as a layer of integration for a holistic performance management. Q. Rev. Bus. Discip. **2**, 165–175 (2015)
29. Nakakawa, A., van Bommel, P., Proper, H.A.: Definition and validation of requirements for collaborative decision-making in enterprise architecture creation. Int. J. Coop. Inf. Syst. **20**, 83–136 (2011)
30. Nakakawa, A., van Bommel, P., Proper, H.A.: Supplementing enterprise architecture approaches with support for executing collaborative tasks—a case of TOGAF ADM. Int. J. Coop. Inf. Syst. **22**, 1350007 (2013)
31. Wagter, R.: Enterprise Coherence Governance Roel Wagter (2013)
32. Plataniotis, G., De Kinderen, S., Proper, H.A.: EA anamnesis: an approach for decision making analysis in enterprise architecture. Int. J. Inf. Syst. Model. Des. **4**, 75–95 (2014)
33. Tang, A., Babar, M.A., Gorton, I., Han, J.: A survey of architecture design rationale. J. Syst. Softw. **79**, 1792–1804 (2006)
34. Frank, U.: Multi-perspective enterprise modeling (MEMO)—conceptual framework and modeling languages. In: Proceedings of the 35 Annual Hawaii International Conference on System Sciences. pp. 1258–1267 (2002)
35. Bock, A.: The concepts of decision making: an analysis of classical approaches and avenues for the field of enterprise modeling. In: The Practice of Enterprise Modeling—8th IFIP WG 8.1. Working Conference, PoEM 2015, Valencia, Spain, 10–12 Nov 2015, Proceedings. pp. 306–321. Springer International Publishing (2015). https://doi.org/10.1007/978-3-319-25897-3_20
36. Goldkuhl, G., Lind, M., Seigerroth, U.: Method integration as a learning process. In: Proceedings of the Fifth International Conference of the British Computer Society Information Systems Methodologies Specialist Group. pp. 15–26. Springer (1997)
37. Jugel, D.: Modeling interactive enterprise architecture visualizations: an extended architecture description. Complex Syst. Informatics Model. Q. 17–35 (2018). https://doi.org/10.7250/csimq.2018-16.02
38. Brim, O.G.J., Glass, D.C., Lavin, D.E., Goodman, N.E.: Personality and Decision Processes, Studies in the Social Psychology of Thinking. Stanford Studies (1962)
39. Jugel, D., Schweda, C.M.: Interactive functions of a Cockpit for enterprise architecture planning. In: 2014 IEEE 18th International Enterprise Distributed Object Computing Conference Workshop Demonstration, pp. 33–40 (2014)

40. Jugel, D., Schweda, C.M.: Interactive functions of a cockpit for enterprise architecture planning. In: Proceedings—IEEE International Enterprise Distributed Object Computing Workshop, EDOCW (2014)
41. Jugel, D., Schweda, C.M., Bauer, C., Zamani, J., Zimmermann, A.: A metamodel to integrate control objectives into viewpoints for EA management. In: CEUR Workshop Proceedings, vol. 2218, pp. 110–119 (2018)

Chapter 16
Transformation and Enactment of Data-Intensive Business Processes Using Advanced Architectural Styles

Jānis Grabis

Abstract Business process redesign is increasingly motivated by analytical requirements, and data intensive activities such as image processing, prediction and classification are increasingly incorporated into business processes. Resulting business processes are referred as to data-intensive business processes. Such processes require data of various types and from different sources as well as analytical data transformations to guide and automate business process execution. Challenges associated with data-intensive business processes are identification and justification of opportunities for using advanced analytical processing methods and selection of appropriate technologies for enactment of these processes. This chapter proposes a method for specifying requirements towards data-intensive activities and uses these requirements to select appropriate implementation and enactment technologies. An example of business process redesign is discussed.

Keywords Data-intensive business processes · Transformation · Microservices

16.1 Introduction

Business processes are subject to continuous improvement not least because of new technologies becoming available. There have been multiple cycles of technology driven business process redesign and the most recent cycle is frequently referred as to digital transformation [1]. «**Digital transformation** is the process of using digital technologies to create something new» [2]. The current cycle of digital transformation has been largely enabled by availability of huge computational resources and large data sets, which can be processed and analyzed. These resources and data can be used to make complex decisions during business process execution. Business processes requiring complex data analytics are referred as to **data-intensive business processes**. Traditional business process management techniques based on BPMN and supported by business process management suites are better geared towards

J. Grabis (✉)
Information Technology Institute, Riga Technical University, Riga, Latvia
e-mail: grabis@rtu.lv

A. Zimmermann et al. (eds.), *Architecting the Digital Transformation*,
Intelligent Systems Reference Library 188,
https://doi.org/10.1007/978-3-030-49640-1_16

dealing with transactional business processes. DMN allows specification of complex business rules assuming that these rules are pre-specified. However, data intensive processes extensively involve analytical capabilities and deal with data analytical and decision-making problems on the fly.

Development and execution of data intensive business is influenced by a number of new requirements. Business process management should be aligned with data analysis life-cycle including support for activities like data preparation, model development, validation and continuous monitoring for changes in data patterns. There is a variety of data and decision-making models and they should be customized and updated for specific business needs. Additionally, data analysis models incorporated in data-intensive business processes have their own lifecycle and often require extensive computational resources at their disposal. That requires integration of core business process activities with activities associated with data analysis and an ability to provision the required computational resources.

A number of technologies are available to address the aforementioned requirements. These technologies range from traditional batch based techniques to data streaming and real-time analysis [3]. Business process architecture [4] is an approach allowing integration of various dimensions of business process redesign and execution. Cloud-computing and service-orientation are two key technologies providing infrastructural services for running computationally intensive operations [5]. This chapter discusses application of these technologies in the framework of digital transformation of data intensive processes.

The objective of this chapter is to propose a multi-dimensional approach to transformation of the data intensive processes. The approach is multi-dimensional in a sense that the business process architecture is used to capture various aspects of developing and operating data intensive business processes and process modeling and implementation layers are considered [6]. Transformation success is measured by cost savings generated taking into account costs associated with data acquisition, development and maintenance of analytical models. The specific research question is what are justifiable investments in analytical capabilities to ensure the transformation success? An example of digitalization of paper receipts is explored to demonstrate the approach.

The rest of the chapter is organized as follows. Section 16.2 provides background information of business process redesign and some of technologies for data intensive business processes. Section 16.3 discusses the transformation approach proposed. An application example of this approach is provided in Sect. 16.4. Section 16.5 concludes.

16.2 Background

As defined above, the data-intensive business processes are ones requiring complex data analytics in terms of either data gathering and processing or computations. This section discusses general aspects of business process redesign and identifies specific requirements characteristic to data intensive business processes.

16.2.1 Business Process Redesign

Business process is a sequence of activities performed to achieve specific business goals. Business processes are continuously improved through their lifecycle [7]. The improvement cycle includes activities of process design, modeling, implementation, monitoring and optimization [8]. Becker et al. [9] provide guidelines for the design of business process. The guidelines include As-Is modeling, analysis of As-Is models, To-Be modeling and optimization. Process documentation, usage of reference models and benchmarking as well as measurement and simulation methods are used for these purposes.

The process improvement methods rely on a set of general principles [8] and are often tailored to specific needs and organizations. Therefore, there is a large variety of methods. Reijers and Mansar [10] define a set of heuristic rules for business process improvement. The qualitative evaluation of the heuristic rules is also provided which helps business analysts to justify their BP improvement decisions. Barros [11] attempts to formalize business improvement guidelines as reusable patterns. These patterns can be used to construct business process from existing best practices. Damij et al. [12] propose a Tabular Activities Development methodology, which addresses both business process improvement and implementation of information system supporting the improved business processes. Process simulation is used to determine the process cycle time and related measurements are one of the key parts of the methodology. Dumas et al. [13] categorize various process improvement alternative and their potential impact on process improvement as well as advocates usage of quantitative models in business process design. The review of BP improvement approaches is provided in [14]. Sidorova and Isik [15] present an extensive review of business process research. They have identified business process design and business process on-going management and control as two of the four cornerstones of core business process research. Decision Model and Notation allows representing decision-making logics in business processes [16, 17]. Although the authors acknowledge importance of quantitative analysis and decision-making in business process design and execution, integration of quantitative models into business processes is rarely explore what is especially important since quantitative and analysis models have their own life-cycle, which should be synchronized with business process life-cycle. It is also desired that models should not be developed for every client using a particular data intensive business processes. The models are pre-packaged or provided as a service and only their configuration is required for particular clients.

The business process architecture is an approach allowing to combine various dimensions of business process redesign [18]. Lapouchnian et al. [19] show that business process architecture is suitable for design of business processes requiring cognitive capabilities. The set of related business processes include processes for analytical model creation, validation and business improvement.

Business processes are implemented and executed using various technologies. ERP systems provide monolithic business process development environment, which is efficient for stable processes while being difficult to modify and to implement custom requirements. Business process management (BPM) suites allow implementing custom processes and jointly with service-oriented architecture create an environment support flexible modification of the processes by selecting appropriate services. However, BPM and to some extent SOA are not well-suited for data analysis purposes, especially, requiring integration of various data processing technologies. Scalability is also achieved at the expense of efficiency. Business intelligence technologies are often used to develop data analysis models though their integration in business processes is often performed in off-line or ad hoc manner [20].

16.2.2 Requirements

Digital transformation of data intensive business processes is driven by business needs:

- Mass-tailoring of business services for specific user and contexts—every user has specific needs and business processes and services should be tailored to these needs with a minimum effort preferable in the autonomous computing mode;
- Automated and transparent decision-making—routine decisions are made by systems and only the most complex and important decisions involve human decision-makers. At the same time, the human decision makers should be able to monitor the decision-making process and to trace decision-making steps;
- Real-time enactment of decisions—decisions should be made in timely manner not to delay the process execution and the most up-to-date data should be used;
- Trust—people involved in decision-making should be confident that decision-making results are accurate and are not compromised.

These business needs lead to technological needs:

- Modification—data analysis models are subject of frequent revision as more data become available and models are refined. The new models should be evaluated and integrated in the business processes without affecting the rest of the process;
- Scalability—data analysis models often require significant computational resources and decisions should be made during business process execution for potentially large number of users. That requires both vertically and horizontally scalable execution environment;
- Security and reliability—data security and reliability issues should be addressed by encryption, access rights management and replication.

There are several other aspects to be considered specifically from the data analytics perspective. That includes support for development and tuning of data analysis models. Business process and data analysis models' life-cycles should be synchronized and model development should be a part of the overall business process redesign and execution. Presentation and understanding of data analytic models is crucial. User of data analytic model should have utmost understanding of models used in business process execution. That can be achieved by explaining levers available for controlling behavior of the models as well as by providing facilities for experimenting with models. The experimentation allows identification of appropriate model usage modes and understanding of model usage consequences. Finally, data analytic models, data used in the models and model solving algorithms come in different forms and the development and execution environment should be flexible to support usage of the most appropriate data processing methods and technologies.

16.2.3 Transformation Technologies

There are number of technologies available for development and redesign of data intensive business processes to meet the aforementioned requirements. This section briefly reviews batch processing, data streaming and microservices as three technologies facilitating data intensive operations.

Batch processing implies that data analytics is performed periodically by analyzing accumulated information. Results of analysis are embedded in business processes. Business rules are a prime example of incorporating batch processing results in business processes. Data warehousing and business intelligence are approaches typically used for batch analysis. Data warehouses are often built on the basis of common architecture [21]. The data storage is designed on the basis of dimensional modeling aimed at capturing essential analytical data characterizing a particular domain. Data are populated following the ETL process, which emphasizes dealing with data sources and ensuring data quality. Different data presentation methods and technologies are used to analyze and to present data extracted from the warehouse. Data warehouses provide a degree of decision-making transparency and trust since data are processes in a controlled environment. Their tailoring to specific needs and modifyability is relatively low and real time enactment of decisions is not the primary application area.

The batch processing is useful if frequency of change in data patterns is relatively slow and decision-making is not strongly dependent of processing live data. However, if data patterns are instantaneous or decision-making strongly depends on live data, batch processing cannot provide timely results. Data streaming allows processing data in real-time and enact data intensive decisions immediately [22]. Lambda architecture is often used to implement data streaming solutions what allows combining batch processing and real-time processing [23]. Data stream is processed using a real-time computation system and decisions are enacted using live data. At the same time, some of the streaming data are stored in the permanent storage to support offline processing. Data streams have flexible structure and data lakes are used to store data.

Microservices [24] is a technology recently gaining prominence as self-contained light-weight containers of business services. They allow for high degree of modularity, containers are created on-demand quickly and with a little overhead as well as various data processing and storage technologies can be utilized. Each microservice has its own technological stack allowing to use implementation and execution technologies most suitable for the data analysis problem at hands. Microservice architecture provides a high degree of modifyability since every container can be updated independently. From the perspective of data intensive business processes, data analytical models are placed in individual containers, which can be updated whenever a new version of the models becomes available. Data intensive operations can be scaled both horizontally and vertically in the cluster of containers or increased resources for a new instance of the container.

16.3 Transformation Approach

The transformation approach consists of three main components: (1) business process architecture; (2) transformation process; and (3) technical architecture. The business process architecture defines an interrelated set of sub-processes required to design and run data-intensive business processes. The transformation process defines steps to be performed to redesign a traditional business process into a data-intensive business process. The technical architecture specifies technologies used to execute data-intensive business processes.

16.3.1 *Business Process Architecture*

Data intensive business processes similarly as other complex processes have many facets what can be captured using business process architecture. BPA names the key organizational processes in a structured manner and defines relationships among them. There are various ways to structure the processes, for example, categorization as primary, supporting and management processes. The processes can also be structure along their life-cycle [13]. In the case of data intensive business processes one can distinguish the following life-cycle processes: setup, operations, and updating (Fig. 16.1).

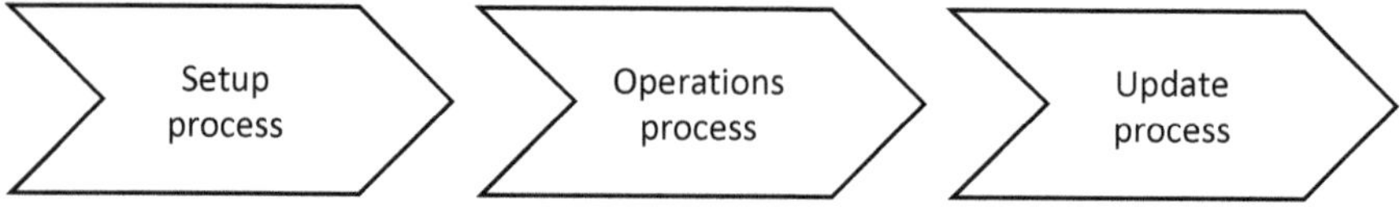

Fig. 16.1 Data-intensive business process architecture

The operations sub-process is the core business process providing the required business functions. Many activities of this process rely on data analysis models, which are invoked during the process execution. These models are setup prior to their usage. The setup sub-process might involve configuration of several models for all data-intensive activities in the business process. Assuming that a number of data analysis models for solving the analytical or decision-making problem are available, typical activities of the setup process are model selection (e.g., exponential smoothing or moving average for forecasting purposes), selection of structural parameters (e.g., input variables for a regression model or layers for neural networks), estimation of models' parameters (e.g., estimation of regression coefficients) and evaluation of modeling performance. The evaluation activities are of particular importance to ensure acceptance of data analysis results and their adoption for business process execution.

The setup sub-process (Fig. 16.2) requires training and test data to estimate and to evaluate models, respectively. The business process architecture represents these as coming from data stores, which physically could belong to different types of storage facilities. It also uses configuration data. The configuration data can be divided in three groups: (1) models' parameters; (2) performance indicators; and (3) design of experiments. The configuration data can be manipulated by business process users. All models' parameters are not included in the configuration data but specifically promoted parameters important for experimenting, tuning and understanding of the models. The performance indicators set targets for both modeling and process performance. That is important because in many situations data analysis models can be used only if reasonably accurate. The design of experiments is necessary for evaluation purposes. That allows business process users to gain understanding of models' behavior and implications of modeling results.

The operations sub-model (Fig. 16.3) is business problem specific. The data analysis models are invoked during execution of its activities. The execution results are stored as transactional records, which could be used for further refinement of the models. The refinement is handled by the update sub-process. The main issues represented in the update process are updating conditions (i.e., when to replace existing models with the updated ones) and handling of model life-cycle issues (e.g., model versioning for long-running processes).

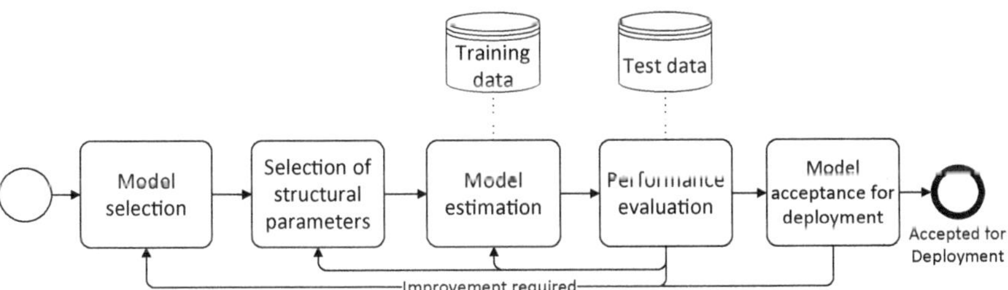

Fig. 16.2 The setup sub-process

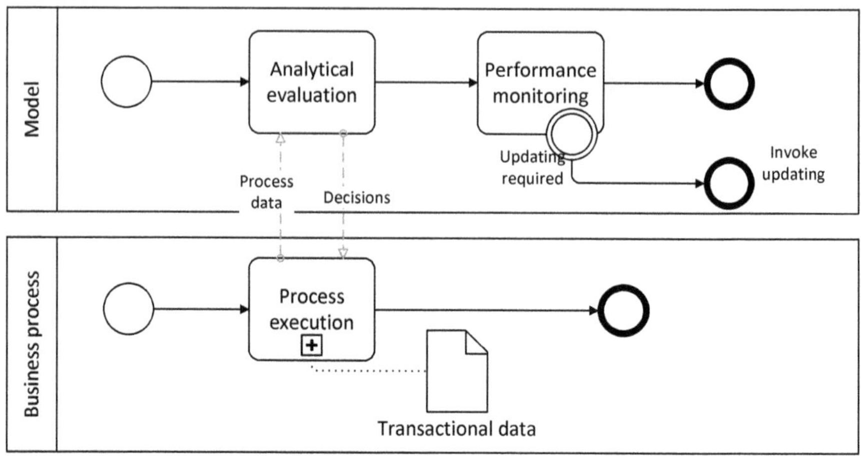

Fig. 16.3 The operations sub-model

16.3.2 Transformation Stages

The transformation process defines steps of process redesign into a data intensive business process and processes execution. The transformation process itself is a part of the overall business process management lifecycle though other stages of this lifecycle are not considered in this chapter.

The following steps are performed to develop a data intensive business processes:

- Define business process—representation of the core business process model is developed assuming that process identification has been performed;
- Identify data intensive activities—the process is vetted to discover opportunities for automation or improved decision-making due to data analytics. A data intensive activity requires data entities, which are not directly associated with the data record currently processed by the activity, and involves analytical transformations;
- Integrate data—data sources for training data are identified and data delivery channels are established;
- Develop data analytical models—data analytical models appropriate for a specific business area are developed. An appropriate type of analytical models is chosen, its parameters are determined on the basis of available training data and performance is evaluated;
- Implement and deploy models—the models are packaged and deployed in the productive environment. The implementation approach is chosen to meet performance requirements concerning scalability and modifiability;
- Redesign process around data intensive activities—introduction of data intensive activities might result in significant changes in the business process to achieve full benefits of digital transformation;

- Integrate models in business process - bind data intensive business process activities with the implemented data analytical models and establish data flows;
- Execute business process—the business process is executed including invocation of the data analytical models for real-time decision-making;
- Update data analysis models—as additional data are accumulated as the result of business process execution, the data analytical models are updated to account for latest tendencies and to improve the accuracy.

Some of these steps are performed in parallel, for example, data integration and development of data analytical models.

16.4 Example

An example of digitalization of paper receipts is explored to demonstrate application of the proposed approach. Digitalization of paper receipts is necessary in many different cases including travel management, expense management and filling out tax returns. This particular example focuses on generic digitalization of paper receipts without elaborating on further processing of the digitalized documents.

The example is elaborated following the proposed approach by exampling redesign of the process including its process architecture and particular attention is devoted to evaluation of benefits of digital transformation.

16.4.1 Case Description

The digitalization of paper receipts comes into play if a person pays for products or services, receives a paper receipt as a proof of payment and afterwards is entitled to some kind of refund. It is assumed that in order to get the refund, a person needs to submit a report (Fig. 16.4). The receipt must be attached to the report, typically, as an image and the person reads data from the receipt and manually fills out a refund form with data from the receipt. The data filled in the form are amount paid, payment data and vendor identification (other data items such as expense category also could be considered). Obviously, this process is tedious, time consuming and error-prone. The submitted report is typically checked by the receiving party as well and she

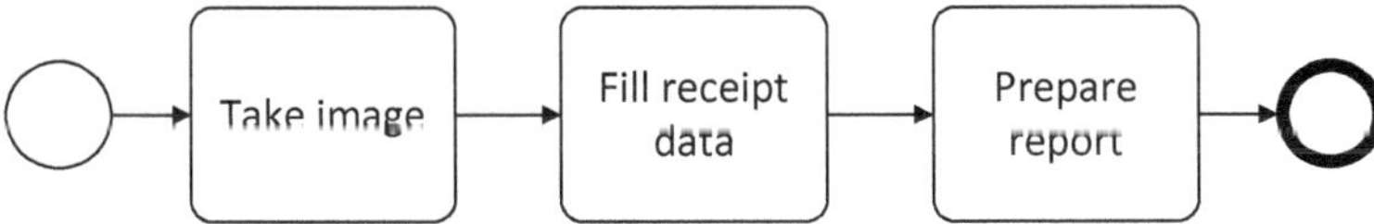

Fig. 16.4 The core business process

needs to repeat steps of comparing the image with data provided in the report. The image of receipt attached to the report potentially also contains a lot of other useful information, which can not be used efficiently.

16.4.2 Process Transformation

It is proposed to redesign the manual processing of receipts with a digitalized process. In the digitalized process, the activity Fill receipt data is replaced by automated data extraction activities (Fig. 16.5). The two added activities are Extract symbols and Identify information. The former activity deals with transforming the image of the receipt into characters and latter activity identifies information about the payment made as encoded in the image. The automated information extraction is not guaranteed to work with 100% accuracy and therefore evaluation and error correction activities are added to the processes. It is mandatory that the extraction accuracy is sufficiently high to outperform the manual processing. In this case the models provide information identification capabilities and support decision-making in the evaluate results by showing the information identification's level of trust.

The Extract symbols and Identify information activities are deemed as data intensive activities because they require data analytical processing. The Evaluate result activity is also potentially data intensive because context data could be used to aid identification and correction of errors. In this example, the scope is restricted to the former two data intensive activities.

In order to extract the symbols, Optical Character Recognition (OCR) is used. Various techniques can be used to identify information about the payment in the extracted symbols. It is proposed to use neural networks because they allow for configuration free tailoring and implementation of the solution [25]. However, their utilization requires large amount of training data. To address peculiarities of using data intensive technologies, the process architecture of the paper receipts digitalization process is defined (Fig. 16.6).

The setup sub-process starts with preparation of input data including training and test data. These data are either previously accumulated images of receipts with known (manually filled) expense data items or specifically generated data sets. Given that neural networks are used to identify expense data items, a structure of the neural

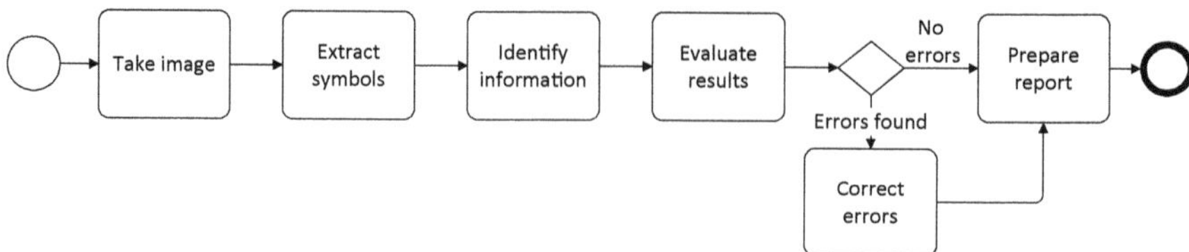

Fig. 16.5 Transformed process with data intensive activities

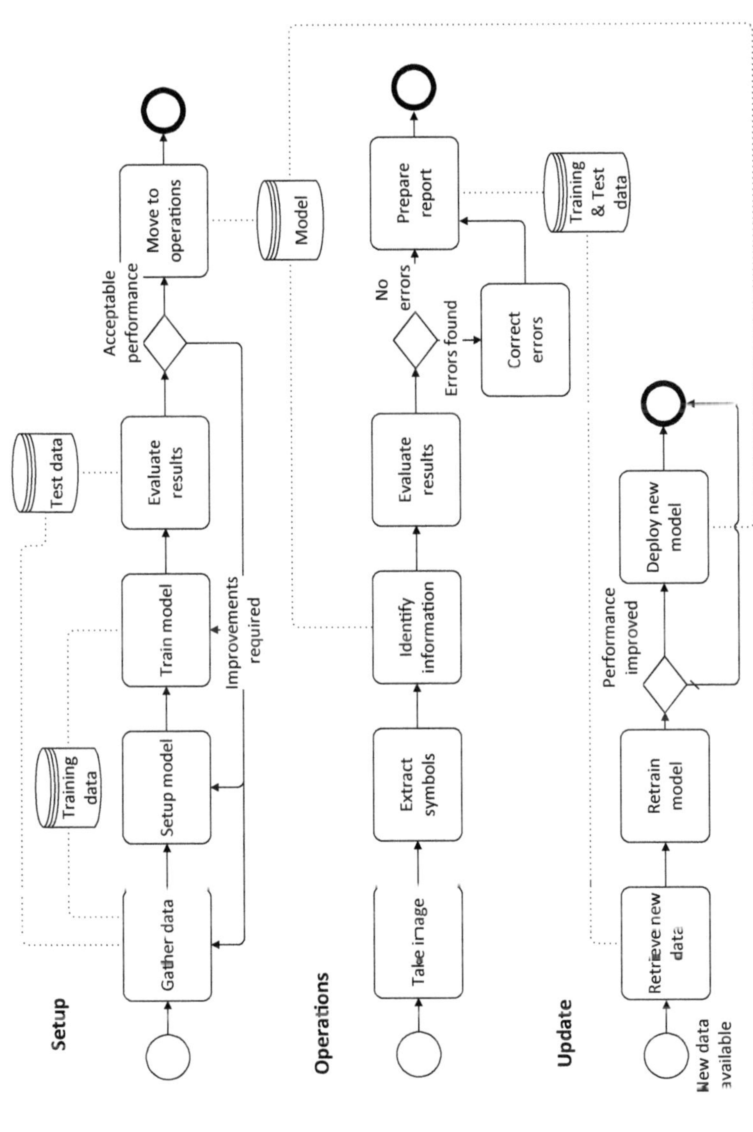

Fig. 16.6 Transformed process with data intensive activities

network is defined and training of the model is performed. The final activity is evaluation of the accuracy of expense data identification. The automated identification is enabled only if this accuracy is satisfactory as defined by performance indicators. Every new enterprise using the receipts processing business process requires tailoring of the model and the setup sub-process provides clear guidance for that and supports automated setup. The configuration parameters and the design of experiments are of particular importance because they allow for understanding of the model and its behavior. In this case, the estimated expenses extraction accuracy threshold is one of the configuration parameters. If the estimated accuracy is below the threshold then the extracted data are not used and manual entry is requested.

The receipt data identification model is load in the model repository and made available to the operations process. The actual processing of receipts takes place in the operations process as described at the beginning of the section. It is important to note that the additional data are gathered during the operations what increases the volume of data available for training and testing. The newly accumulated data are used in the updating process. The model is retrained to improve accuracy or to identify changes in patterns. The updating is potentially resource consuming process and an organization needs to decide on frequency of updating. The updated models are deployed only if they provide performance improvements. The updating process is also automated and run in the background of operations.

16.4.3 Technology

The technologies supporting data intensive processes are used to implement the redesigned receipts processing process. The main attention is devoted to microservices while batch processing and streaming could be considered for further improvement of the solution.

The microservices are used to support scalability and modifiability of the Extract symbols and Identify information activities. The data processing services and models are implemented as containers, which can be replicated in the case of increasing computational load and can be replaced without affecting the other parts of the solution once an updated model becomes available. The batch processing could be used to analyze typical errors encounter during the process execution. The analysis would yield post-processing rules for the Identify information activity and recommendation of error correction. The data streaming could be used to provide real time support to users including application usage support and contextual information for correcting errors.

Figure 16.7 show the technical architecture of the solution supporting the data intensive receipts digitalization business process. The Take image activity creates an image processing job and places it in the queue (implement using RabbitMQ[1]).

[1] https://www.rabbitmq.com/.

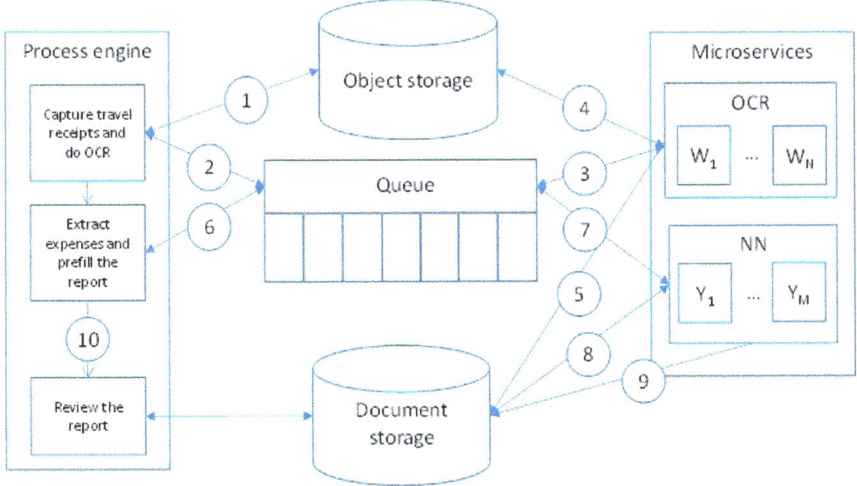

Fig. 16.7 Technical architecture (the numbers indicate the sequence of interactions)

The images captured are also stored in object storage implemented using Swift.[2] The image to text conversion is realized as a microservice using the Tesseract[3] library. Incoming image conversion messages are pulled from the queue by microservice instances and the instances read image data from the object storage. The recognized text is stored in the document-oriented database (implemented using MongoDB[4]) and the queue is notified that the text is available for further processing. That triggers the expense data extraction using neural networks (realized using TensorFlow[5] and dedicated neural networks (NN) are created for each expense item). The neural network was trained in the model setup component and production neural networks are again implemented as microservices. The extraction results are stored in the document-oriented database and the data intensive activity captures events indicating that extraction results are available for verification.

Following the business process architecture allows to redesign the business process as well as to establish supporting processes for model setup and update. The technical solution is highly scalable and the components are decoupled one from another. That allows to modify the components, to use specialized technological components as necessary and to integrate the solution in the enterprise's existing technological landscape. In this case, the expense data extraction algorithm is regularly modified to achieve better accuracy. Once the new version of the algorithm is approved for application it is deployed as a microservice without a delay.

[2]https://docs.openstack.org/swift/latest/.

[3]https://github.com/tesseract-ocr.

[4]https://www.mongodb.com/.

[5]https://www.tensorflow.org/.

16.4.4 Evaluation

In order to justify investment in digital transformation, process redesign benefits and expenses are evaluated. The evaluation should be performed taking into account the whole business process architecture. Additionally, investments in technology and data as well as operating expenses of the data intensive solution should be considered. Typically, these include the cost of using computational resources (e.g., cloud computing resources) and the cost of acquiring data.

The evaluation compares manual processing of receipts and digitalized processing of receipts. The cost of manual processing of receipts C_M is calculated as

$$C_M = r * t_M * N * H,$$

where t_M is the time of manual processing per receipt, r is processing rate in EUR per time unit, N is the number of receipts per period and H is number of periods in the planning horizon.

The cost of digitalized processing of receipts C_D incorporated cost of setup C_S, operations C_O and update C_U processes

$$C_D = C_S + C_O + C_U,$$

where

$$C_S = (c_1 + c_2)V + C_T$$

$$C_O = \sum_{i=1}^{H} r * t_D * N + r * t_E * (1 - \gamma_i) + (c_3 + c_4) * N$$

$$C_U = \sum_{i=1}^{H} c_2(V + \Delta V_i)Y_i$$

C_S is a combination of data acquisition cost c_1 per GB and model training cost c_2 what also depends on data volume V in GB and C_T on-time initial model development cost. C_O is a sum of cost of digitalized processing, cost of error correction and fees for using services and licenses needed for data intensive processing, where t_D is time of digitalized processing (presumably smaller than t_M), t_E is time of error correction, γ_i is the receipt digitalization accuracy in the ith period, c_3 is the costs of using external services and c_4 is licensing cost per using protected algorithms if any. C_U is the costs of retraining taking into account the additional data volume ΔV_i available in the ith period and Y_i is equal to 1 if updating is performed in the ith period and 0 if no updating is performed. It is assumed the paper digitalization accuracy depends on data volume available for training. If data volume increases, the accuracy also increases. In this case, logistics regression function is used to model dependence of the accuracy on the data volume.

Table 16.1 defines values of parameters used to illustrate the evaluation of process digitalization. The training cost is calculated per GB of data processes and characterizes the amount paid for computational resources consumed (for instance, the fee for using resources from a cloud provider; the value used in the example is a rough estimate and much more refined evaluation of the cost of the computational resources could be performed). It is assumed that new data are obtained every time period and are accumulated for updating. The update process in not necessary performed every time period.

Initially, the manual processes and digitalized processes are compared (Fig. 16.8). The numerical results are obtained over 24 periods. It is assumed that there are no setup and update costs associated with the manual process (discounting the costs associated with the base information system). The total cost of the digitalized process

Table 16.1 Parameters of the receipt digitalization process

Group	Parameter	Value	Unit	Description
Setup	C_T	10,000	EUR	Data analytics development cost
	V	500	GB	Initial data volume
	c_1	1	EUR/GB	Data acquisition cost
	c_2	1	EUR/GB	Data analytics model training cost
Operations	c_3	0.05	EUR/receipt	OCR service cost
	c_4	0.1	EUR/receipt	Licensing cost
	t_M	1.7	Minutes	Manual processing time
	t_D	0.7	Minutes	Digitalized processing time
	t_E	0.5	Minutes	Error correction time
	M	50,000	Receipts/month	Receipts to be processed
	r	20	EUR/h	Receipts processing labor rate
Updating	c_2	1	EUR/GB	Data analytics model updating cost

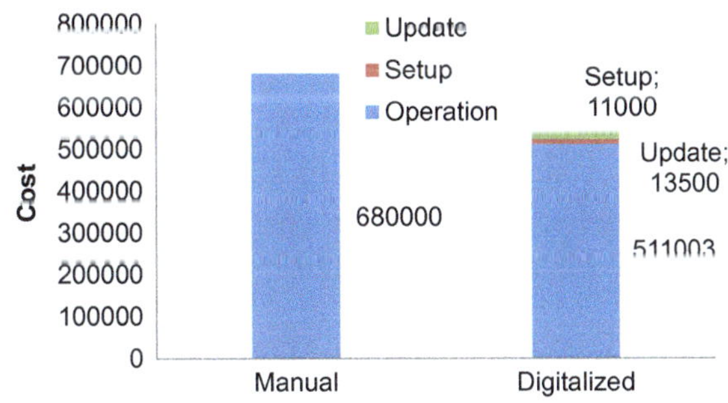

Fig. 16.8 Comparison of total cost for the manual and digitalized processes

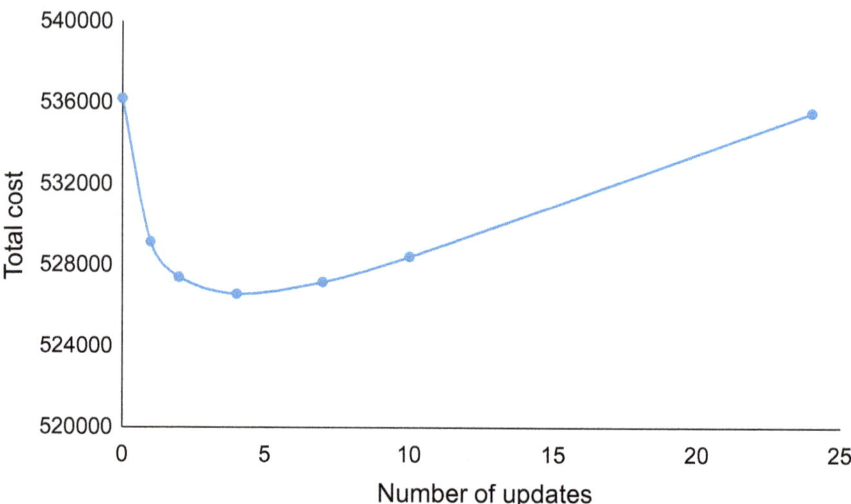

Fig. 16.9 The total cost of the digitalized process according to the number of updates

is by 25% lower than in the case of the manual process. In the given example, the setup and update costs are relatively negligible though they are highly case specific, especially, initial outlay on model development.

The update cost depends on the frequency of updating. The frequency is chosen to balance update cost and business process improvements gained by increased accuracy of information retrieval. It is assumed that having more data helps to improve accuracy of the information retrieval. Figure 16.9 shows the dependency of the total cost on updating frequency. If updating is made every time period (i.e., as soon as new data become available), the update cost increase outweigh benefits of accuracy improvement. If no updates are made then the solution does not benefit from accumulation of the additional data and the total cost increases. The optimal number of updates in this case is 4 thought timing of updates is also should be considered. An optimization model could be constructed to optimize timing of updates to minimize the total cost.

16.5 Conclusion

The paper proposes a framework for design and execution of data-intensive business processes. Integration of setup, operations and update sub-processes allows to implement data intensive business processes as one package rather than implementing the key process and its supporting processes separately as it is often a case. At the current stage of the proposed research, a proof-of-concept has been developed and the further elaboration of framework is required.

The business process architecture approach shows that various sub-processes are interlinked and share training, test, transactional and model data. However, from the technological perspective it is essential the sub-processes can be modified independently without interfering with normal business process operations. Otherwise cost of developing and running of data-intensive business processes exceeds benefits of automation and analytical capabilities. That is illustrated using the example. It shows that the justifiable investment in the analytical capabilities is a trade-off between process execution efficiency gains and the total cost of implementing and running the business process architecture. The model updating in particular should be performed only if efficiency gains outweight the updating cost. In the example considered, the updating cost was due to using computational resources for model retraining.

The example described is currently implemented as a prototype and experiments are conducted with this prototype to tune data analysis models and to evaluate efficiency of the redesigned processes. The solution meets the requirements of digital transformation by ensuring high level of scalability and modifiability. New workers to perform data intensive tasks can be created on-demand. Whenever the update process is completed a new data analytical model can be deployed using microservices without affecting the rest of the solution.

References

1. Zimmermann, A., Schmidt, R., Jugel, D., Möhring, M.: Adaptive enterprise architecture for digital transformation. Commun. Comput. Inf. Sci. **567**, 308–319 (2016)
2. Salesforce: What Is Digital Transformation? (2018) https://www.salesforce.com/products/platform/what-is-digital-transformation/
3. Abawajy, J.: Comprehensive analysis of big data variety landscape. Int. J. Parallel Emergent Distrib. Syst. **30**(1), 5–14 (2015)
4. Weske, M.: Business Process Management: Concepts, Languages, Architectures in Business Process Management: Concepts, Languages, Architectures, Springer (2007)
5. Moreno-Vozmediano, R., Montero, R.S., Llorente, I.M.: Key challenges in cloud computing: enabling the future internet of services. IEEE Internet Comput. **17**(4), 18–25 (2013)
6. Grabis, J., Kampars, J.: Application of microservices for digital transformation of data-intensive business processes. In: ICEIS 2018—Proceedings of the 20th International Conference on Enterprise Information Systems, pp. 736–742 (2018)
7. Van der Aalst, W.M.P., Ter Hofstede, A.H.M., Weske, M.: Business process management: A survey. In: van der Aalst, W.M.P., Weske, M. (eds.) Business Process Management. BPM 2003. Lecture Notes in Computer Science, vol. 2678. Springer, Berlin, Heidelberg (2003)
8. De Morais, R.M., Kazan, S., de Pádua, S.I.D., Costa, A.L.: An analysis of BPM lifecycles: from a literature review to a framework proposal. Bus. Process Manag. J. **20**(3), 412–432 (2014)
9. Becker, J., Kugeler, M., Rosemann, M.: Process Management: A Guide for the Design of Business Process. Springer, New York (2011)
10. Reijers, H.A., Liman Mansar, S.: Best practices in BP redesign: an overview and qualitative evaluation of successful redesign heuristics. Omega **33**, 283–306 (2005)
11. Barros, O.: Business process patterns and frameworks: reusing knowledge in process innovation. Bus. Process Manag. J. **13**, 47–69 (2007)
12. Damij, N., Damij, T., Grad, J., Jelenc, F.: A methodology for BP improvement and IS development. Inf. Softw. Technol. **50**(112), 7–1141 (2008)

13. Dumas, M., La Rosa, M., Mendling, J., Reijers, H.A.: Fundamentals of Business Process Management. Springer, Berlin, Heidelberg (2018)
14. Zellner, G.: A structured evaluation of business process improvement approaches. Bus. Process Manag. J. **17**(2), 203–237 (2011)
15. Sidorova, A., Isik, O.: BP research: a cross-disciplinary review. Bus. Process Manag. J. **16**(4), 566–597 (2010)
16. Biard, T., Mauff, A.L., Bigand, M., Bourey, J.: Separation of decision modeling from business process modeling using new decision model and notation (DMN) for automating operational decision-making. IFIP Adv. Inf. Commun. Technol. **463**, 489–496 (2015)
17. Mertens, S., Gailly, F., Poels, G.: Towards a decision-aware declarative process modeling language for knowledge-intensive processes. Expert Syst. Appl. **87**, 316–334 (2017)
18. Lapouchnian, A., Yu, E., Sturm, A.: Design dimensions for business process architecture. Lect. Notes Comput. Sci. **9381**, 276–284 (2015)
19. Lapouchnian, A., Babar, Z., Yu, E., Chan, A., Carbajales, S.: Designing process architectures for user engagement with enterprise cognitive systems. Lect. Notes Bus. Inf. Process. **305**, 141–155 (2017)
20. Chou, D.C., Bindu Tripuramallu, H., Chou, A.Y.: BI and ERP integration. Inf. Manag., Comput. Secur. **13**(5), 340–349 (2005)
21. Kimball, R., Ross, M., Thornthwaite, W., Mundy, J., Becker, B., Caserta, J.: Kimball's Data Warehouse Toolkit Classics: 3. Wiley (2014)
22. Cugola, G., Margara, A.: Processing flows of information: from data stream to complex event processing. ACM Comput. Surv. **44**(3), 1–70 (2012)
23. Marz, N., Warren, J.: Big Data: Principles and Best Practices of Scalable Real-Time Data Systems. O'Reilly Media, Newton (2013)
24. Dragoni N., et al.: Microservices: yesterday, today, and tomorrow. In: Mazzara, M., Meyer, B. (eds.) Present and Ulterior Software Engineering. Springer (2017)
25. Schuster, D., Muthmann, K., Esser, D., Schill, A., Berger, M., Weidling, C., Aliyev, K., Hofmeier, A.: Intellix—End-user trained information extraction for document archiving. In: Proceedings of the International Conference on Document Analysis and Recognition, ICDAR, pp. 101–105 (2013)

Chapter 17
A Tool Supporting Architecture Principles and Guidelines in Large-Scale Agile Development

Ömer Uludağ, Sascha Nägele, Matheus Hauder, and Florian Matthes

Abstract In today's business environments, organizations are confronted with technological advancements, regulatory uncertainties, and time-to-market pressures. The ability to detect relevant changes and to react timely and effectively becomes an important determinant for business survival. As a result, enterprises apply agile methods to larger projects as a part of their digital transformation. The adoption of agile methods at scale poses new challenges such as establishing effective knowledge networks or coordinating various development activities to produce desirable enterprise-wide effects. The latter can be addressed by applying architecture principles. However, there is a lack of academic research on how architecture principles can be created and applied in large-scale agile development. Against this backdrop, we propose a prototypical web application called "Architecture Belt" that supports the establishment of architecture principles. It uses social design principles and the analogy of belts in martial arts to enforce the application of architecture principles by exerting institutional pressures on agile teams.

Keywords Large-scale agile development · Architecture principles · Architecture Belt

17.1 Introduction

Nowadays, enterprises are exposed to turbulent market conditions due to factors such as increasing customer demands, globalization, novel technological advancements,

---

Ö. Uludağ (✉) · S. Nägele · M. Hauder · F. Matthes
Department of Informatics, Technical University of Munich, Munich, Germany
e-mail: oemer.uludag@tum.de

S. Nägele
e-mail: sascha.naegele@tum.de

M. Hauder
e-mail: hauder@outlook.de

F. Matthes
e-mail: matthes@tum.de

A. Zimmermann et al. (eds.), *Architecting the Digital Transformation*,
Intelligent Systems Reference Library 188,
https://doi.org/10.1007/978-3-030-49640-1_17

and intense time-to-market-pressures [1–3]. The problem of how companies can successfully deal with unpredictable and dynamic environments has been a prevailing topic both in industry and academia for a few decades [4]. One of the most popular options to deal with uncertain and unpredictable environments is to foster organizational agility by introducing agile methods such as Scrum or eXtreme Programming (XP) [4–6]. Agile methods were originally designed for small, co-located, self-organizations teams working closely with customers in a single-project context [1, 7]. The success of agile methods has made them attractive also outside this context, especially for larger software development projects in large companies [8]. As the fundamental assumptions of agile methods are challenged in large-scale endeavors [9, 10], organizations face unprecedented challenges such as (1) dealing with internal silos [11], (2) establishing inter-team communication and coordination [9], and (3) aligning multiple development activities to achieve desirable enterprise-wide effects and agility [1, 12].

In the last few years, enterprise architecture management (EAM) has established itself as valuable management function to drive large-scale transformations by providing guidance for necessary coordination efforts [13, 14]. A powerful instrument of EAM initiatives to effectively guide and steer large-scale transformations is the formulation and usage of architecture principles [14, 15]. Architecture principles safeguard transformation processes by restricting design freedom in a purposeful manner [14, 16]. Notwithstanding the significance of architecture principles for coordinating large-scale agile endeavors, extant literature disregards (1) how architecture principles are established in general, (2) how they should be enforced without encountering the resistance of agile teams, (3) and how this process can be supported (cf. [14, 17, 18]). In this chapter, we aim to fill this research gap by proposing a prototypical web application called "Architecture Belt" that supports the establishment of architecture principles in large-scale agile development.

The remainder of this chapter is organized as follows. In Sect. 17.2, we lay out the state of research on architecture principles for steering large-scale agile transformations. In Sect. 17.3, we present an approach to establish architecture principles and guidelines in large-scale agile development. In Sect. 17.4, we present the Architecture Belt along with its core features. We discuss our main findings and present the evaluation results of the prototype in Sect. 17.5 before concluding the paper with a summary of our results a conclusion and remarks on future research in Sect. 17.6.

17.2 Background and Related Work

Agile methods such as Scrum, XP, and Crystal Clear, which are in line with the values of the Agile Manifesto [19], share some common characteristics such as iterative and incremental development life cycles, co-located teams focusing on small releases, and a planning strategy based on a release plan or feature backlog [20] where the architectural design of the developed product is not very important [21]. Agile methods imply that the architecture should evolve incrementally and constantly rather than being imposed by some direct structuring force (emergent design) [22].

However, the sole practice of emergent design is insufficient when applying agile on a larger scale as it may cause excessive redesign efforts, architectural divergence, and functional redundancy increasing the complexity of application architectures [22, 23]. In large-scale agile development, some amount of architectural planning and governance becomes even more important [22] as they ensure the alignment of multiple agile teams to produce desirable organization-wide effects [12] and provide a common target vision by connecting strategic considerations to the execution of the agile projects [14]. Thus, for large-scale software-development endeavors, agility is enabled by architecture, and vice versa [24]. Similar to [14], we believe that the mutual alignment and coordination of large-scale agile transformations can be achieved by using architecture principles.

According to the online Merriam Webster Dictionary, principles denote "comprehensive and fundamental laws, doctrines, or assumptions" [25] and provide insights into the causes of certain effects rooted in natural laws, facts or beliefs [14]. There is no single, widely applicable definition of (enterprise) architecture principles, however, the general intend of architecture principles is to set a constraint on the design space of enterprises and (information) systems and guide design decisions (cf. [26–29]).

Regardless of the importance of architecture principles for the coordination of large-scale agile endeavors, organizations face a number of challenges in their formulation and usage. First, architecture principles are usually created and enforced top-down by EAM initiatives and thus are sometimes ignored by agile teams as they usually prefer independence and self-responsible decisions [14, 30, 31]. Second, because there are no structured feedback mechanisms between agile teams and enterprise architects, enterprise architects have no insight into the extent to which the defined principles are helpful to the work of agile teams [18, 32]. Third, as architectural works in large-scale agile projects are more continuous [9], enterprise architects are missing suitable tools to check whether agile teams comply with specified architecture principles [33]. Hence, enterprise architects cannot do justice to the task of proactively supporting agile teams during the application of architecture principles. Last but not least, since organizations usually do not have a central place for the management and documentation of principles, and the communication between enterprise architects and agile teams exhibits some major gaps [18, 32], the relevance of architecture principles for agile teams is not immediately apparent [18].

17.3 Establishing Architecture Principles and Guidelines in Large-Scale Agile Development

In a previous work [18], a collaborative approach for establishing architecture principles has been proposed (see Fig. 17.1) based on insights from a single-case study of a global insurance company. This approach adapts the generic architecture principle process by [14] and complements the top-down enforcement [30] of architecture

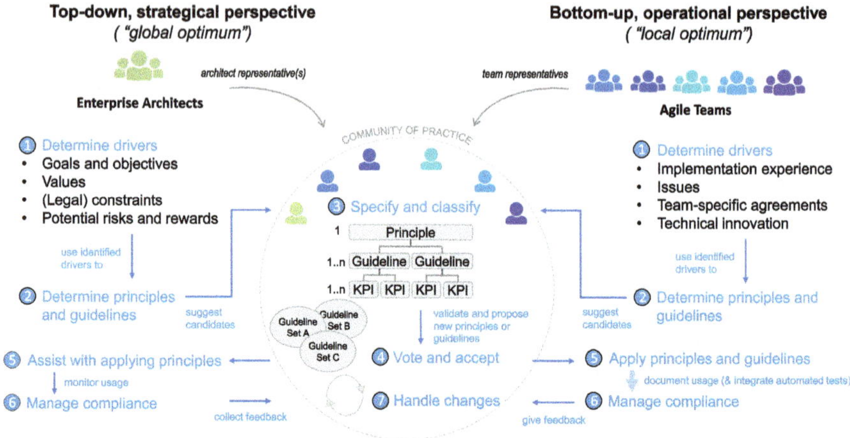

Fig. 17.1 Collaborative approach for establishing architecture principles and guidelines

principles and guidelines with an influence-centric view. This adaption is necessary because traditional top-down methods lack the consideration of bottom-up processes and put the main emphasis on "*command and control*", which we see as contradicting with agile and lean values. To effectively govern agile teams, a lean approach based on collaboration and motivation is required [34].

The heart of the collaborative approach constitutes the community of practice (CoP) consisting of enterprise architects and agile developers. The general idea of CoP is not limited to principles and guidelines, but also to make high-level design and architectural decisions together with all agile teams, similarly to Feature Design CoPs as described by [35]. The proposed approach envisages that enterprise architects contribute to the CoP with their strategic perspective to achieve a global, organization-wide optimum and that agile teams take an operational perspective that aims for a local optimum [36]. To keep the CoP relatively small in size and the decision-making ability efficient, we recommend that both enterprise architects and agile teams send representatives without necessitating everyone to participate. As suggested by [35], one main success factor for CoPs is that they are open to anyone who wants to participate.

Step 1—Determine drivers

Enterprise architects and agile teams determine their main drivers for specifying architecture principles. As enterprise architects take the strategical perspective, their drivers are mainly based on goals, objectives, and values defined as part of the IT strategy or legal constraints [14]. Agile teams use their implementation experience, actual implementation issues, team-specific agreements, and technical innovations for deriving principles [18].

Step 2—Determine principles and guidelines

Based on the identified drivers, both groups create a list of possible principles, select relevant principles, and formulate principle statements [14]. Subsequently, these principles are suggested as candidates for the CoP.

Step 3—Specify and classify

The proposed principles are specified into one or more guidelines that are more specific than the principles. In addition, we recommend to establish one or more metrics to ease the fulfillment evaluations [18].

Step 4—Vote and accept

Using a basic democratic consensus decision, the CoP votes on the adoption, refinement or rejection of a principle or guideline. In our opinion, a democratic vote requires a certain degree of organizational maturity. We believe that principles or guidelines driven by legal requirements should not be voted on.

Step 5—Apply principles and guidelines

Agile teams apply architecture principles and guidelines that are accepted by the CoP and enterprise architects assist them during the realization [18]. Our collaborative approach complements the enforcement-centric view of EAM by an influence-centric view by exerting normative and mimetic pressures on agile teams (see Fig. 17.2). First, the approach exerts normative pressure by using the analogy of belts from martial arts that awards agile teams that are complying with principles. By making the respective belts transparent to all agile teams, we aim to create influence based on mimetic pressures. Main advantages of these two social pressures are that the success stories of agile teams are made visible to others and that agile teams with higher ranks act as a role model and support lower belt teams, thus enhancing the overall compliance with principles and guidelines.

Step 6—Manage compliance

Acceptance issues of agile teams towards governance efforts [31] are solved by providing agile teams both the responsibility to adhere and the freedom to neglect a particular guideline. In both cases, they must document these decisions as the CoP needs an overview of guidelines with a low level of compliance. Low compliance can indicate implementation or specification issues. Architecture Belt (see Sect. 17.4) supports this phase by providing an overview of which guidelines are applied by which teams.

Step 7—Handle changes

Enterprise architects collect feedback from agile teams that have a genuine impression on the actual consequences of governance activities [37]. Our approach incorporates the experience of agile teams by involving them in the change process and provides a feedback mechanism where people can comment on principles, request changes, or discuss with peers on specific experiences [18].

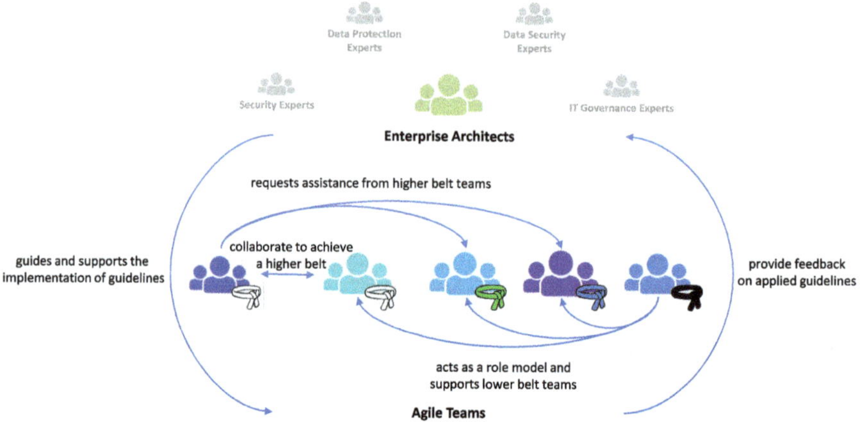

Fig. 17.2 Facilitating the application of architecture principles and guidelines

17.4 Architecture Belt

17.4.1 Core Features and Main Views

Architecture Belt[1] is designed to support the process of establishing architecture principles and guidelines in large-scale agile development (see Sect. 17.3). Architecture Belt currently focuses on assisting enterprise architects and agile teams applying and managing compliance with principles and guidelines. The system architecture of Architecture Belt is kept very simple. It consists of the Angular single-page-application serving as the frontend, which is communicating with the Spring Boot backend and its exposed REST-API via HTTP calls. The Spring Boot backend accesses a MySQL database with the use of JPA[2] and Hibernate ORM.[3] In the following, we present the core features of Architecture Belt based on the main views of the application.

Figure 17.3 shows the main feature of Architecture Belt, namely an overview of guidelines that are relevant to an agile team. This view provides a quick summary of team-related guidelines and their respective fulfillment status. For example, Fig. 17.3 shows that the *12factor* team has two team-specific white belt guidelines, namely *Codebase* and *Dependencies*. Currently, *12factor* only complies with the *Dependencies* guideline, but does not fulfill the *Codebase* guideline. Therefore, *12factor* needs to rework its current codebase by explicitly declaring dependencies to other libraries to change its compliance status of *Dependencies* to *fulfilled*. Architecture Belt supports *12factor* by providing a description of the *Dependencies* guideline

[1]https://github.com/mhauder/belt.

[2]http://spring.io/projects/spring-data-jpa.

[3]http://hibernate.org/orm/.

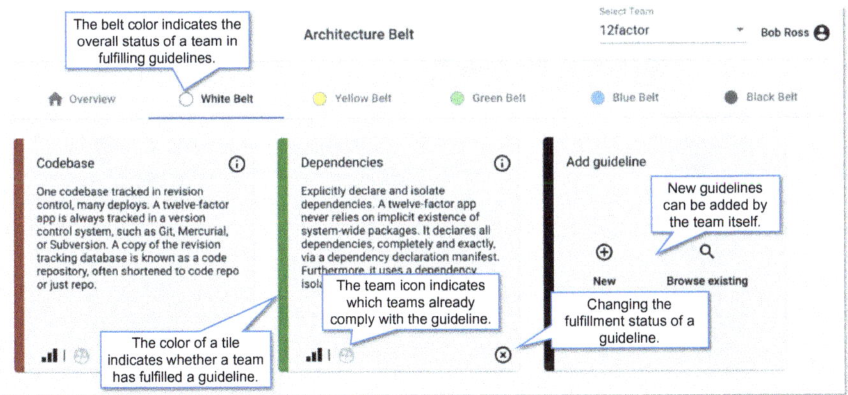

Fig. 17.3 Team-specific guideline overview with belt category selection and fulfillment status of each guideline in the selected belt

and indicating which teams already comply with it (see Fig. 17.3). *12factor* can use this information to contact other teams and ask for assistance. The information icon button in the top right corner of each guideline leads to the guideline detail screen which is described later on. After both guidelines are fulfilled, *12factor* ranks up to the next belt level, i.e., yellow belt, which constitutes new guidelines that have to be considered by *12factor*. In addition, the navigation bar on the top can also be used to switch between belts and access guidelines.

Some guidelines may also require agile teams to demonstrate their compliance with a guideline by linking to existing artifacts. Figure 17.4 illustrates a case in which a link towards an evidence artifact must be provided to proof the compliance with the *Customer Journey* guideline, e.g., a link to the Customer Journey Map. After providing the link, the compliance status of the *Customer Journey* guideline is set to *fulfilled*.

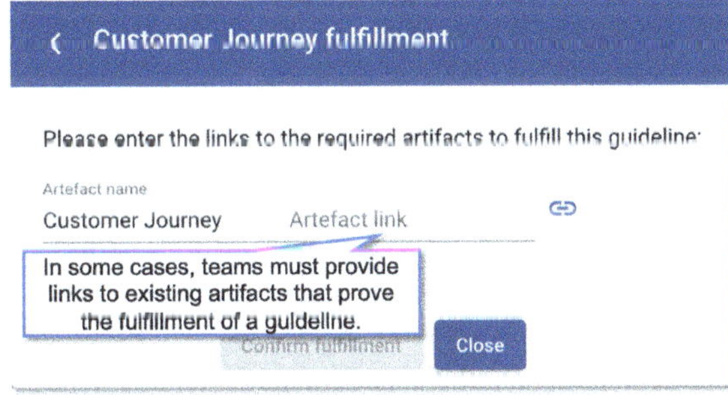

Fig. 17.4 Guideline fulfillment dialog that requires a link to an existing artifact to prove the compliance

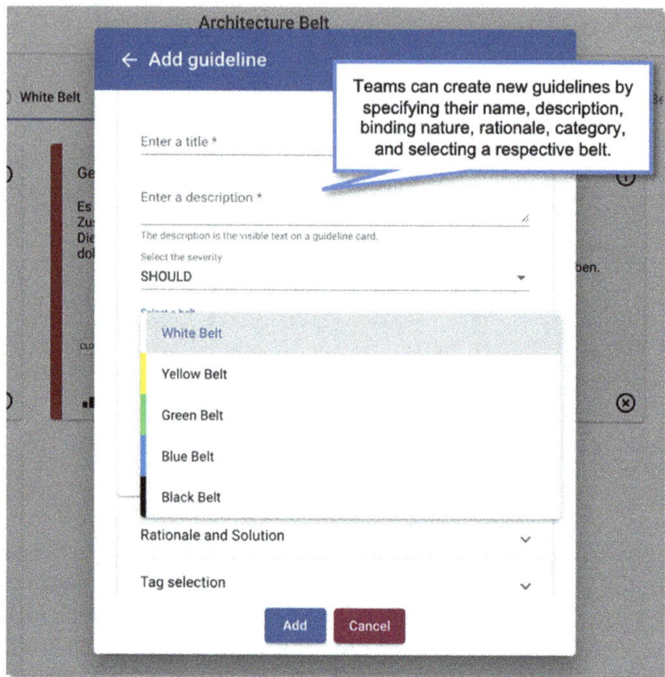

Fig. 17.5 Dialog for adding a new guideline

Another core feature of Architecture Belt is the feature to create entirely new guidelines or to add existing ones from the organization's general guideline pool that are relevant for a team. To add a new guideline, a new dialog opens (see Fig. 17.5). It prompts the team to enter the name and description of a guideline. In addition, the team must specify its binding nature, i.e., *should* or *must*, and the respective belt. Last but not least, the team is asked to specify the rationale and category of a guideline.

Architecture Belt offers a feedback feature to document experiences with existing guidelines (see Fig. 17.6). Each agile team can document their perception of a guideline, here *Customer Journey*, and provide potential improvements. Architecture Belt notifies responsible enterprise architects about new feedback so that they can revise guidelines with the help of the CoP if needed.

Architecture Belt also assists agile teams by providing a guideline detail screen with title, description, and fulfillment criteria. Figure 17.7 shows a detail screen for the *Codeanalyse* guideline. The rationale provides a better understanding by elaborating the context and motives. Additional resources may refer to further information, e.g., wiki pages, sample code, etc. The expert teams section refers to agile teams that comply with the *Codeanalyse* guideline and could help others to implement it. The button with the clock icon leads to the guideline history, i.e., it shows when *Codeanalyse* was set to fulfilled by which user. The comment button allows *CloudOps* to define team-specific solution and rationale for the *Codeanalyse* guideline.

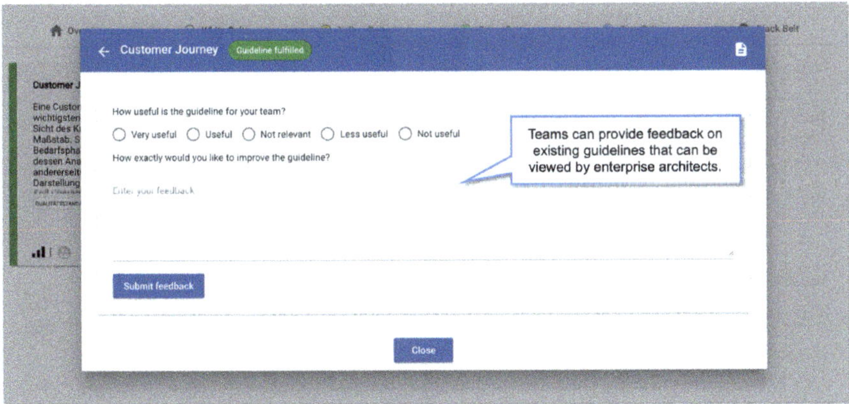

Fig. 17.6 Dialog for providing feedback on a specific guideline

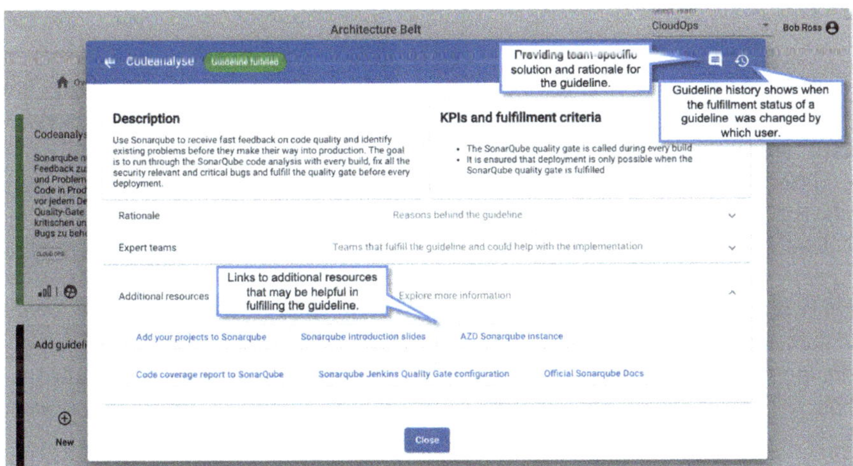

Fig. 17.7 Guideline detail view dialog of a specific guideline

Another core element of Architecture Belt is the team dashboard (see Fig. 17.8). It displays the home screen of all users and shows the current belt of a team and the percentage of progress to the next belt level. In this example, the *12factor* team currently has a white belt and needs to fulfill one more guideline to get the yellow belt. On the right, the home screen displays other teams, namely *CloudOps* and *Test Team A*, working on the same belt as *12factor* and their current progress. This function exerts normative and mimetic pressures (see Sect. 17.3) by motivating teams to act as a role model and by facilitating mutual support between teams facing similar challenges. Hence, Architecture Belt aims to spark greater collaboration to achieve higher compliance with guidelines.

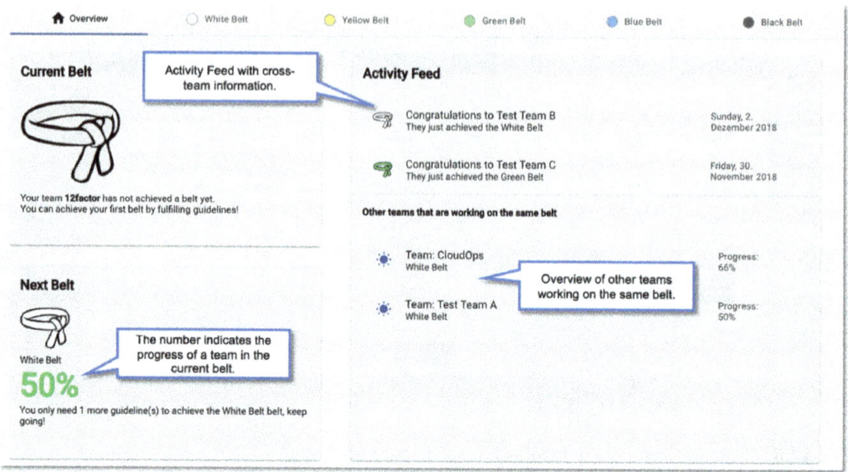

Fig. 17.8 Team dashboard displaying the current belt, activity feed, and progress of other teams

The team creation feature of Architecture Belt allows users to easily add new teams or assign users to specific agile teams. During the creation process, users can select predefined project or team characteristics as illustrated in Fig. 17.9.

Based on the specified characteristics, Architecture Belt recommends a set of relevant guidelines (see Fig. 17.10).

Last but not least, Architecture Belt is again exerting normative and mimetic pressures by providing an overview of teams and guidelines and their respective levels of compliance (see Fig. 17.11). This view allows users to get an overview of their teams and to set their progress in relation with others. Figure 17.11 shows that the *12factor Newbies* and *Partly 12factor* teams have the white belt, while team *12factor Ninjas* has reached the blue belt. Figure 17.11 also illustrates that *12factor*

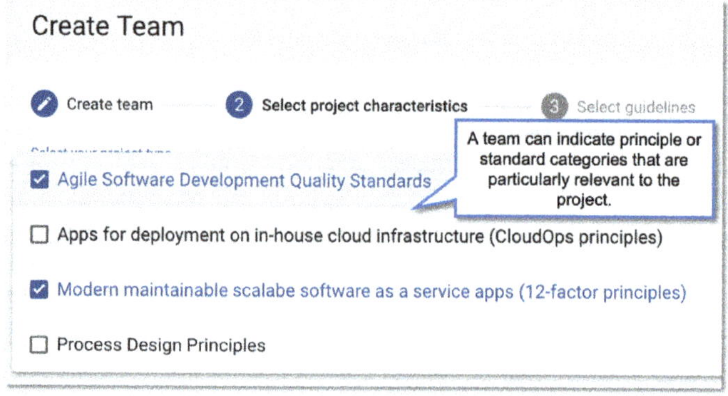

Fig. 17.9 Dialog for selecting project characteristics when adding a new team

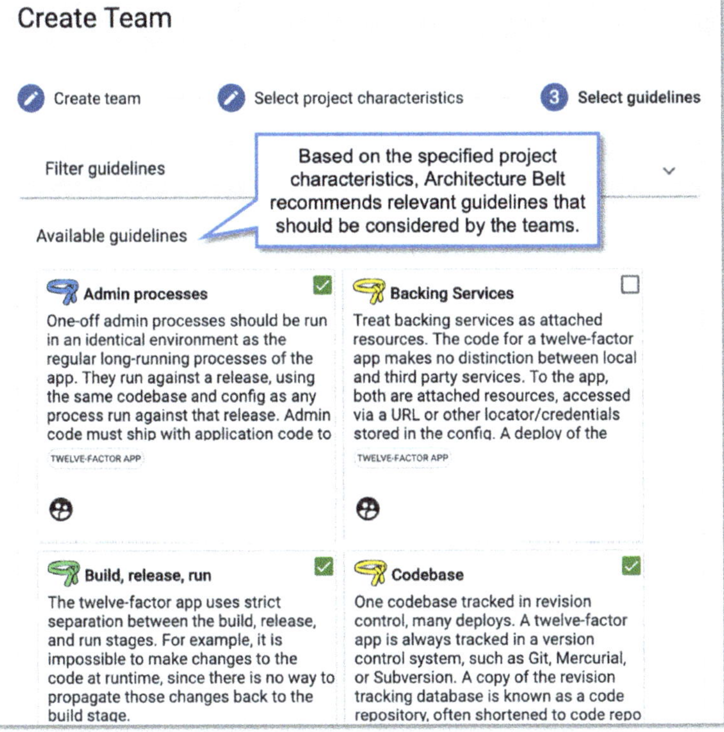

Fig. 17.10 Recommendation of relevant guidelines based on specified project characteristics

Fig. 17.11 Overview of teams and guidelines and their respective fulfillment status

Ninjas must fulfill the *Logs* and *Concurrency* guidelines to reach the black belt. As this team is very advanced, the *12factor Newbies* and *Partly 12factor* teams can ask *12factor Ninjas* for assistance. Since not all guidelines are relevant to a team, Fig. 17.11 shows that the *Partly 12factor* team must not consider the four guidelines: *Dev/prod parity*, *Dependencies*, *Logs*, and *Concurrency*.

17.4.2 Applied Social Design Principles

We incorporated insights from social science into the design of Architecture Belt as the value of the tool is highly dependent on the active usage and contribution of its users. In the following, we list selected design claims (DC) based on the research by Resnick and Kraut [38] that cannot only improve the level of participation and contribution, but also lead to better compliance with guidelines:

> *DC #1: "Making the list of needed contributions easily visible increases the likelihood that the community will provide them."*

The guideline overview for a team (see Fig. 17.3) is designed to quickly and easily show the members of a team which guidelines have already been met and where an action is needed within the team.

> *DC #12: "People are more likely to comply with a request when they see that other people have also complied."*

This design claim fits well with the news feed and belt-specific progress leaderboard (see Fig. 17.8) displayed on the team dashboard page, showing teams achieving new belts and how far they are progressing in the belt that is relevant for the current team. In addition, the expert team feature shows teams who have been complying with a guideline.

> *DC #13: "Providing members with specific and highly challenging goals, whether self-set or system-suggested, increases contribution."*

Based on the belt analogy from martial arts, Architecture Belt challenges teams to reach a higher level until they reach the black belt. Teams can set their own goals by adding team-specific guidelines.

> *DC #15: "Goals have greater effects when people receive frequent feedback about their performance with respect to the goals."*

For this purpose, the team dashboard (see Fig. 17.8) always indicates the current progress in the current belt on the way to the belt that can be achieved next.

> *DC #16: "Combining contribution with social contact with other contributors causes members to contribute more."*

This design claim highlights the necessity to integrate the tool into a broader context. By including Architecture Belt in existing processes and collaboration, especially the ones focused on personal, face-to-face interaction, the awareness for the tool and actual usage could increase sharply. The tool could be included in existing CoPs, sprint retrospectives or during the assessment of minimal viable product iterations.

> *DC #18: "Performance feedback—especially positive feedback—can enhance motivation to perform tasks."*

Similar to DC #15, we implemented this design claim by showing the teams their progress in a current belt and encouraging them to achieve compliance with further guidelines (see Fig. 17.8).

> *DC #19: "Site designs that encourage systematic, quantitative feedback generate more verbal feedback as well."*

By promoting to give feedback on a Likert scale on how valuable the guideline is for a specific team (see Fig. 17.6), the goal is to increase the amount of more detailed feedback and reasoning behind the rating as well.

> *DC #21: "Comparative performance feedback can enhance motivation, as long as high performance is viewed as desirable and potentially obtainable."*

The progress indicator on the dashboard (see Fig. 17.8) presenting other teams working on the same belt aims to provide comparative performance feedback as described in DC #21 and can be seen as a sort of leaderboard.

> *DC #23: "Rewards, whether in the form of status, privileges or material benefits—motivate contributions."*

Achieving a higher belt can be seen as a sort of status (see Fig. 17.8).

17.5 Discussion

17.5.1 Key Findings

As enterprises frequently run into challenges when formulating and using architecture principles in large-scale agile development [18], Architecture Belt provides adequate means for encountering them. First of all, Architecture Belt encourages agile teams to participate in the creation of principles by allowing them to create new principles or providing feedback on existing ones. In line with [8, 30], we believe that sole top-down driven EAM initiatives are insufficient at enterprise scale and have a lower acceptance in agile environments. One possible way to address this challenge is to involve local stakeholders, i.e., agile teams, in respective governance efforts. Correspondingly, Architecture Belt aims to complement the enforcement-centric view of

EAM by an influence-centric view through the use of institutional pressures [39] and social design principles [38]. Thereby, Architecture Belt aims to convince agile teams that their social status will be raising if they comply with architecture principles, i.e., by using the belt analogy. In addition, by stating the rationale of principles, agile teams get a better understanding of the benefits of complying with principles, thus enhancing EAM governance. By providing a central catalog of all principles that unites all efforts regarding architecture principles that are relevant for agile teams, Architecture Belt makes EAM deployments more transparent [30]. Since agile teams are pressured to deliver business value frequently, they often underprioritize architectural works, resulting in the accumulation of technical debts [9]. Architecture Belt addresses this issue by minimizing the effort agile teams have to invest when applying architecture principles. Currently, agile teams have to manually specify whether they comply with a principle. However, in the future, Architecture Belt can be integrated by code analysis tools for automatizing the compliance check with principles. Thus, we are consonant with [9] that architectural work in large-scale agile projects is more continuous and needs to be automatized as much as possible. We also agree with [18, 36, 40, 41] that enterprise architects have to take a more supportive and consulting role for agile teams by assisting them during the application of architecture principles. The support of Architecture Belt for this process is two-fold: First, by providing an overview of the compliance status of agile teams with principles, enterprise architects can identify teams that need assistance more quickly. Second, by getting feedback from agile teams on existing principles, enterprise architects can continuously improve principles.

17.5.2 Evaluation

We conducted semi-structured interviews with fifteen experts to evaluate the prototype. The interviewees were seven agile developers, seven enterprise architects, and one solution architect of an insurance company. Figure 17.12 shows the evaluation results of the tool support where we asked participants to decide on six statements based on a five-point Likert scale. First, Architecture Belt's team-specific guideline overview (see Fig. 17.3) and guideline selection feature (see Fig. 17.10) provide a better overview of guidelines that agile teams must or should comply with. Second, Architecture Belt's feedback feature (see Fig. 17.5) provides significant value add for agile teams by enabling them to feedback on existing guidelines. Third, the team-specific overview of guidelines and fulfillment indicators (see Fig. 17.3) save the effort of checking which guidelines a team needs to fulfill. Fourth, the recommendation feature of Architecture Belt helps to identify relevant guidelines based on specified project characteristics (see Fig. 17.10) very helpful in identifying relevant guidelines that fit the use cases of agile teams. Fifth, Architecture Belt's view that provides an overview of teams and guidelines and their respective fulfillment status (see Fig. 17.11) was considered very useful. Last but not least, the team-specific

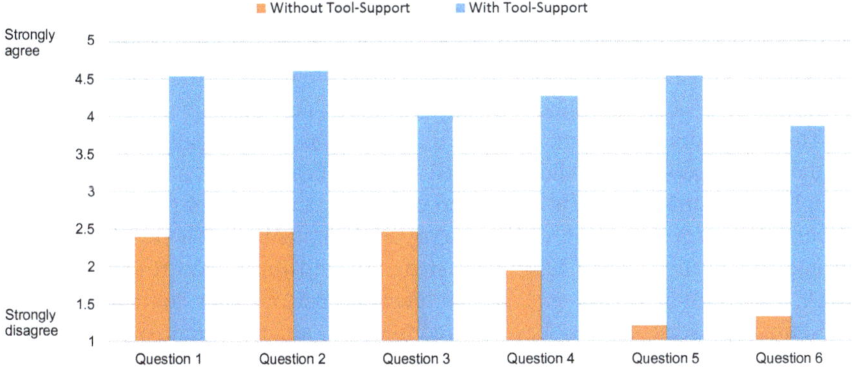

Question 1: Agile teams and enterprise architects have a good overview on which guidelines agile teams must or should adhere to.

Question 2: It is easy for agile teams and enterprise architects to give feedback on guidelines.

Question 3: It takes little effort to identify which guidelines a specific agile team is fulfilling and which ones it is not fulfilling.

Question 4: It is easy for agile teams to look through existing guidelines and select the guidelines that fit their use cases.

Question 5: It is simple to get an overview of which agile teams are compliant to which guidelines.

Question 6: It is clear why an agile team decided against using a specific guideline.

Fig. 17.12 Evaluation results of Architecture Belt (y-axis represents the mean of the statements provided)

solution and rationale descriptions (see Fig. 17.7) facilitate the process to identify the reasons why an agile team decided against a guideline.

17.6 Conclusion, Limitations, and Future Work

Architecture principles constitute powerful simple-rule-like instruments of EAM initiatives to drive large-scale agile transformations by providing guidance for necessary coordination efforts and restricting design freedom in a purposeful manner [13–16]. Despite their raison d'être to guide large-scale transformations, extant studies on architecture principles still lack insights into how principles can be established in large-scale agile development without encountering the resistance of agile teams and how this process can be supported. This chapter makes a contribution to the above-mentioned research gap by proposing a prototypical web application called Architecture Belt that supports the establishment of architecture principles in large-scale agile development.

The benefits of Architecture Belt are threefold: First, it motivates agile teams in the formulation and usage of architecture principles and guidelines by the use of gamification and social design principles. By increasing their intrinsic motivation, the acceptance issues of agile teams towards principles and guidelines can be improved. Second, Architecture Belt provides a central place for managing and documenting principles and guidelines as well as related information. As a consequence, agile teams need less effort and time for identifying existing guidelines and relevant contact persons. Third, Architecture Belt leads to better compliance with principles and guidelines, as it encourages agile teams to contribute to this governance effort, e.g., providing feedback on applied principles or adding new guidelines.

This study does not come without its limitations. First of all, from a research methodology perspective, the presented web application was used and evaluated by a limited number of agile teams and enterprise architects of the case company. Second, the current version of the prototypical implementation supports only the manual verification of the fulfillment status of guidelines. Accordingly, in future work, we will deploy and evaluate Architecture Belt in other companies that are pursuing large-scale agile endeavors. In addition, we aim to implement an integration service that collects data from various sources which are relevant to the guideline fulfillment such as GitHub,[4] Xray[5] or SonarQube.[6] We will also extend it with a service that performs automated tests based on the collected data to determine the level of compliance with guidelines. Based on these two services, Architecture Belt can be integrated into existing continuous delivery pipelines so that the application can act as a quality gate and check the fulfillment status of certain guidelines before a deployment to production is possible. Hitherto, Architecture Belt's main focus was on architecture principles, however, it can also be used in other domains, e.g., for data security, data protection or process design principles.

References

1. Kettunen, P.: Extending software project agility with new product development enterprise agility. Softw. Process Improv. Pract. **12**, 541–548 (2007)
2. Sambamurthy, V., Bharadwaj, A., Grover, V.: Shaping agility through digital options: reconceptualizing the role of information technology in contemporary firms. MIS Q. **27**, 237–263 (2003)
3. Overby, E., Bharadwaj, A., Sambamurthy, V.: Enterprise agility and the enabling role of information technology. Eur. J. Inf. Syst. **15**, 120–131 (2006)
4. Sherehiy, B., Karwowski, W., Layer, J.: A review of enterprise agility: concepts, frameworks, and attributes. Int. J. Ind. Ergon. **37**, 445–460 (2007)
5. Lee, G., Xia, W.: Toward agile: an integrated analysis of quantitative and qualitative field data on software development agility. Mis Q. **34** (2010)
6. Bharadwaj, A., El Sawy, O.A., Pavlou, P.A., Venkatraman, N.: Digital business strategy: toward a next generation of insights. MIS Q. 471–482 (2013)
7. Dingsøyr, T., Falessi, D., Power, K.: Agile development at scale: the next frontier. IEEE Softw. **36**, 30–38 (2019). https://doi.org/10.1109/MS.2018.2884884
8. Dikert, K., Paasivaara, M., Lassenius, C.: Challenges and success factors for large-scale agile transformations: a systematic literature review. J. Syst. Softw. **119**, 87–108 (2016)
9. Dingsøyr, T., Moe, N., Fægri, T., Seim, E.: Exploring software development at the very large-scale: a revelatory case study and research agenda for agile method adaptation. Empir. Softw. Eng. **23**, 490–520 (2018)

[4]https://github.com/.

[5]https://www.getxray.app/.

[6]https://www.sonarqube.org/.

10. Dybå, T., Dingsøyr, T.: What do we know about agile software development? IEEE Softw. **26**, 6–9 (2009)
11. Paasivaara M (2017) Adopting SAFe to scale agile in a globally distributed organization. In: Global Software Engineering (ICGSE), 2017 IEEE 12th International Conference on. pp 36–40
12. Ovaska, P., Rossi, M., Marttiin, P.: Architecture as a coordination tool in multi-site software development. Softw Process Improv Pract **8**, 233–247 (2003)
13. Lange, M., Mendling, J., Recker, J.C.: Realizing benefits from enterprise architecture: a measurement model. In: Proceedings of the 20th European Conference on Information Systems (ECIS) (2012)
14. Greefhorst D, Proper E (2011) Architecture Principles: The Cornerstones of Enterprise Architecture. Springer
15. Korhonen, J.J., Halén, M.: Enterprise architecture for digital transformation. In: 2017 IEEE 19th Conference on Business Informatics (CBI), pp. 349–358 (2017)
16. Hoogervorst, J.A.P.: Enterprise Governance and Enterprise Engineering. Springer Science & Business Media (2009)
17. Greefhorst, D., Proper, E.: A practical approach to the formulation and use of architecture principles. In: 2011 IEEE 15th International Enterprise Distributed Object Computing Conference Workshops. IEEE, pp. 330–339 (2011)
18. Uludağ, Ö., Nägele, S., Hauder, M., Matthes, F.: Establishing architecture guidelines in large-scale agile development through institutional pressures: a single-case study. In: 25th Americas Conference on Information Systems (2019)
19. Beck K, Beedle M, Van Bennekum A, et al (2001) Manifesto for Agile Software Development
20. Augustine, S.: Managing Agile Projects. Prentice Hall, Upper Saddle River (2005)
21. Babar, M.A.: An exploratory study of architectural practices and challenges in using agile software development approaches. In: 2009 Joint Working IEEE/IFIP Conference on Software Architecture European Conference on Software Architecture, pp. 81–90 (2009)
22. Leffingwell, D., Martens, R., Zamora, M.: Principles of Agile Architecture. Leffingwell, LLC Rally Softw Dev Corp. (2008)
23. Mocker, M.: What is complex about 273 applications? Untangling application architecture complexity in a case of European investment banking. In: 2009 42nd Hawaii International Conference on System Sciences, pp. 1–14 (2009)
24. Nord, R.L., Ozkaya, I., Kruchten, P.: Agile in distress: architecture to the rescue. In: International Conference on Agile Software Development, pp 43–57 (2014)
25. Merriam Webster Online: Principle (2019). https://www.merriamwebster.com/dictionary/pri nciple
26. van Bommel, P., Buitenhuis, P.M., Proper, S.J.B.A., Hoppenbrouwers, E.H.A.: Architecture principles—a regulative perspective on enterprise architecture. In: Enterprise modelling and information systems architectures: concepts and applications. In: Proceedings of the 2nd International Workshop on Enterprise Modelling and Information Systems Architectures, St. Goar, Germany, 8–9 Oct (2007)
27. Haki MK, Legner C (2013) Enterprise Architecture Principles in Research and Practice: Insights from an Exploratory Analysis. In: Proceedings of the 21st European Conference on Information Systems ECIS
28. Op't Land, M., Proper, E.: Impact of principles on enterprise engineering. In: ECIS, pp. 1965–1976 (2007)
29. Stelzer, D.: Enterprise architecture principles: literature review and research directions. In: Proceedings of the 2009 International Conference on Service-Oriented Computing (2010)
30. Winter R (2016) Establishing "Architectural Thinking" in Organizations. In: Horkoff J, Jeusfeld MA, Persson A (eds) The Practice of Enterprise Modeling. Springer, pp 3–8
31. Kulak, D., Li, H.: The Journey to Enterprise Agility: Systems Thinking and Organizational Legacy. Springer International Publishing, Cham, Switzerland (2017)
32. Uludağ, Ö., Reiter, N., Matthes, F.: What to expect from enterprise architects in large-scale agile development? A multiple-case study. In: 25th Americas Conference on Information Systems (2019)

33. Uludağ, Ö., Reiter, N., Matthes, F.: Improving the collaboration between enterprise architects and agile teams: a multiple-case study. In: Zimmermann, A, Schmidt, R, Lakhmi, J. (eds.) Architecting the Digital Transformation. Springer (2020)
34. Ambler, S.W.: Scaling agile software development through lean governance. In: 2009 ICSE Workshop on Software Development Governance, pp. 1–2 (2009)
35. Paasivaara, M., Lassenius, C.: Communities of practice in a large distributed agile software development organization—case Ericsson. Inf. Softw. Technol. **56**, 1556–1577 (2014)
36. Uludağ, Ö., Hauder, M., Kleehaus, M., et al.: Supporting large-scale agile development with domain-driven design. In: International Conference on Agile Software Development, pp. 232–247 (2018)
37. Bente, S., Bombosch, U., Langade, S.: Collaborative Enterprise Architecture (2012)
38. Resnick, P., Kraut, R.E.: Building Successful Online Communities: Evidence-Based Social Design (2009)
39. Brosius, M., Aier, S., Haki, M.K., Winter, R.: The institutional logic of harmonization: local versus global perspectives. In: Aveiro, D., Guizzardi, G., Guerreiro, S., Guédria, W. (eds.) Advances in Enterprise Engineering XII, pp. 3–17. Springer International Publishing, Cham (2019)
40. Canat, M., Català, N.P., Jourkovski, A., et al.: Enterprise architecture and agile development—friends or foes? In: IEEE 22nd International Enterprise Distributed Object Computing Workshop (2018)
41. Drews, P., Schirmer, I., Horlach, B., Tekaat, C.: Bimodal enterprise architecture management: the emergence of a new EAM function for a BizDevOps-based fast IT. In: IEEE 21st International Enterprise Distributed Object Computing Workshop, pp. 57–64 (2017)

Part V
Digital Applications

Chapter 18
Improving the Collaboration Between Enterprise Architects and Agile Teams: A Multiple-Case Study

Ömer Uludağ, Niklas Reiter, and Florian Matthes

Abstract In an increasingly digital world, traditional enterprises are confronted with rapidly changing customer demands, increasing market dynamics, and continuous emergence of new technological advancements. Confronted with the imperatives of a digital world, companies are striving to adopt agile methods on a larger scale to meet these requirements. In recent years, enterprise architecture management has established itself as a valuable governance mechanism for coordinating large-scale agile transformations by connecting strategic considerations to the execution of transformation projects. However, the collaboration between enterprise architects and agile teams is not always frictionless as both exhibit antithetic mindsets. Against this backdrop, we present a multiple-case study of five leading German companies that aims to shed light on this field of tension. Based on our results from 68 semi-structured interviews with 25 industry experts, we present a set of tactics for improving the collaboration between enterprise architects and agile teams.

Keywords Enterprise architecture management · Large-scale agile transformation · Multiple-case study

18.1 Introduction

In increasingly hypercompetitive environments, organizations face a multitude of challenges such as volatile customer demands, rapid technological advancements, and tensing regulatory uncertainties [1, 2]. As a consequence, organizations are urged to undergo organizational transformations to respond readily to environmental

Ö. Uludağ (✉) · N. Reiter · F. Matthes
Department of Informatics, Technical University of Munich, Munich, Germany
e-mail: oemer.uludag@tum.de

N. Reiter
e-mail: niklas.reiter@tum.de

F. Matthes
e-mail: matthes@tum.de

A. Zimmermann et al. (eds.), *Architecting the Digital Transformation*, Intelligent Systems Reference Library 188,
https://doi.org/10.1007/978-3-030-49640-1_18

changes [3–5]. To master their respective digital transformation, organizations are extensively adopting agile practices which often necessitate large-scale agile transformations [1, 6, 7]. However, the process of large-scale agile transformations entails key managerial challenges [1] such as the enablement of effective knowledge networks [8], the change from a life-cycle model towards iterative and feature centric models [9], and the effective coordination and alignment of multiple large-scale agile activities [10].

In recent years, enterprise architecture management (EAM) has established itself as a valuable governance mechanism for coordinating enterprise transformations by connecting strategic considerations to the execution of large-scale agile projects and fostering the mutual alignment of business and IT [11–13]. Within this context, key responsibilities of enterprise architects include (1) maintaining a holistic vision of development initiatives, (2) harmonizing governance requirements across large-scale agile projects, and (3) guiding agile teams through business and technical roadmaps [14].

However, the collaboration between enterprise architects and agile teams is not always frictionless as both exhibit antithetic ways of thinking and mindsets [15]. First and foremost, enterprise architects take a top-down perspective focusing on long-term goals and strategies, thus, conflicting with agile teams' short-term ambitions to satisfy business representatives [16]. Agile teams' pressure to deliver frequently business value may lead to the negligence of long-term architectural improvements [16]. Concomitant with this tension, advocates of agile approaches fear architecture as a remnant of some ugly past, decry it as "*Big Design Upfront*" [17, 18], whereas architects argue that "*[refactoring] is not a replacement for sound upfront design; if an architecture is decent you can improve it, but re-factored junk is still junk*" [19]. Moreover, the "*command and control*" culture of enterprise architects opposes the "*servant leadership*" culture of agile teams [15]. Hence, agile teams prophesy that the era of "*PowerPoint architects*" has come to an end and expect enterprise architects "*to keep themselves connected to the code being created*" [15].

Although this field of tension is a prevalent phenomenon in a multitude of traditional organizations, only a few pertinent articles on this topic are available (cf. [11, 20–22]) and a call for empirical research on this topic has been issued (cf. [21, 22]). Therefore, we aim to respond to the call for research and discuss the collaboration between enterprise architects and agile teams in large-scale agile transformations. To achieve this, we conducted a multiple-case study with five leading German enterprises from different industries by capturing the interplay between both disciplines, identifying related challenges and observing a set of good practices. Against this backdrop, we formulate the following research questions:

RQ 1 *Which typical challenges do enterprise architects face when collaborating with agile teams?*

RQ 2 *What tactics exist to address the related challenges of enterprise architects?*

The remainder of this chapter is structured as follows. Section 18.2 begins with a discussion of relevant literature on the interplay between large-scale agile development and EAM. Section 18.3 portrays our qualitative study approach, the selection as well as a compilation of our sample and our data analysis approach. Section 18.4 presents the results of the case studies. In Sect. 18.5, we discuss the main findings our study before concluding our study with a summary of our results and remarks on future work in Sect. 18.6.

18.2 Background and Related Work

Agile development practices such as eXtreme Programming (XP) [23], Scrum [24], and Crystal Clear [25] focus on delivering working software systems in small iterations and increments [26] where architectural design issues are neglected [27]. For example, Scrum does not place any emphasis on architecture-related practices. In Scrum, the architecture of one-project application can always be redesigned and repackaged to achieve a higher degree of reuse (emergent design) [11]. In addition, local project-oriented agile practices often lose sight of overarching goals and enterprise-wide objectives as well as overlook the need for holistic enterprise architecture (EA) [20, 28]. The practice of emergent design and the lack of EA in large-scale agile environments may lead to many additional challenges such as technical debts, architectural divergence, and functional redundancy increasing the complexity of the overall EA [28–30].

Based on [31], we define EA as the fundamental organization of an enterprise, embodied in its components, their relationships to each other and to the environment, and the principles governing its design and evolution. We are consonant with [18, 28] and argue that EA is a strategic capability enabling agility in large-scale software development endeavors. A sustainable EA is critical for developing and realizing enterprise strategy and roadmap, and responding to always changing business landscape [32]. This claim is affirmed by [33] who argue that organizations traverse four stages of EA, namely *Business Silos*, *Standardized Technology*, *Optimized Core*, and *Business Modularity* to benefit from greater organizational agility. As firms migrate through the EA stages, they shift from a focus on local optimization to global optimization with important implications for global flexibility (see Fig. 18.1). Within this process, companies limit local flexibility and enforce standardization which later enables a much higher organizational flexibility. This phenomenon is also called "*Agility Paradox*".

Our findings are consonant with those of [11, 20, 28, 34] and argue that extant literature still lacks insights into the link between EA and large-scale agile development. Nevertheless, there are some important contributions that need to be mentioned here.

A structured literature review by [14] revealed that some scaling agile frameworks, which were proposed by practitioners to resolve issues associated with team size, customer involvement, and project constraints [35], highlight the importance of EA in large-scale agile development. These frameworks include the Disciplined Agile Delivery [36], Enterprise Agile Delivery and Agile Governance [37], Scaled Agile Framework [38], and Gill Framework [39]. A status-quo of agile practices applied

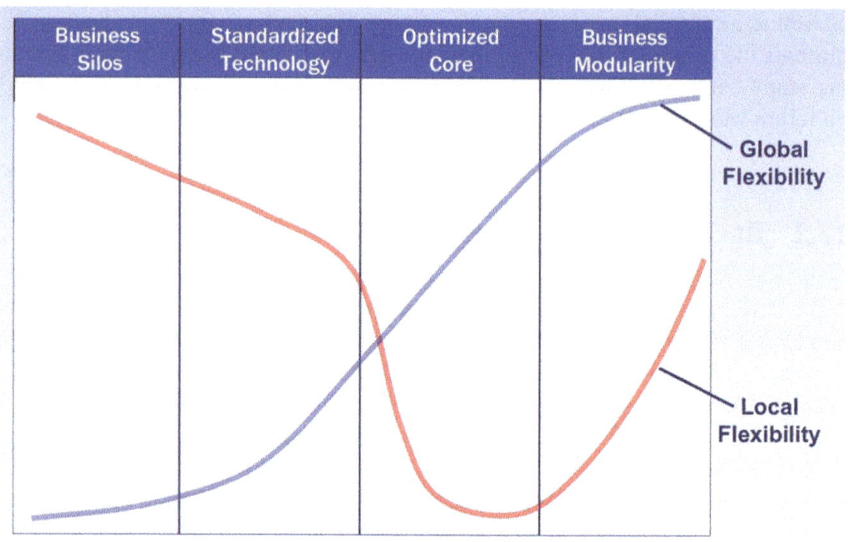

Fig. 18.1 Changes in organizational flexibility through the EA stages [33]

to EAM is presented by [11]. The authors elaborate a design of an organization specific agile EAM practice based on theoretical concepts in EA frameworks which are extended by agile principles. In addition, the authors conducted an online survey with 105 industry experts to identify frequently applied agile practices for EAM [11]. This revealed that most EAM initiatives apply an iterative and incremental approach, near all organizations apply EAM in a self-organized manner, and that their EAM initiatives act cross-functionally. Based on ten expert interviews with three German companies [20], describe how agile methods can be used to create EA deliverables and how enterprise architects can collaborate with agile teams. The approaches, i.e., TOGAF [40] and Scrum [41], which were analyzed by [20] put a strong focus on EAM or agile software development, which is then extended by aspects of the other. However [20], claim that these approaches lack specific details on how the interaction between EAM and agile teams should work in practice. An integrated adaptive EA driven approach for large agile environments is proposed by [28], which aims to guide and align the agile project level architecture and design with the holistic EA. The author of [28] argues that the solution architecture represents an integration point between EA and agile development and can be used to guide the evolving architecture and emergent design in agile projects. A single-case study from a global insurance company describes how domain-driven design can be used to support large-scale agile development [42]. The findings of [42] show that agile teams need some kind of architectural guidance from enterprise architects who have a holistic overview of the teams and applications they develop. In addition, the observed stakeholders within the large-scale agile endeavor appreciated that enterprise architects not only coach the teams on new methods but also support them in their application and exploitation. Based on twelve expert interviews with four companies, [34] revealed that the surveyed organizations do not regard agile and architecture as opposites, which confirms the statement of [17] that agile and architecture can coexist and complement each other well. In addition [34], claim that extensive use of modeling and EA at the team level is considered as over-architecting as the lower levels are more dynamic and change frequently. Another common issue mentioned by [34] is that architects, agile teams, and product owners have communication gaps and that architects have no difficulties when working with agile teams, but agile teams tend to disagree with the working methods of architects.

18.3 Research Methodology

Since our research is motivated by a practical problem and we aim to study a nascent field such as large-scale agile software development, we apply case study research as it provides an in-depth overview of real-life situations and uncovers the reasons for the occurrence of certain contemporary phenomena [43]. In the following, we outline the design of this multiple-case study according to the guidelines of [44].

Case study design: The goal of this study is to explore the challenges faced by enterprise architects collaborating with agile teams and a set of related tactics to address them. Our case study is therefore mainly exploratory but also bears a descriptive or explanatory character [44]. We investigated five organizations and four units of analysis that allows for cross-case analysis [45]. The four units of analysis include (1) the formulation and application of architecture principles, (2) the decision-making process of architecture-related topics in EA boards, (3) the new role of the EAM in assisting agile teams, and (4) the value contribution of EAM in supporting large-scale agile development endeavors. The cases were purposefully selected because the companies under investigation are currently performing large-scale agile transformations and their well-established EAM functions face unprecedented challenges within this context. We selected cases from different industries to avoid potential industry bias. An overview of the case organizations and interviews is shown in Table 18.1.

Data collection: We focused on first and third data collection techniques in line with [46]. First-degree data collection techniques were applied through semi-structured interviews and workshops. A total of 68 individual and groups interviews were conducted. In almost all companies, at least one senior executive, one enterprise architect, and one agile team member were interviewed to gain a diverse perspective on this field of tension and further triangulate our results. The interviews followed a semi-structured structure and were conversational to allow interviewees to explore their experiences and views in detail [45]. Each interview was primarily conducted in face-to-face meetings and at least two researchers were present in the interviews to facilitate observer triangulation [44]. The case organizations were also invited to our focus group that met four times between September 2017 and February 2019 [47]. Extensive discussions within this workshop series deepened the existing knowledge base and validated the interview results. We supplemented our findings with internal

Table 18.1 Overview of case organizations and conducted interviews

Industry and code name of case organization	Head-quarter location	Company size [employees]	No. of inter-views	Position of interviewees
Global insurance company ("GlobalInsureCo")	Germany	140,000+	12	Agile developer; enterprise architect; chapter lead agile coaching
Car manufacturer ("CarCo")	Germany	130,000+	16	Chief technology officer; enterprise architect; group lead IT; product owner; requirements engineer; scrum master
Information technology company ("ITCo")	Germany	7000+	8	Enterprise architect; product owner
Retail company ("RetailCo")	Germany	50,000+	16	Chapter lead business process architecture; chief scrum master; enterprise architect; product owner; scrum master
Public sector insurance company ("PublicInsureCo")	Germany	6700+	16	Agile developer; enterprise architect; head of IT governance department; head of EAM

documents from the case companies representing a third-degree data collection technique. The purposeful selection of different data sources and roles from different case organizations enabled the triangulation of data sources [48].

Data analysis: We used open coding for analyzing documents, workshop protocols, and transcribed interviews as suggested by [49]. Preliminary codes were merged and checked for consistency and completeness by another researcher. Subsequently, groups of code phrases were consolidated into concepts that were later referred to the formulated research questions.

18.4 Results

18.4.1 *Case Descriptions*

Table 18.2 gives an overview of the investigated organizations and their ongoing large-scale agile transformations.

18.4.2 *Establishing Architecture Principles*

As architecture principles constitute a powerful simple-rule-like instrument of enterprise architects to effectively guide and steer large-scale transformations and agile teams [13, 50], we were curious about the challenges the organizations perceived in introducing or implementing principles of which the most important were:

- **CA-1**: Low commitment of agile teams regards implementing architecture principles due to low intrinsic motivation, acceptance issues or conflicting goals to deliver business value frequently.
- **CA-2**: Lack of appropriate tools and (automated) mechanisms for verifying and controlling the compliance of agile teams with architecture principles.
- **CA-3**: Lack of knowledge and understanding why principles are important and why they should be considered by agile teams.
- **CA-4**: Poor communication of existing and new architecture principles towards agile teams.

18.4.3 *Enterprise Architecture Boards*

According to [51], a key element in a successful EA governance strategy is to build a cross-organizational architecture board that oversees the implementation of the EA strategy. Thus, we aimed to explore within this unit of analysis which typical challenges arise when EA governance boards are scheduled that aim to coordinate large-scale agile transformations of which the most important were:

- **CB-1**: EA governance bodies tend to decide over the heads of agile teams without knowing their concerns.
- **CB-2**: Lack of involvement of agile teams in relevant architecture processes and discussions.
- **CB-3**: Rigid and tedious EA governance procedures slow down the development speed of agile teams.

Table 18.2 Overview of the large-scale agile transformations of case organizations

	GlobalInsureCo	CarCo	ITCo	RetailCo	PublicInsureCo
Transformation begin	Beginning of 2016	Beginning of 2017	Beginning of 2015	End of 2017	Mid of 2016
Reasons for the transformation	Effectiveness, quality, and efficiency Time-to-market Modern working environments for employees Customer-centricity	Development speed End-to-end responsibilities	Finding young and motivated employees Short response times	Time-to-market Flexibility to changing conditions Customer experience Collaboration between business & IT	Changes in business model Time-to-market Customer satisfaction Flexibility in software development
Applied scaling frameworks	LeSS Scaled Scrum Spotify Model	LeSS SAFe Scaled Scrum	LeSS SAFe Scaled Scrum	SAFe Scaled Scrum Spotify Model	eScrum Nexus SAFe Scaled Scrum
Transformation approach	Factory approach with dedicated co-locations	Top-down driven by the CIO and senior executives Line of business piloting	Steered by a central transformation team	Piloting with large projects Line of business piloting Driven by management board	Piloting with large projects Line of business piloting
Transformation scale	Enterprise-wide transformation	Enterprise-wide transformation	Enterprise-wide transformation	Enterprise-wide transformation	Enterprise-wide transformation

18.4.4 Role of Enterprise Architects

As the role of enterprise architects in large-scale agile development is demanding due to the reason that it requires continuous coordination and negotiation between many stakeholders [14, 16, 52], we asked the interviewees the challenges enterprise architects face when supporting large-scale agile endeavors. We identified the following important challenges:

- **CC-1:** Lack of capacity and high workload of enterprise architects' make it difficult for them to deliver their services on time and in appropriate quality.
- **CC-2:** Lack of time availability of enterprise architects does not allow them to adequately accompany and support agile teams.
- **CC-3:** Enterprise architects cannot cope with the speed of the decision-making process of agile teams.
- **CC-4:** Enterprise architects encounter acceptance problems and are perceived as a disturbance by agile teams.

18.4.5 Value Contribution of Enterprise Architects

Previous studies revealed that the added value of enterprise architects has not yet reached the team level, thus, encountering the resistance of agile teams [15, 34, 52]. Thus, we were curious about the main reasons for this phenomenon of which the most important were:

- **CD-1**: Lack of technical-know of enterprise architects make architecture artifacts provided by enterprise architects less relevant to agile teams.
- **CD-2:** Lack of regular feedback mechanisms between enterprise architects and agile teams.
- **CD-3**: Communication between agile teams and enterprise architects is facilitated via third persons leading to loss of information, dissatisfactions by agile teams, and their unwillingness to communicate with enterprise architects.

18.4.6 Tactics to Improve the Collaboration

In the following, we will present five tactics that aim to address the challenges faced by enterprise architects when collaborating with agile teams. Table 1.3 provides an overview of the identified challenges and tactics to address them.

Tactic #1—Defining architecture principles as Quality Gate policies

This tactic addresses the challenge of lack of appropriate tools and (automated) mechanisms for verifying and controlling the compliance of agile teams with architecture principles (CA-2) by proposing an automated process which is integrated into the

Table 18.3 Overview of which challenges can be addressed by which tactics

Tactics	Challenges														
	CA-1	CA-2	CA-3	CA-4	CB-1	CB-2	CB-3	CC-1	CC-2	CC-3	CC-4	CD-1	CD-2	CD-3	CD-4
Tactic #1—Defining architecture principles as quality gate policies	–	✓	–	–	–	–	–	–	–	–	–	–	–	–	–
Tactic #2—Empowered communities of practices for architecture	–	–	–	–	✓	✓	–	–	–	–	–	–	–	–	–
Tactic #3—Agile architecture decision-making model	–	–	–	–	–	–	✓	–	–	–	–	–	–	–	–
Tactic #4—Applying ScrumBan for optimizing workflow and improving delivered services	–	–	–	–	–	–	–	✓	–	✓	–	–	–	–	–
Tactic #5—Conducting architecture spikes for assisting agile teams	–	–	–	–	–	–	–	–	–	–	–	✓	–	–	–

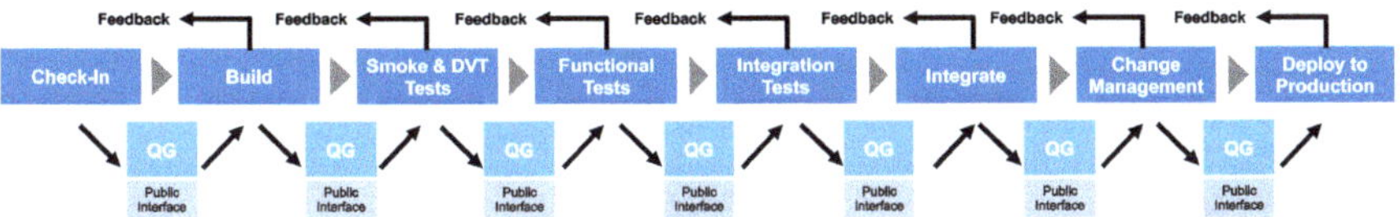

Fig. 18.2 Automated process for ensuring compliance with architecture principles by agile teams

Continuous Integration and Continuous Delivery (CI/CD) pipeline of agile teams. This tactic was by applied by RetailCo to enforce quality and architecture policies, i.e., principles and guidelines, based on the Quality Gate (QG) concept. QGs represent milestones during the software delivery process where special quality requirements are checked [53] (see Fig. 18.2). At each QG, the application to be deployed is evaluated against predefined quality- and architecture-related criteria. Based on the fulfillment of these criteria, gatekeepers, e.g., enterprise architects, make a decision about whether an application may proceed or not. The CI/CD pipeline checks architectural conformity of all future application and gives feedback to agile teams in case of violating predefined rules. For each QG, a public API interface is provided that allows direct injections of rules and tests so that the conformance of agile teams regards architecture can be automated.

Tactic #2—Empowered Communities of Practices for Architecture

This tactic addresses the challenge of EA governance bodies decide over the heads of agile teams without knowing their concerns (CB-1) and lack of involvement of agile teams in relevant architecture processes and discussions (CB-2) by proposing an empowered Community of Practice for Architecture (CoPA) consisting of enterprise architect and agile team representatives. In all companies, CoPAs were regularly organized by enterprise architects to facilitate the exchange of architecture-related topics across the organization. Usually, CoPAs are open to all interested parties and do not force participation. At RetailCo, the participation was mandatory for all architects as the CoPA served not only for the exchange of information but also had some decision-making power for coordinating architecture topics as well as for making architecture decisions. In nearly all interviews, the acceptance of agile teams and enterprise architects regards CoPAs was very high. Agile team members specifically stated that CoPAs offer them a platform to actively participate in architectural decisions, decide on the selection of new tools or contribute to the creation of architecture principles. RetailCo and GlobalInsureCo defined a collaborative decision-making process for their CoPAs (see Fig. 18.3) that makes architectural decisions based on a basic democratic consensus decision, i.e., the participating members vote and decide on decisions based on a simple majority.

Aside from discussing architecture-related topics, participants regularly propose new tools or architecture principles which they would like to use in their daily work but require the approval of an EA governance body as this decision might also affect other teams, and thus have to be included in the Book of Standards (BoS). During a CoPA meeting, one participant, e.g., agile developer, presents the problem of the

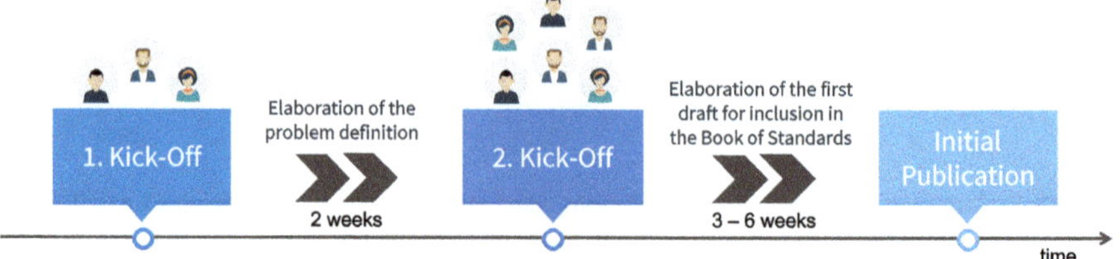

Fig. 18.3 Collaborative decision-making process for architecture-related topics within CoPAs

topic (first kick-off), i.e., he/she illustrates why the company should define MongoDB as a new standard, what the advantages of MongoDB are compared to traditional relational databases, and what disadvantages can occur after its introduction. After the presentation, a small group of experts is formed that elaborates the problem definition until the next CoPA meeting. Within this period, the expert group collaboratively elaborates solutions regarding the identified and presents the prepared information at the next CoPA meeting (second kick-off). Based on the presentation, the CoPA members vote whether the proposed solution should be included in the BoS, rejected or reworked. If the former, the expert group gets additional three to six weeks to elaborate a first draft for including the solution in the BoS (initial publication).

Tactic #3—Agile Architecture Decision-Making Model

This tactic addresses the challenge of rigid and tedious EA governance procedures that slow down the development pace of agile teams (CB-3) by proposing a decision-making model for classifying architecture decisions based on their importance and impact. We found out that case organizations categorized decisions based on different criteria such as urgency, financial or organizational scope, and strategic importance, and introduced several EA governance bodies empowered with different levels of authority to increase the decision-making speed (see Fig. 18.4).

Architecture decisions that affect only a single agile team and have a low strategic importance for the organization, e.g., software architecture, tool stack or external libraries, agile teams are allowed to make the decision themselves without having to include further EA governance bodies (self-governance). As soon as other agile teams are affected by this decision, an agile team has to escalate it to a Center of Excellence for Architecture (CoEA) representing an expert group for controlling development activities across the organization and consisting of different types of architects, e.g., domain, enterprise or system architects, and some managing roles of the IT department. This expert group evaluates the strategic importance of the incoming architectural decision and determines the most suitable persons in the company to make this decision, e.g., the CoEA itself, a CoPA or an agile team (expert group governance). The idea for this escalation mechanism is to mitigate the risk of empowering CoPAs or agile teams with people who (1) do not have a view of the global-optimum of the organization, (2) block decisions in order not to deal with new issues or (3) do not have to bear the consequences of bad decisions. An empowered CoPA consisting of

Fig. 18.4 Agile EA governance quadrant

enterprise architect and agile team representatives decides on architectural decision which have no strategic importance and are important for the common collaboration of the agile teams working on the same large-scale agile development program, e.g., recommended architecture principles, dependencies or interface design of the sub-products that need to communicate with each other (community governance). Based on a basic democratic consensus decision, the CoPA makes the architectural decision (see Tactic #2). Our proposed agile architecture-decision making model highly relates to the Enterprise Agile Delivery and Agile Governance framework proposed by [37] that aims to protect agile teams from the rigidness of traditional EA governance models by proposing a so called "*Agile Governance Buffer Zone*".

Tactic #4—Applying ScrumBan for optimizing workflow and improving delivered services

This tactic addresses the challenges of enterprise architects to deliver their services on time and in appropriate quality (CC-1) and to cope with the speed of the decision-making process of agile teams (CC-3) by applying ScrumBan as an agile management methodology to deliver services just-in-time and to easily adapt to quickly changing stakeholder needs. This tactic was applied by the EA team of PublicInsureCo that plans and controls the topics and tasks by using practices of Kanban and Scrum without being a real Scrum team, i.e., developing software. The EA team works in small iterations, i.e., two-weekly Sprints, to ensure that the EA team can easily adapt to changing environments. The EA team follows the basic practices of Scrum, i.e., participating in common Scrum ceremonies such as Backlog Refinement, Sprint Planning, Daily Scrum, Sprint Review, and Sprint Retrospective, exercising Scrum roles such as the Product Owner or Scrum Master (taken over by the head of EAM), and using Scrum artifacts such as Product and Sprint Backlog as well as a Sprint Burn-Down Chart. Within the Backlog Refinement, the EA team prioritizes incoming requests based on their strategic importance. At the Sprint Planning, the tasks to be performed are distributed to the members of the EA team in a self-organizing way. Daily Scrums are organized by the EA team to gain an understanding of what work has been done and what work remains. At Sprint Review meetings, the EA team requests feedback from its "customers" and performs inspect and adapt events, i.e., Sprint Review, to improve its work and delivered services. The EA team complements Scrum by using a Kanban board to visualize work and limit work-in-progress so that the EA team can plan and execute its work. Thereby, the EA team uses a team collaboration tool to manage its backlog, to track tasks, and to visualize their progress using a digital ScrumBan board (see Fig. 18.5). At the Backlog Refinement, the EA team prioritizes incoming requests by their stakeholders as user stories and refines them by creating sub tasks for each user story ("*To Do*"; first column of the ScrumBan board). At the Sprint Planning, the user stories and sub tasks are assigned to members of the EA team and the team members decide which user stories and tasks should

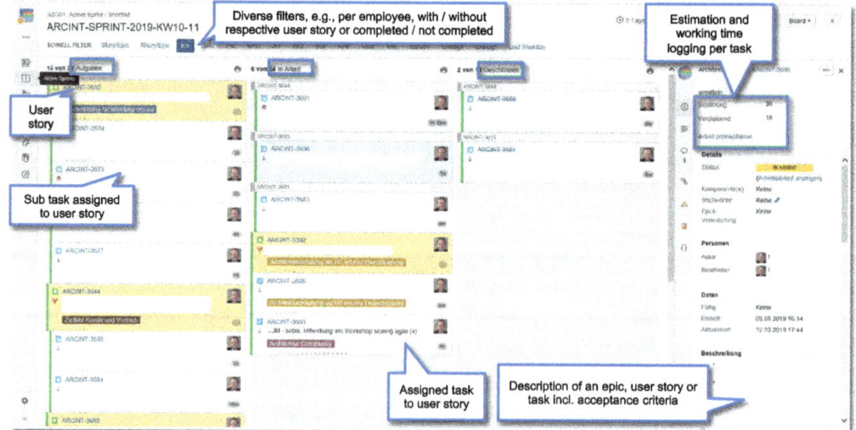

Fig. 18.5 ScrumBan board visually representing the progress of the EA team

be performed in the next Sprint ("*In Progress*"; second column of the ScrumBan board). At the Sprint Review, the EA team members present their performed tasks and user stories, e.g., delivered service or created EA deliverable, and change their status to "*Completed*" (see third column of the ScrumBan board).

The EA team uses also a digital Burn-Down Chart provided by the team collaboration tool to track the actual and estimated amount of work to be done in the next Sprint (see Fig. 18.6). The horizontal of the Burn-Down Chart in Fig. 18.6 indicates the time and the vertical y-axis indicates the pending work. By tracking the remaining work throughout the Sprint, the EA team can manage its progress and respond accordingly. The green line (burn-up line) indicates the time spent for tasks in the Sprint whereas the red line (burn-down line) indicates the remaining time in the Sprint.

In addition, the EA team uses a digital Cumulative Flow Diagram (CFD) that shows the various statuses of work items for the Sprint (see Fig. 18.7). The horizontal x-axis of the CFD indicates the time while the vertical y-axis indicates the tasks to do. Each colored area of the CFD equates to a workflow status defined at the ScrumBan board, i.e., "*To Do*", "*In Progress*" or "*Completed*".

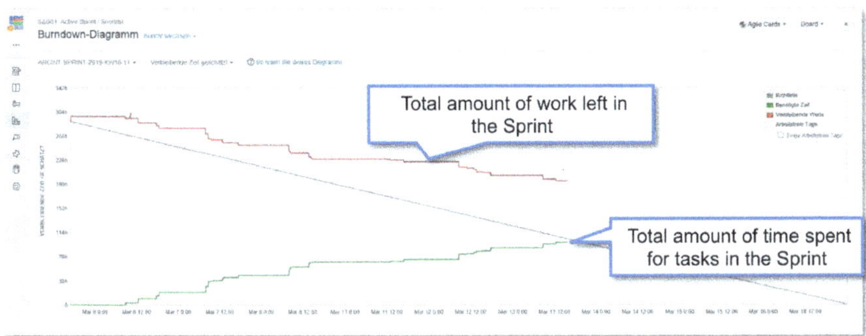

Fig. 18.6 Sprint Burn-Down Chart demonstrating the amount of work that has been completed by enterprise architects in a Sprint and the total work remaining

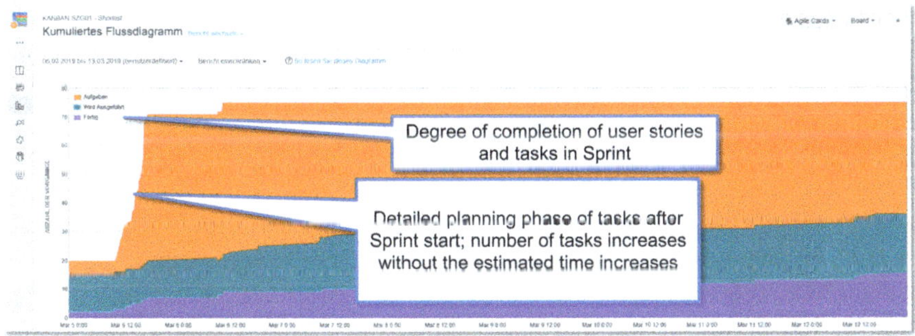

Fig. 18.7 Cumulative Flow Diagram illustrating the degree to which enterprise architects perform their tasks in the Sprint

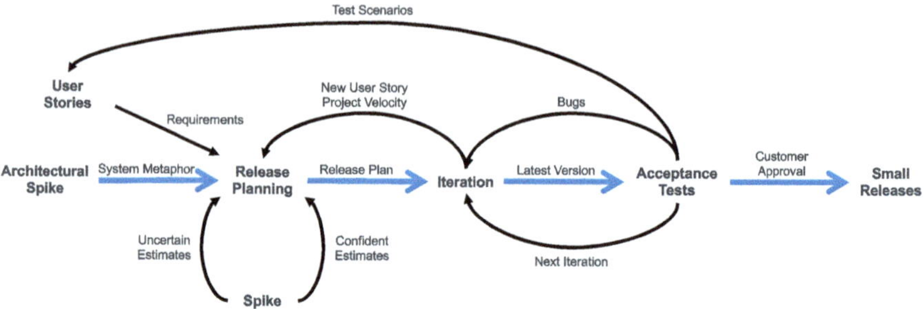

Fig. 18.8 Overview of the XP development process [54]

Tactic #5—Conducting Architecture Spikes for assisting agile teams

This tactic addresses the challenge of providing relevant architecture artifacts to the agile teams (CD-1) by conducting Architecture Spikes for agile teams. A common practice to deal with unknown requirements in user stories, e.g., a new technology, is conducting so-called Architecture Spikes [54]. An architecture Spike is an experiment that aims to reduce the risk of architectural problems and improve the technical understanding of the used frameworks and tools. Although XP recommends that agile teams should conduct Architecture Spikes (see Fig. 18.8), we observed that enterprise architects of CarCo, ITCo, and PublicInsureCo also conducted those to provide agile teams technical support. Thereby, enterprise architects build small minimal viable products and provide them as architecture patterns which are later used by the agile teams to estimate unknown user stories. This XP practice is also sometimes called Sprint 0 as its goal is to set an initial architectural skeleton for agile teams.

18.5 Discussion

As "*agile breaks everything*" [15], adopting agile methods has far-reaching implications on a company's work procedures, structures, and cultures [55, 56]. As a consequence, agile also breaks the EAM function wherefore many practitioners wonder whether it still has its raison d'être in the future or should be abolished. A similar phenomenon was previously experienced by the IT, strikingly evolving from a supporting function to a strategic weapon for driving new business models [57]. Based on our observations, we expect the EAM function to play an even more strategic role for companies in the future, moving from a cost center to a profit center by creating value for the organization. However, these expectations make the role of enterprise architects even more demanding and necessitate a fundamental shift in their traditional ways of working and thinking.

To coordinate large-scale agile transformations, enterprise architects have to closely collaborate with agile teams [21, 52]. To this end, enterprise architects must leave the "*ivory tower*" and attend agile events of the teams, e.g., weekly plannings or demos. On the one hand, agile teams' perception of enterprise architects will improve significantly, which is currently a common issue [52]. On the other hand, enterprise architects can offer advice and guidance based on the real product rather than a document that the teams may or may not follow [15]. We observed that agile teams complain about the lack of technical know-how of enterprise architects. We believe, like [15], that the days of the "*PowerPoint architects*" have come to an end. Although we think that in the near future, not all enterprise architects will regularly do pair programming side by side with agile developers (cf. [58]), we believe that enterprise architects will increasingly conduct architecture spikes [23] to address architectural issues of agile teams. The majority of the interviewed enterprise architects claimed that they were unable to respond to requests from agile teams in time and with the appropriate quality. One observed countermeasure to this issue is that enterprise architects are increasingly using agile and lean practices (cf. [11, 20]), e.g., ScrumBan [59]. We recognized that tedious EA governance procedures slowed down the development speed of agile teams who therefore strive to stay "off the radar". We agree with [37] that organizations must protect agile teams from traditional EA governance boards by empowering agile teams (self-governance) or CoPAs (community governance) to make architectural decisions as long as they are entitled to do so. We believe that the traditional chief architects [60] who have a "*command and control*" mentality will not survive the fundamental shift of the EAM function. Instead, we observed that both enterprise architects and agile teams are increasingly opting for supporting architects [60]. The supporting enterprise architect consults and assists agile teams during the realization of architectural standards, is able to code for short demonstrations, and acts as a facilitator for cross-team architecture exchange (cf. [22]). A drawback of this role is its reduced decision-making power, which is why the supporting architect must convince agile teams argumentatively, otherwise architectural improvements might be neglected by the teams. Last but not least, as architectural work in large-scale agile projects is more continuous [16], enterprise architects are confronted with high frequency of changes in IT systems and dynamic EAs [22]. To proactively support agile teams with appropriate "*up-to-date*" models, the tools of enterprise architects have to be integrated into the development and deployment toolchain of agile teams [22], e.g., CI/CD pipeline, and the documentation of the EA must be automatized as much as possible (cf. [61]).

18.6 Conclusion, Limitations, and Future Research

Confronted with the imperatives of hypercompetitive business environments, firms are urged to undergo large-scale agile transformations to respond to environmental changes [1, 6]. The EAM function provides appropriate governance mechanisms for coordinating and aligning multiple large-scale agile activities so that desirable organization-wide effects and agility can be achieved [10, 13]. In this context, enterprise architects and agile teams need to work closely, which is not always frictionless as both exhibit antithetic ways of thinking and mindsets [15]. Our research is motivated by the prevalence of this phenomenon in a multitude of traditional organizations as well as by the lack of empirical studies. We conducted a case study with five cases from different industries to get an in-depth understanding of this field of tension.

Nonetheless, we acknowledge the limitations of our study. First of all, we identified five cases to be representative of enterprises striving to undergo large-scale agile transformations. These companies may not be fully representative of all industries. Thus, we encourage researchers to perform case studies in other industries, e.g., aerospace and oil industries or state-owned enterprises. Second, we did not investigate young companies being agile "*from day one*" that might have a different collaboration culture between enterprise architects and agile teams. Therefore, we call for research on cross-case analysis to compare the obstacles and practices of enterprise architects collaborating with agile teams in different organizational environments to validate or revise our findings and test our provided set of tactics in other organizational settings. Third, since large-scale agile transformations can be interpreted as episodic, socio-technical change processes [1], we acknowledge that we do not offer a longitudinal study on how the collaboration between enterprise architects and agile teams may change during the transformation process. Thus, a longitudinal perspective might provide important insights how organizations mastered the identified challenges.

References

1. Fuchs, C., Hess, T.: Becoming agile in the digital transformation: the process of a large-scale agile transformation. In: 39th International Conference on Information Systems. San Francisco (2018)
2. Overby, E., Bharadwaj, A., Sambamurthy, V.: Enterprise agility and the enabling role of information technology. Eur. J. Inf. Syst. **15**, 120–131 (2006)
3. Sambamurthy, V., Bharadwaj, A., Grover, V.: Shaping agility through digital options: reconceptualizing the role of information technology in contemporary firms. MIS Q. **27**, 237–263 (2003)
4. Highsmith, J.R.: Agile Project Management: Creating Innovative Products. Pearson Education (2009)
5. Besson, P., Rowe, F.: Strategizing information systems-enabled organizational transformation: a transdisciplinary review and new directions. J. Strateg. Inf. Syst. **21**, 103–124 (2012)
6. Gerster, D., Dremel, C., Kelker, P.: "Agile meets non-agile": implications of adopting agile practices at enterprises. In: 24th Americas Conference on Information Systems
7. Dikert, K., Paasivaara, M., Lassenius, C.: Challenges and success factors for large-scale agile transformations: a systematic literature review. J. Syst. Softw. **119**, 87–108 (2016)
8. Šmite, D., Moe, N.B., Šāblis, A., Wohlin, C.: Software teams and their knowledge networks in large-scale software development. Inf. Softw. Technol. **86**, 71–86 (2017)
9. Nerur, S., Mahapatra, R., Mangalaraj, G.: Challenges of migrating to agile methodologies. Commun. ACM **48**, 72–78 (2005)
10. Ovaska, P., Rossi, M., Marttiin, P.: Architecture as a coordination tool in multi-site software development. Softw. Process Improv. Pract. **8**, 233–247 (2003)
11. Hauder, M., Roth, S., Schulz, C., Matthes, F.: Agile enterprise architecture management: an analysis on the application of agile principles. In: International Symposium on Business Modeling and Software Design (2014)
12. Weill, P., Ross, J.W.: IT Savvy: What Top Executives Must Know to Go from Pain to Gain. Harvard Business Press (2009)

13. Greefhorst, D., Proper, E.: Architecture Principles: The Cornerstones of Enterprise Architecture. Springer, Berlin (2011)
14. Uludağ, Ö., Kleehaus, M., Xu, X., Matthes, F.: Investigating the role of architects in scaling agile frameworks. In: 21st International Conference on Enterprise Distributed Object Computing Conference. pp. 123–132 (2017)
15. Kulak, D., Li, H.: The Journey to Enterprise Agility: Systems Thinking and Organizational Legacy. Springer International Publishing, Cham (2017)
16. Dingsøyr, T., Moe, N., Fægri, T., Seim, E.: Exploring software development at the very large-scale: a revelatory case study and research agenda for agile method adaptation. Empir. Softw. Eng. **23**, 490–520 (2018)
17. Abrahamsson, P., Babar, M.A., Kruchten, P.: Agility and architecture: can they coexist? IEEE Softw. **27** (2010)
18. Nord, R.L., Ozkaya, I., Kruchten, P.: Agile in distress: architecture to the rescue. In: International Conference on Agile Software Development, pp. 43–57 (2014)
19. Meyer, B.: Agile! The Good, the Hype and the Ugly. Springer Science & Business Media (2014)
20. Hanschke, S., Ernsting, J., Kuchen, H.: Integrating agile software development and enterprise architecture management. In: 48th Hawaii International Conference on System Sciences (2015)
21. Canat, M., Català, N.P., Jourkovski, A., et al.: Enterprise architecture and agile development: friends or foes? In: 22nd International Enterprise Distributed Object Computing Workshop (EDOCW), pp. 176–183 (2018)
22. Drews, P., Schirmer, I., Horlach, B., Tekaat, C.: Bimodal enterprise architecture management: the emergence of a new EAM function for a BizDevOps-based Fast IT. In: 21st International Enterprise Distributed Object Computing Workshop, pp. 57–64 (2017)
23. Beck, K.: Extreme Programming Explained: Embrace Change. Addison-Wesley Professional
24. Schwaber, K., Beedle, M.: Agile Software Development with Scrum, 1st edn. Prentice Hall, Upper Saddle River (2001)
25. Cockburn, A.: Crystal Clear: A Human-Powered Methodology for Small Teams. Pearson Education (2004)
26. Augustine, S.: Managing Agile Projects. Prentice Hall, Upper Saddle River (2005)
27. Babar, M.A.: An exploratory study of architectural practices and challenges in using agile software development approaches. In: 3rd European Conference on Software Architecture, pp. 81–90 (2009)
28. Gill, A.Q.: Adaptive enterprise architecture driven agile development. In: International Conference on Information Systems Development, ISD 2015
29. Leffingwell, D., Martens, R., Zamora, M.: Principles of Agile Architecture. Leffingwell, LLC Rally Softw Dev Corp. (2008)
30. Mocker, M.: What is complex about 273 applications? Untangling application architecture complexity in a case of European investment banking. In: 2009 42nd Hawaii International Conference on System Sciences, pp. 1–14 (2009)
31. ISO/IEC/IEEE 42010:2011(E) Systems and Software Engineering—Architecture Description (2011)
32. Doucet, G., Gøtze, J., Saha, P., Bernard, S.A.: Coherency management: Using enterprise architecture for alignment, agility, and assurance. J. Enterp. Archit. 4:
33. Ross, J.W., Weil.l P., Robertson, D.: Enterprise Architecture as Strategy: Creating a Foundation for Business Execution. Harvard Business Press (2006)
34. Canat, M., Català, N.P., Jourkovski, A., et al.: Enterprise architecture and agile development friends or foes? In: 22nd International Enterprise Distributed Object Computing Workshop (2018)
35. Vaidya, A.: Does dad know best, is it better to do less or just be safe? Adapting scaling agile practices into the enterprise. PNSQC ORG 1–18 (2014)
36. Ambler, S., Lines, M.: Scaling Agile Software Development: Disciplined Agility at Scale (2014)

37. Marks, E.: Governing Enterprise Agile Development Without slowing it Down: Achieving Friction-Free Scaled Agile Governance via Event- Driven Governance (2014)
38. Scaled Agile: SAFe®for Lean Enterprises. http://www.scaledagileframework.com/. Accessed 6 Apr 2019
39. Gill, A.Q.: The Gill Framework®V 3.0. http://www.adaptinn.com/the-gill-framework/. Accessed 6 Apr 2019
40. The Open Group: TOGAF Version 9.2 (2019)
41. Schwaber, K., Sutherland, J.: The Scrum Guide (2017)
42. Uludağ, Ö., Hauder, M., Kleehaus, M., et al.: supporting large-scale agile development with domain-driven design. In: International Conference on Agile Software Development. pp. 232–247 (2018)
43. Easterbrook, S., Singer, J., Storey, M.-A., Damian, D.: Selecting empirical methods for software engineering research. In: Shull, F., Singer, J., Sjøberg, D. (eds.) Guide to Advanced Empirical Software Engineering, pp. 285–311. Springer, London (2008)
44. Runeson, P., Höst, M.: Guidelines for conducting and reporting case study research in software engineering. Empir. Softw. Eng. **14**, 131 (2008)
45. Yin, R.: Case Study Research: Design and Methods, 5th edn. Sage Publications, Thousand Oaks (1994)
46. Lethbridge, T.C., Sim, S.E., Singer, J.: Studying software engineers: data collection techniques for software field studies. Empir. Softw. Eng. **10**, 311–341 (2005)
47. Uludağ, Ö.: Scaling Agile Practices Workshops (2019). https://wwwmatthes.in.tum.de/pages/1lihu1sjq8jpk/Scaling-Agile-Practices-Workshops. Accessed 11 Mar 2019
48. Stake, R.: The Art of Case Study Research. Sage Publications, Thousand Oaks (1995)
49. Miles, M., Hubermann, M., Saldana, J.: Qualitative Data Analysis: A Methods Sourcebook. Sage Publications, Thousand Oaks (2014)
50. Korhonen, J.J., Halén, M.: Enterprise architecture for digital transformation. In: 2017 IEEE 19th Conference on Business Informatics (CBI), pp. 349–358 (2017)
51. The Open Group: The TOGAF® Standard, Version 9.2: Architecture Board (2018). http://pubs.opengroup.org/architecture/togaf9-doc/arch/. Accessed 21 Mar 2019
52. Uludağ, Ö., Reiter, N., Matthes, F.: What to expect from enterprise architects in large-scale agile development? A multiple-case study. In: 25th Americas Conference on Information Systems (2019)
53. Hüttermann, M.: Quality and Testing. DevOps for Developers, pp. 51–64. Apress, Berkeley, CA (2012)
54. Marchesi, M., Succi, G., Wells, D., et al.: Extreme Programming Perspectives. Addison-Wesley (2003)
55. Zheng, Y., Venters, W., Cornford, T.: Collective agility, paradox and organizational improvisation: the development of a particle physics grid. Inf. Syst. J. **21**, 303–333 (2011)
56. Bostrom, R.P., Heinen, J.S.: MIS problems and failures: a socio-technical perspective, part II: the application of socio-technical theory. MIS Q 11–28 (1977)
57. Melville, N., Kraemer, K., Gurbaxani, V.: Information technology and organizational performance: an integrative model of IT business value. MIS Q. **28**, 283–322 (2004)
58. Williams, L., Kessler, R.: Pair Programming Illuminated. Addison-Wesley Longman Publishing Co., Inc. (2002)
59. Ladas, C.: Scrum-ban. In: Lean Softw. Eng. Essays Contin. Deliv. High Qual. Inf. Syst(2008). http://leansoftwareengineering.com/ksse/scrum-ban. Accessed 30 Mar 2019
60. Toth, S.: Vorgehensmuster für Softwarearchitektur: Kombinierbare Praktiken in Zeiten von Agile und Lean. Carl Hanser Verlag, Munich, Germany (2015)
61. Kleehaus, M., Uludağ, Ö., Matthes, F.: Towards a continuous feedback loop for microservice-based environments. In: 11th International Conference on the Quality of Information and Communications Technology (2018)

Chapter 19
Digitalization in Retail Banking

Oleg Svatoš

Abstract Traditional banks keep information about their clients just for themselves and do most of the business with their clients on a direct basis, without showing them all options on the market they have. Home banks have an information advantage over competing financial institutions but the digital transformation of the banking industry may completely change that. Introduction of PSD2 and related regulations officially opens the financial market to third parties with whom the clients may share online their transaction history from different banks providing them so with behavioral data the third parties can build their individualized offers on. Innovation boom slowly starts as currently only transactions of payment accounts are in the regulation involved but an extension into the credit business and others are already in the works. This chapter discusses possible business models, applications, business risks and new opportunities for retail banks and their customers enabled by the upcoming digital transformation.

Keywords PSD2 · Retail banking · Digital transformation

19.1 Introduction

Digital transformation is in the banking industry developing at its own speed. The banking industry is in its nature very conservative industry and the application of new technologies is relatively slow. Trust and reliability is a very important factor in this. The clients put their trust in banks. This is why the banks usually adopt only those technologies which are already somewhere working for some reasonable time and most risks of such technologies are already known and resolved. This creates a picture of the banking industry that gives the idea that the banking industry is always behind the technological wave but in their case, they just have to be very careful not harm the trust they possess. Moreover, if we look at the innovations produced by financial markets, which usually cause tremendous troubles like the subprime bonds,

O. Svatoš (✉)
Faculty of Informatics and Statistics, University of Economics in Prague, Prague, Czech Republic
e-mail: svatoso@vse.cz

A. Zimmermann et al. (eds.), *Architecting the Digital Transformation*,
Intelligent Systems Reference Library 188,
https://doi.org/10.1007/978-3-030-49640-1_19

maybe it is better that it is so. Banks are built on the trust of their clients and if one of these respected institutions fails this trust significantly, it puts in danger the whole financial sector.

Digitalization in established banks is not then represented by some radical change of their business models or significant product innovations which we would call as digital transformation. The main focus is mainly just on the transformation of current business processes into paperless processes. It does not mean that the documents are not necessary anymore, they are just now electronic form. The same goes for retail services. The direct contact with the client is only extended by internet banking so that one does not have to go physically to the bank. Most of the things, one can do remotely, is just work that the clerk in the office would have to do. So the result of digitalization in banking so far is that the actual work, which was done by the banks, has been shifted to the customers. Processes stay similar, one just has to do it by oneself now.

The digitalization process did not lead to some kind of banks' openness so far either. Banks protect clients' data as they are expected but the banks also protect data from the clients themselves. The banks consider the clients' data as their own property. It is their know-how their business is based on and gives them an advantage over the other competing banks and financial institutions. Without this data, it is hard for the competition to propose comparable offers as they do not know the client's behavioral history. Clients can deliver partially these data in paper form but this is still the only fraction of the data, the home bank has available, and it requires some effort from the client too. In contrast with that, the home bank's offer is just a single click away… Internet banking opens the access to transactional history a little bit, as one can access one's transaction history and generate banks statements for individual needs, but usually it is not digitally signed (trusted) and some kind of metadata, which would allow immediate automated processing followed by categorization and behavioral scoring, is in the retail business missing. There are attempts to overcome these difficulties like screen scraping [1] but that is a specific solution for specific clients, who are brave enough to share their access credentials with somebody else, and with the introduction of PSD2 also illegal [2].

Another factor in this is the speed at which a retail bank can provide an offer. Nice examples are the unsecured loans for which only the home bank can provide the individualized offer at the time of purchase. Other competitors have to create offers with some buffer in their interest rate or they have to help themselves with tricks like that a client gets at the beginning higher interest rate but if the client pays well, the last installments for the loan will not be required or will be paid back.

This status quo of the banking industry may change due to the open banking initiative [3] which resulted in the Payment Services Directive (PSD2) [4]. Third parties defined in the regulation, which would be allowed by a client to step between the client and client's bank(s), are able to mediate directly some of the banks' services and the banks are required to provide them with APIs to be able to do so. Third parties can be not only the banks but also other institutions that acquire a license from a national regulator for one of the services specified by the PSD2 directive. This is eagerly awaited by technological companies, so-called fintechs [5], represented not only by start-up companies but also by big social media enterprises like Google or Facebook, which plan to build new services around the required APIs. The PSD2 can be the start of real digital transformation in the banking industry.

Bank conservatism is in very contrast with technological companies. Banks try to avoid any failure, technological companies grow on failure. They are able to introduce innovations in a very short time and they either succeed or fail. If they fail, they just try doing something else and their customers are aware of that. Putting these two worlds together may seem like suicide but there are significant benefits for the retail clients and also for the banks on the horizon originating in this cooperation if all this is managed and regulated properly.

The approach taken by the PSD2 allows the combination of banking and fintech so that the commonly present failure of fintechs does not cause any significant harm to the financial industry. The very basic financial services (directly connected to money) stay in the hands of the banks but coordination and data transfers between these services and the client may go through the third-party applications the clients give permission to. Regulation forces the banks not only to open up through APIs but it also forces them to build complete infrastructure which allows the clients to share their transactional data with other institutions online. If a client gives the institution(s) approval, the data may be then used for far more services then just for payment services and account information aggregation. It creates a new platform on top of which the digital transformation of the retail financial market may take place.

19.2 Payment Service Directive 2

The Payment Service Directive (PSD2) [4] was approved in 2015 and the implementation deadline into local legislation for the EU states was January 13th, 2018. Until this date European Banking Authority (EBA) was required to propose technical standards, which would specify in detail the requirements (RTS) for technical solutions of APIs, the banks are required to implement and open to the third parties. The RTSs were delayed [6] and, since they come into effect after 18 months after their approval, the whole process was slowed down. The delay created a so-called interim period of ca. 20 months in which the banks slowly started developing the APIs but the actual granting of access to the third parties was on a voluntary basis as the technical requirements specified in RTSs were not in effect yet.

As far as the jurisdiction is concerned, the PSD2 only applies to those parts of the payment transaction that are carried out within the EU.

The PSD2 specifies the following participants on the retail financial market the regulation focuses on:

- *Third-Party Payment Service Providers (TPP)*—institutions that have approval or license from a national regulator that they can provide the payment initiation service, account information service or balance validation service (see below) and have liability insurance. The traditional banks are in this category by definition since they fall under much stricter regulation but a TPP can be also another non-banking company or even small start-up if it passes the process of licensing for one of the services at the national regulator.

- *Payment Initiation Service Providers (PISP)*—TPPs which initiate a payment from payment account in the client's home bank on the client's request.
- *Account Information Service Providers (AISP)*—TPPs which retrieve transactional history from payment account in the client's home bank on the client's request.
- *Account Servicing Payment Service Providers (ASPSP)*—the traditional banks which provide the customers with payment accounts and payment services.
- *Payment Service User (PSU)*—end-user using the payment services either directly through one's bank or indirectly through a TPP.

The PSD2 specifies three services for which the banks (ASPSPs) have to publish their APIs and allow the TPPs to call upon the client's request and approval. The banks have to enable access to these APIs for free and cannot force the TPPs into some contractual relationship with them.

In this chapter we go into detail over the two main services: payment initiation service and account information service. The third service, the confirmation on the availability of funds, is just a simple service that answers yes or no and covers specific needs of card business, which impact on digital transformation we find as minor.

19.2.1 Payment Initiation Service

The payment initiation service is provided by licensed TPPs, which upon the client's request initiate a payment from the payment account in the client's bank (Fig. 19.1).

The payment initiation process illustrated in Fig. 19.1 involves the following steps:

1. A PSU agrees with the terms and conditions of a PISP,
2. The PSU mandates the PISP to initialize a payment,

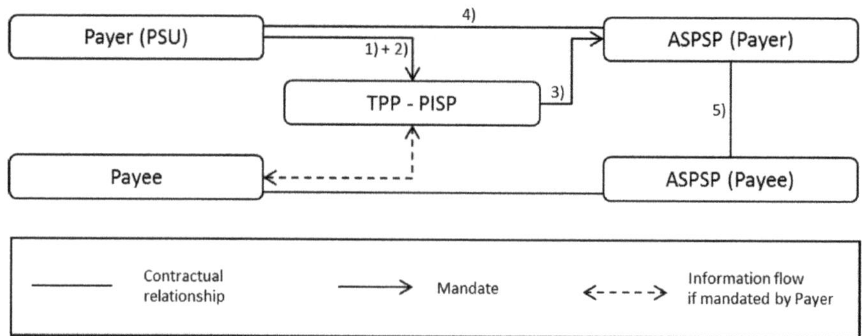

Fig. 19.1 General scheme of PISP service [7]

3. The PISP communicates the payment order to the PSU's ASPSP,
4. The PSU authorizes the payment to ASPSP,
5. The ASPSP executes the payment transaction initiated by the PISP on behalf of the PSU.

For the sake of clarity, we keep the schemes and step descriptions at a conceptual level so that a reader does not get overwhelmed by implementation detail. This detail is described in different specifications, which differ from each other, as the PSD2 nor RTS's go into such implementation detail. The complete detail can be found for instance in [8] or [9].

Figure 19.1 illustrates not only the necessary information flow but also contractual relationships. A TPP, in this case, acts as a coordinator and does not have to have a contract with any of the parties involved. That is very unusual in the banking business and makes the relations flexible and opened for innovations. This is also a reason why there is liability insurance for TPPs required and direct approval of the payment initiation to the bank by the client implemented. The ASPSPs cannot charge any additional fees, other than they would charge in case the client was using the ASPSP's own internet banking application. These services have to be equal in their conditions.

19.2.2 Account Information Service

The purpose of account information service is to allow aggregation of one's transaction history from different payment accounts in different banks into one overview provided by a TPP's application (Fig. 19.2).

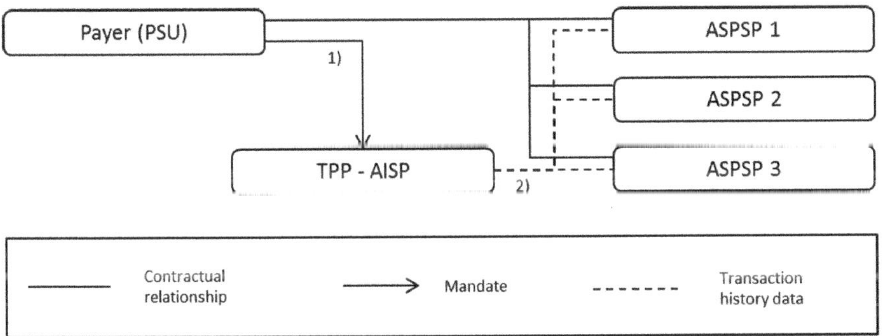

Fig. 19.2 General scheme of AISP service [7]

The account information acquisition process illustrated in Fig. 19.2 involves the following steps:

1. A PSU grants AISP access to the information on his accounts held by one or several ASPSPs for next 90 days [10],
2. Acting on behalf of the PSU, the AISP reports the financial situation.

From the point of view of digital transformation, this is the groundbreaking service. The idea of aggregating one's transaction history, so that one can see one's balances and transaction history in one unified overview is nice but the more important thing, in this case, is what one can do with this data next. The transaction data represents very private data, to which only the clients' banks had full online access. The account information service changes that and now a client can work with this data online and provide it to other parties if one agrees to. Again, there is not necessary for a TPP to have a contractual relationship with either the client or the ASPSP. The client's identity is again verified by the authentication and authorization process provided by the bank (ASPSP) when granting access to the AISP to the client's payment account.

A TPP is by definition forbidden to use the acquired data another way then just presenting it to the client but this can be extended if the client voluntarily agrees. At the same time, this permission must not be a condition for AISP service execution.

Implementation of PSD2 makes also alternative access to transaction data connected to payment accounts, such as screen scraping, illegal and forces so the third parties to use the APIs [2]. Access to other services stays intact by this regulation and so the screen scraping still stays alive an in a grey zone [11].

19.3 Impact of PSD2 on Digital Transformation of the Banking Industry

At first sight, it may seem that the PSD2 brings only online access to simple services that have been here already for some time. But when looking at it in detail, it not only forces the banks to implement the required APIs, but it also forces them to implement the whole necessary infrastructure that goes along with it: authentication and authorization of clients, fraud detection systems, licenses for the third parties and also necessary liability insurance, etc. This way the retail financial market may transform into a new market with services that have not been seen yet.

The important factor for digital transformation is that the PSD2 takes into account the need for flexibility. When we look at Figs. 19.1 and 19.2, we can see that the contractual relationships stay as they are, there is no extension required, and the services defined in PSD2 only build upon them, so that trust is secured by the banks and the TPPs can flexibly provide the AISP, PISP services.

Figure 19.3 illustrates the new platform the PSD2 implementation actually constitutes and the services which are available. The infrastructure is provided by the banks which in this case, due to the heavy regulation of the banking industry, represent

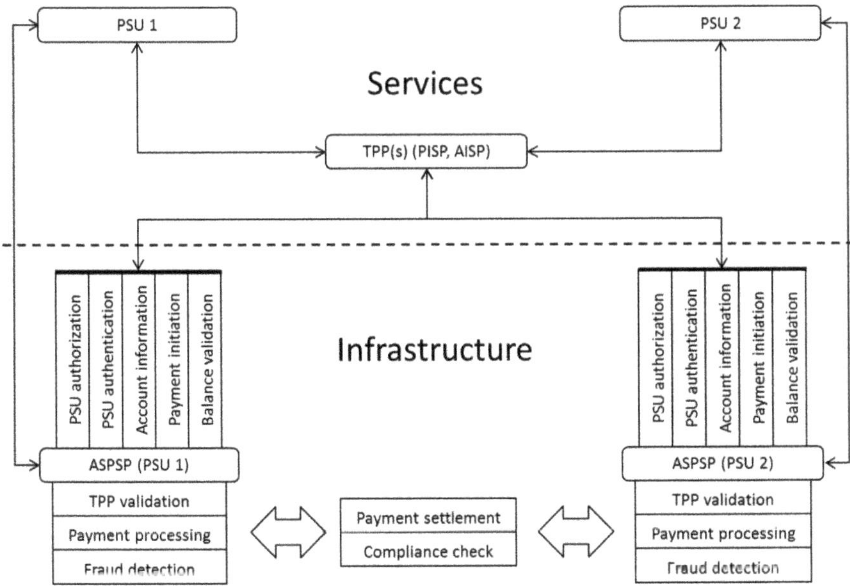

Fig. 19.3 The new platform constituted by the PSD2 implementation

trusted entities that can provide one with trusted data of a verified client. The PSD2 with its services brings the possibility for TPPs not only to use PISP and AISP services but also utilize authorization and authentication services provided by the ASPSPs, which eventually allow one to identify a user by some trusted authority online. The client's home bank identifies the client upon service request (AISP/PISP/Fund sufficiency) and the other providers can, thanks to the bank's authentication/authorization process, rely on it [12].

The inner processes of the banks and the infrastructure for the payment execution are already established by the banks and the PSD2 does to change that, on the contrary, it takes advantage of that. PSD2 extends the duties of the banks to validate whether TPPs have a proper license, and it also extends requirements put on the fraud detection system that banks have to implement and execute. With the help of required liability insurance of TPPs, this should minimize the risks arising from the admittance of TPPs into the banking ecosystem and secure the trust of banks' and TPPs' clients in the services provided by TPPs.

Not only the infrastructure of the banks has to be ready but also clients and their technical equipment have to be ready too. Deloitte's analysis [13] shows this premise is fulfilled. The penetration of the internet and mobile phones grows (already 80% smartphone penetration in the UK in 2016 with stable year to year growth since 2012 [13], p. 12). Preference for making payments and checking balance via internet banking is also much higher than any other channel the banks provide [13], p. 14.

Figure 19.3 illustrates what all services the TPPs have available and it is upon their innovation what kind of new digital services they can provide to their potential clients. We assume that the result in total can be either small digital evolution or real digital transformation of the retail finance business.

The small digital evolution involves:

- *Parallel payment services.* These services are already here (for instance PayPal or Revolut) but they are usually dependent on credit card operators or the payers' bank internet bank implementation. PSD2 offers independence if a TPP fulfills the requirements specified for AISP/PISP. In parallel with the traditional card payments and bank transfers, the PSD2 makes possible for the TPPs to constitute their own payment networks. Due to the usual bank fee policy in case of international payments probably only within the particular state but with some form of intelligent clearing (for instance like TransferWise or Revolut have) it is possible to create even international payment network within the EU where the PSD2 is applicable.
- *Online reinvesting robots for conservative consumers.* This service would be based on moving money from one savings account to another according to individual conditions provided by the banks and liquidity preference of the client. A robot would always make a suggestion to the client how to optimize the earning on his portfolio, and the client would either approve the transaction or not.
- *Upgrade of current loan processing.* The banks will modify their credit approval processes so that they can take advantage of possible access to transaction data of the applicant in different banks. This allows online credit approval in a short time based on proper risk-based pricing so that a client gets the best rate fitting his risk profile. This is in contrast with a common case when a bank does not have access to the client's transaction history and it has to ask the client to deliver one's bank statements so that they can evaluate one's riskiness. There were developed different workarounds on how to do it online but the PSD 2 makes all this much simpler if the client, of course, agrees to give access to his transaction account. PSD2 provides an option but the banks cannot deny one the loan if one only does not agree with the usage of AISP service.

Deloitte study [13], p. 14 shows that in the UK the space for the small steps has already been created. Over 70% of consumers and 80% of SMEs prefer making domestic payments and checking their account balance over the internet. But the digital transformation does not have to stop at this point. According to the study the digital distribution channel is preferred in cases of applications for unsecured loans by ca. 35% of consumers and around by 50% of SMEs. This is the space, where the banks make money and it will most probably become a new hotspot for the digital transformation of the banking industry. Supporting data for potential growth in the digital sales channel we can find in the credit cards, which are a form of pre-approved unsecured loans, where we can according to the survey [13], p. 14 see that for over 50% of consumers is the digital way the preferred way for credit card applications. In the case of SMEs, it is even over 70%.

When we start to develop the ideas on how the services provided by infrastructure constituted by PSD2 implementation could be used, we come also to rather disruptive scenarios that would represent the real digital transformation of the retail financial business (at least in the unsecured loans area). The key parameter in this analysis is the idea, who would be in the role of a TPP. In general, there are considered small fintech's and banks, which would eventually engage in the digital evolution described above. But if we extend this set by other parties interested in joining in, we get to different interesting business models that could really start the digital transformation of the retail finance business.

19.3.1 *Broker as a TPP*

If a client would provide through PSD2 APIs one's transaction history to a broker, it would allow the broker not only to do better financial analysis and advisory for the individual client but also to provide clients with loans for the lowest price possible.

With the approval of the client, the broker would send the client's transaction history to all the banks, it cooperates with, and the banks would be immediately able to calculate their offer fitting to the client's risk profile. This can be extended by the brokers by one more step—e-auctions. Since the main parameters for calculation of a loan would be provided by the broker to the contenders, it is only a matter of the level of interest margin the loan providers are ready to sacrifice. This can apply not only to retail unsecured loans but also to secured ones if the collateral quality can be evaluated online. This would significantly change the business for the banks. They would have to compete hard for every deal—the nature of their relations with the brokers would get less important and the ability to provide the best rate in comparison to the whole retail market more important.

Another opportunity in the retail business is the deposits. If a broker was allowed to administer deposits on saving accounts by its clients, it would eventually collect a relatively big volume of deposits the banks would compete for by providing individual conditions. The broker would then just ask its clients for approval for new account conditions and the suggested money transfer. All would be administered online over PISP services.

19.3.2 *Social Media as a TPP*

For precise scoring of the client's creditworthiness, it is necessary to have information about the client's behavior. The transaction history from payment accounts is one of the most important sources for the client's riskiness evaluation but not the only one. In the era of digitalization, one leaves traceable digital track in the digital space, and there are some companies that are able to trace it. Most apparent are Google and Facebook. With their data, data from registers (credit, insolvency, etc.) and the payment transaction history from banks, they are able to deliver the most precise risk profile scoring possible which can be updated it on daily basis—something, for what others, including the banks, have considerably less data available. Big social media may become then as TPPs providers of independent risk evaluation of clients for the financial institutions. The banks are bound by the regulation to check the risk profile of their clients [14], so in this case, they still would have to do supervision, but eventually, they would get more actual and precise scoring results than they would get from their own risk engine. The digital transformation would, in this case, allow shifting one of the key banking functions outside the banks.

The idea described above can be extended further. The social media may act as comparison engines and mediators of loans for their clients. They would not only be able to evaluate the client's creditworthiness but also through an e-auction would provide their clients with the best offers for a loan. The main difference here, when comparing with brokers, is that the social media may not need the commissions from banks, the broker business is built on. The social media needs information about their clients so that they can sell well targeted adverting services and the loan mediation would be just an appealing way to get to very interesting clients' data including their payment transaction history. The loan itself could be in this case just a byproduct. If the social media enter the retail loan market with the focus on data, as described above, this would transform the broker business significantly and could put them out of the retail banking business completely.

19.3.3 Online Retail Store as a TPP

Online retailers sell their goods not only for cash but they also allow their customers to finance their purchase by a loan. This is done online and it has to deal with all difficulties described in the sections above. A retailer usually has a contract with one loan provider, and again, the PSD2 may completely change this.

The same way as the brokers may engage in e-auctions for loan products, the online retailers could do the same. It would allow them not only provide the customers with the goods or services they want but also with the best financing options, the retail financial market can provide. For retailers, it could mean higher business and for the clients best loan offers that fit their risk profile. This would also mean that a client could get in many cases better loan conditions than one's home bank would currently offer them in form of unsecured loan, credit card or an overdraft account. This would significantly change the business model of retail banks. On one hand, they would have to deal with a decrease in their margins but on the other, this would give the banks an opportunity to be at the right place at the right time i.e. at the shop at the time of sale. This can eventually lead to a significant increase in the potential volume of loans a bank would have a chance to provide.

19.3.4 Further Options

The list of possible ways in which direction the digital transformation may set out presented in this chapter is not definitely complete, as the reality has not got to that point yet. We have elaborated only the ones which are currently most apparent and are derived from what worries current daily business of the retail financial market members most—clients to get best rate for loan or deposit, banks to get the most precise risk score so that they are able to provide the best conditions for particular client with reasonable profit margin. We will eventually see chaining of services provided over the ASPSPs' APIs, forming together complex products, in which the banking products would represent just a "module" in somebody else's product.

Further development also depends on what other APIs the banks will open. The basic infrastructure is ready, and it is on the market members how deep they will dive into digital transformation. The business models described above show that it includes not only the technologies but also the transformation of the current business into a relatively different one with different rules and players. On the one hand, the banks will lose their privilege over the clients' data and this may harm their business as it is set up today. But on the other, the digital transformation also brings an opportunity for the individual banks to gain access to a much larger market which can significantly increase the volume of their business if they adapt to the new conditions and take advantage of them.

19.4 Risks and Threats to Digital Transformation in the Banking Industry

From a technological point of view, the business models described above are feasible, the technologies for their implementation including security management either exist or are in the production already [15, 16]. The digital transformation in the retail banking industry is not then mainly about the banks and their ability to implement the technologies and to secure them (they have relatively unlimited the resources for that) but it is about the transformation that has to take place in the hearts and minds of their customers. In our case, consumers and SMEs have to be able and willing to use all these innovations. As noted above, the main principle, the banking is based on, is the trust and it took a long time and endless amount of regulation for the banks to gain the trust of the clients they now possess. And even now, their clients are still suspicious. The studies [17, 18] show that people use the digital channels but they are still reserved to the idea of open banking it is something new they do not have experience with. On the other hand, Deloitte [13] shows that that the consumers and SMEs are able to do things digitally without the supervision of a bank clerk and that the motivation to change one's bank when one gets a better interest rate in another bank is very high [13], p. 15.

What is the biggest threat that threatens the digital business described above? Loss of trust. Any scam, security breach, private data leakage may significantly slow down the adaption process of new technologies (business transformation) or even stop it. The PSD2 lets into the financial industry new flexible entities which are used to learn from their mistakes. This *learn by doing* approach is not what would secure the trust of their customers whose money and private information would be involved. Source of risk, in this case, are mainly those TPPs, which do not realize that the industry they are entering does not appreciate usual start-up set up—failure in 9 cases out of 10 [10]. If they fail to realize that, although the regulation tries to avoid it [19], the market would remain small as the clients will not trust small start-ups and they will not share their private data with them. Clients would probably consider only sharing the data with entities they have already put their trust in—the banks but also the big online players like Google and Facebook. The oligopoly, in this case, may slow down the digital transformation too, although, it may not be so severe since the technological companies follow different goals than the banks. Their goal is to get the information to be able to sell the most precise targeted advertising, and not to be the best credit mediator.

Another case is the necessity of strong security and also a little bit more bureaucracy than the fintech start-up companies are used to. One has to make the client aware that one is doing an important decision when approving something in one's application. Nice examples are the approvals of rights, an application needs to be granted when installing on an android phone. People usually do not pay attention to this and approve whatever the application suggests. The lack of attention is not such big issue with the payment service as the individual payments have to be approved directly by the clients to the home banks, but in case of client transaction history and how one will share it, that may become an issue as there are wide options limited just by what a client agrees with. If the TPPs would fail in making the clients pay attention to what they approve in the TPPs' applications, it would again harm the trust of the actual and potential customers and would slow down the whole digital transformation process.

The threats come not only from the individual participants but also from the legislation which has currently enabled this innovation. The regulation in its pursuit of clients' protection has the ability to impose such requirements on the TPPs that would make the services very inconvenient for the customers. A nice example is the SCA [20] or the 90-day validity of the client's approval for payment account when using the AISP service [21]. If a client has to look for one's credential data for the payment accounts in individual banks, one has integrated with TPP's application, every time when one wants to use the PISP/AISP service, it harms the whole PSD2 concept. One usually does not remember one's credentials and if every 90 days one's approval of access to all payment accounts in different banks expires, it has to be followed by looking for the credentials and putting them again in. This creates an inconvenience a client does not like at all and makes him wonder whether it is worth it. At the same time, for TPPs it is a matter of survival to be able to keep their customers logged in their platform. If customers stop logging in, all their accesses would expire and the customers may not come back again. Since the applications and the clients' data are what the TPPs will usually make money on, it would kill their business. Their mission is to have the clients using the application on a daily basis and the regulation, which would not take this into consideration, may harm development in this segment significantly.

Another trouble, that has the origin on the regulation side, is the looseness of technical standards for APIs the banks are required to publish. As the RTSs do not go into full technological detail, we can see a rising number of different standards for the particular APIs implementation [22, 23]. These are compliant with RTSs but different in functions and data structures they use as they are developed independently on each other. This is most visible in differences between the individual EU member states. Almost every country has its own standard [22] that undergoes its separate continuous development. This is costly for TPPs to implement and maintain as the individual standards change continuously. Only financially strong companies will be able to keep up with that. This again limits the number of contenders in the TPPs' services market.

19.5 Conclusions

The introduction of PSD2 forces retail banks to create access to payment accounts through APIs. These can be then accessed by third parties which are required to have approval from the bank's client and a license from the national regulator. PSD2 allows not only for the clients to be able to control their payment accounts in different banks from one application but also opens completely new world of possibilities as it sets up a new digital environment which solves the fundamental issues with client identity and transaction data verification and gives a client opportunity to share online with other institutions one's behavioral data stored in transactions of payment accounts in one's bank(s).

We can see some innovations already appearing in the form of integrating web and mobile applications but the real digital transformation of the retail financial industry, this regulation may bring, is still in ideas.

Based on the capabilities of already deployed technologies, market surveys and analysis we assume that a client will be eventually able to use one's transaction data from one's bank wherever one needs and agrees to, allowing so to the new platforms to be able to provide the client with individualized financial products from different banks. One will be given the possibility to compare them with competition and to select the most fitting ones.

As we have discussed in this chapter, the innovations do not have to stop at the level of payments, since the main potential, the PSD2 brings, is the ability to share online one's behavioral history in form of payment account transaction data, which then makes possible for the individual parties from retail financial market to offer individualized credit offers to clients no matter what their home bank is. The motivation is on both sides, since for the banks the unsecured credits are what the profit brings and the clients want, of course, to have the lowers interest rate of the credit possible. The game gets even more interesting when we consider what kind of third party providers may enter the retail financial market and what their motivations are. Eventually, it may happen that the new fintech companies take over the market of arranging of credits as many of them have additional information about their clients, the banks do not, and the data delivered over PSD2 APIs would just add the last piece of the puzzle they are now missing for precise credit scoring. Motivation on the fintech side might not only be in information integration and credit arrangement since the biggest ones have completely different core business—targeted advertising. Eventually, these fintechs may arrange the credits for clients for free and use the data gained from the transactions for more precise targeting of their advertising. This would make both, clients and fintechs, happy and possibly put a large number of brokers out of the business. This digital transformation may then significantly change the way how retail financial products, especially unsecured loans, are arranged and distributed today.

As with any other innovation, this comes also with its risks and threats which are also discussed. The trust is essential in the banking industry and there has to be done the maximum to maintain it. The introduction of new players to the retail financial market, who come from outside of the financial market, requires careful setup and regulation so that their innovation potential does not change into destruction. The same can be stated about the regulation which tries to maintain the trust of all participants. If it is too loose, the potential frauds may destroy the trust in innovations arising from the implementation of PSD2 if it is too strict, it would make usage of this innovation too complicated and would not be accepted by the potential AISP/PISP users. The fintech companies should not underestimate the importance of the regulation and the effect of lobbing. The financial business is heavily regulated sector and lobbing is an inseparable part of the financial business. The fintech should never give up on that, otherwise, the banks and their lobby may influence the regulation in their favor such way, that the requirements imposed on fintech make their business hardly practicable.

Enterprises, which consider entering this new market, should think beyond the technology. Money is just a medium of exchange—a way how one actually gets to the final product or service one wants or needs. The biggest potential benefit of PSD2 lies in the combination of AISP and PISP with other services, not the money transfers itself. The banks will inevitably implement the basic services required by the regulation but they will be very careful with experimenting. This is where the newcomers have the chance to succeed and to introduce groundbreaking innovations that would shift the digital transformation forward.

Enterprises in the race for most innovative and user-friendly service must not underestimate the security. They can and should rely on security measures implemented by the banks (authorization/authentication) and very strongly protect the clients' data, one either agrees to share with them or uses/requires for transaction execution. Minimalization of stored data and having the GDPR in mind is a good start.

It is also important to keep in mind that not everybody who uses digital devices is self-serviceable. The broker business shows that the people when dealing with finances (life insurance, mortgages, etc.) in a lot of cases need assurance that they are doing the right thing. This has to be seriously considered when designing a new service and the applications involved otherwise a large number of potential customers would be left out.

The potential arising from letting the new players into the retail financial market is huge. How well this opportunity will be used depends now on how the individual participants will approach it and how the regulation will be successful in maintaining trust among individual parties. We hope that we are at the very beginning of a small revolution in the retail financial industry.

References

1. Wood, C.: Why do FinTechs want to save screen scraping? In: Nordic APIs (2017). Retrieved from https://nordicapis.com/fintechs-want-save-screen-scraping/
2. European Commission. Payment Services Directive (PSD2): Regulatory technical standards (RTS) enabling consumers to benefit from safer and more innovative electronic payments (2017). Retrieved from http://europa.eu/rapid/press-release_MEMO-17-4961_en.htm
3. Manthorpe, R.: What is Open Banking and PSD2? In: Wired (2018). Retrieved from https://www.wired.co.uk/article/open-banking-cma-psd2-explained
4. Directive (EU) 2015/2366 of the European Parliament and of the Council of 25 November 2015 on payment services in the internal market, amending Directives 2002/65/EC, 2009/110/EC and 2013/36/EU and Regulation (EU) No 1093/2010, and repealing Directive 2007/64/EC. Retrieved from https://eur-lex.europa.eu/legal-content/EN/TXT/?uri=CELEX:32015L2366
5. Kagan, J.: Fintech. In: Investopedia (2019). Retrieved from https://www.investopedia.com/terms/f/fintech.asp. Accessed 15 Jan 2019
6. Papamichael, P.G., Evagorou, C.: PSD2 Finalised Standard on SCA and CSC: The Wait is Over, but Questions Remain. In: Mondaq (2018). Retrieved from http://www.mondaq.com/uk/x/689054/Financial+Services/PSD2+finalised+standard+on+SCA+and+CSC+the+wait+is+over+but+questions+remain. Accessed 15 Jan 2019
7. Deutsche Bank AG.: Payment Services Directive 2: Directive on Payment Services in the Internal Market "(EU) 2015/2366" (2016). Retrieved from http://cib.db.com/docs_new/White_Paper_Payments_Services_Directive_2.pdf
8. Open Banking Ltd.: Open Banking Standard (2019). Retrieved from https://www.openbanking.org.uk/providers/standards/
9. Česká Bankovní Asociace: Czech Standard for Open Banking v1.2 (2018). Retrieved from https://www.czech-ba.cz/cs/aktivity/standardy/cesky-standard-pro-open-banking
10. Patel, N.: 90% Of Startups Fail: Here's What You Need To Know About The 10% In: Forbes (2015). https://www.forbes.com/sites/neilpatel/2015/01/16/90-of-startups-will-fail-heres-what-you-need-to-know-about-the-10/
11. Mac Dowel, A.G.: Screen scraping is dead, long live screen scraping (2017). In: Finextra. Retrieved from https://www.finextra.com/blogposting/14793/screen-scraping-is-dead-long-live-screen-scraping
12. Hamish T., Jones M., Torpey M.: Trust and Digital Identification in an Open Banking World. In: International Banker (2018). Retrieved from https://internationalbanker.com/banking/trust-and-digital-identification-in-an-open-banking-world/
13. Deloitte LLP: How to flourish in an uncertain future: Open banking and PSD2 (2017). Retrieved from https://www2.deloitte.com/content/dam/Deloitte/cz/Documents/financial-services/cz-open-banking-and-psd2.pdf

14. European Banking Authority: Guidelines on Creditworthiness Assessment EBA/GL/2015/11 (2015). Retrieved from https://eba.europa.eu/documents/10180/1092161/EBA-GL-2015-11+Guidelines+on+creditworthiness+assessment.pdf/
15. Accenture: PSD2 & Open Banking Security and Fraud Impacts on Banks (2016). Retrieved from https://www.accenture.com/_acnmedia/PDF-40/Accenture-PSD2-Open-Banking-Security-Fraud-Impacts.pdf
16. Masuda, et al.: Risk management for digital transformation and big data in architecture board: a case study on global enterprise **4**(1) (2018)
17. Murray-West, R.: Less than a third of customers trust open banking? In: FT Advisor (2018). Retrieved from https://www.ftadviser.com/mortgages/2018/03/05/less-than-a-third-of-customers-trust-open-banking/
18. Palframan, M.: Three quarters of Britons haven't heard of open banking (2018). Retrieved from https://yougov.co.uk/topics/finance/articles-reports/2018/08/01/three-quarters-britons-havent-heard-open-banking
19. European Banking Authority: Guidelines on Security Measures for Operational and Security Risks under PSD2 (2018). Retrieved from https://eba.europa.eu/documents/10180/2081899/Guidelines+on+the+security+measures+under+PSD2+%28EBA-GL-2017-17%29_EN.pdf/c63cfcbf-7412-4cfb-8e07-47a05d016417
20. European Banking Authority: Opinion of the European Banking Authority on the implementation of the RTS on SCA and CSC (2018). Retrieved from https://eba.europa.eu/documents/10180/2137845/Opinion+on+the+implementation+of+the+RTS+on+SCA+and+CSC+%28EBA-2018-Op-04%29.pdf
21. European Commission: Regulatory Technical Standards on Strong Customer Authentication and common and secure communication under Article 98 of Directive 2015/2366 (PSD2) (2018). Retrieved from https://eba.europa.eu/documents/10180/1761863/Final+draft+RTS+on+SCA+and+CSC+under+PSD2+%28EBA-RTS-2017-02%29.pdf
22. European Payments Council.: Summary of the API Evaluation Workshop (2018). Retrieved from https://www.europeanpaymentscouncil.eu/sites/default/files/kb/file/2018-04/API%20EG%20019-18%20v1.1%20Summary%20of%20the%20API%20Evaluation%20Workshop%20-%2020180228.pdf
23. Hines P.: Open Banking API Standards: What are the Options? How Do I Choose? In: Celent (2017). Retrieved from https://www.celent.com/insights/532743958/

Chapter 20
E-Healthcare Service Design Using Model Based Jobs Theory

Shuichiro Yamamoto

Abstract The demand on the innovation of enterprise services are rapidly increased. There are various modeling methods that can be applied to design digital transformation of services. However, there are few visual modeling methods to design innovations. Without consistent visual modeling methods, it is difficult to integrate service innovations and digital transformations. In this chapter, we propose a Model Based Jobs Theory (MBJT) which can visually model the Jobs Theory of Christensen. Moreover, we examine the applicability of MBJT by designing an e-Healthcare service.

Keywords Jobs theory · E-Healthcare · Enterprise architecture · ArchiMate · Case study

20.1 Introduction

Christensen proposed the Jobs Theory [1] to effectively design the innovation. This chapter proposes a Model Based Jobs Theory (MBJT) based on the Jobs Theory of Christensen by using Goal Oriented Requirements Engineering Approaches.

At first the chapter explains the basic concepts of Jobs Theory. Then, the job analysis table is proposed by the consideration on the relationship between Jobs Theory and Requirements Engineering. Moreover, Model Based Jobs Theory (MBJT) is proposed by mapping the jobs analysis table into goal models using ArchiMate. ArchiMate is the standard Enterprise Architecture modeling language of The Open Group (http://www.opengroup.org/). Then, an e-Healthcare service is designed by MBJT. MBJT is also compared with similar approaches such as BMC (Business Model Canvas).

S. Yamamoto (✉)
Graduate School of Informatics, Nagoya University, Furo-Cho Chikusa-Ku, Nagaoya, Japan
e-mail: syamamoto@acm.org

A. Zimmermann et al. (eds.), *Architecting the Digital Transformation*,
Intelligent Systems Reference Library 188,
https://doi.org/10.1007/978-3-030-49640-1_20

The rest of this chapter are organized as follows. Section 20.1.2 describes related work. Section 20.1.3 proposes MBJT. An example of e-Healthcare service design is also provided by using MBJT. The applicability of MBJT is discussed in Sect. 20.1.4. Finally, Sect. 20.1.5 summarizes and provides the future work.

20.2 Related Work

20.2.1 Jobs Theory

Jobs Theory [1] focuses the consumer jobs to be done. There are the cause effect relationships between the jobs to be done and the solutions, because consumers accomplish jobs by using the product solutions which companies provide.

If the jobs as activities to resolve the causes of problematic situations of consumers were not clear, then the products to be developed as solutions never be consumed.

The jobs to be done are defined as the process that derives progress which consumers tried to accomplish. The specific context where jobs are achieved is the situation of jobs. The situation shows something that consumer takes pains.

Jobs continuously repeats when consumers need progresses. The hire is defined as that consumers use products to resolve jobs.

The purchasing and using products are the two types of hire. Consumers purchase products only once. The purchased product is using repeatedly for each jobs.

The big hire means that consumer purchases products. The little hire means that consumer uses products.

Christensen pointed the necessity of emotional and social aspects in addition to the functional aspect. Aspects are the qualities which derive better solutions. Aspects are decomposed as inevitable elements of jobs, and integrated into the total story that customers experience in the course of jobs. Solutions are the means to achieve jobs. It is necessary to discover means for jobs that only have insufficient solutions so far. Table 20.1 shows elements of the jobs theory.

Table 20.1 Elements of jobs theory

Elements	Explanations
Customer	Who requires to resolve problems
Job	Process that derives progress which consumers tried to accomplish
Progress	What customers achieve by jobs
Hire	Use of products by customers to solve jobs
Situation	Specific context where jobs occur
Aspect	Qualities which derive better solution
Solution	Means to achieve jobs

20.2.2 E-Health Service Design Approaches

e-Health is expected to realize the wellness of personal healthcare using Information communication technology(ICT) for healthcare services [2]. It is necessary to design business model to the success of e-Health services. Schiltz and others proposed use cases for the business models of e-Health service [3]. For example, there was a use case of cloud based image archive service named PACS (Picture Archiving and Communication System) utilizing Radiology Information System (RIS).

Concept of Operation [4] for the National personally controlled electronic health record (PCEHR) system [5] was created for the purpose of delivering a high level picture of the system and how it works in a moderate level of detail in order to easily communicate with stakeholders. The version of ConOps is created following the framework for a national electronic health record system agreed by the Australian Health Ministers Conference in April 2010.

Business Model Canvas (BMC) [6] had been used to describe e-Health business models. The business process modeling approaches are also proposed [7, 8]. Basyman et al. [9] proposed an Activity-based Process Integration approach that incrementally integrates activities of a new process to current processes. They use $i*$ [10] and URN (User Requirements Notation) [11] to describe goal models and business process integration for the emergency process in the hospital. Goal oriented approaches [12] are applied to analyze healthcare processes [13, 14].

Enterprise Architecture (EA) [15, 16] is also necessary to realize the business transformation using ICT. EA has been used to model healthcare services. For example, Sharaf et al. [17] discussed EA for the mobile healthcare cloud service. Ahsan et al. [18] proposed a visual approach using ArchiMate to describe healthcare activities. ArchiMate [19–22] is a language to model EA. As ArchiMate has rich features to represent A, there are many researches to design business models and business values. Meertens et al. [23] proposed a mapping between ArchiMate and BMC by using business ontology. Iacob et al. [24] proposed a mutual transformation method between EA and business model. Caetano et al. [25] proposed an enterprise modeling method to integrate ArchiMate and BMC by using a semantic model. Yamamoto [26] proposed a design method to derive business values in ArchiMate based on actor relationships.

Yamamoto [27] surveyed approaches to visualize EA models. Yamamoto and others [28] proposed a method to visualize Jobs Theory [1] based on ArchiMate. The paper proposed the MBJT by integrating the Jobs Theory and the goal oriented requirements model by using ArchiMate. Moreover, a case study for applying MBJT to an e-Healthcare use case was described. The meta-model of MBJT has also been proposed.

Muller et al. [29] proposed to use BPMN for describing Healthcare processes. Although BPMN can be used to describe business processes, it is impossible to describe an application and technology components in BPMN. It is also impossible to describe interrelationships among business processes, application components and technology devices. Therefore, BPMN is not appropriate for representing business models including application and technology elements.

Although Silva et al. [30] proposed automatic evolution operations for EA models and meta-models, these operations did not focus on business model evolution of the healthcare domain.

Bork et al. [31] identified meta-model specification techniques and compared the expressive power of the techniques. As the paper focused on meta-model specification techniques they did not mention business model development processes. Santana et al. [32] proposed a method to derive EA view point models from existing models at organizations. Smith et al. [33] extracted functional requirements themes from Dutch government agencies to construct BRM (Business Rule Management). BRM focuses to transform legal requirements into business decisions.

There are many approaches to represent healthcare business processes as follows.

Mertens et al. [34] proposed a recommendation-based robust business process engine to realize flexible healthcare processes. Behnam et al. [35] proposed a Care Process Metamodel (CPM) to describe healthcare processes and to implement business intelligence care process monitoring solutions. The CPM did not include healthcare devices and application services. Dogac et al. [36] proposed a business process specification language to support collaborative eHealth processes. Vilasdechanon et al. [37] used the function modelling method (IDEF0) to present saline management processes. Brown et al. [7] introduced a Healthcare Business Process Reference Model to standardize business process roles and interactions between them. They also showed the Healthcare reference model concepts. Antonacci et al. [38] have introduced a model-driven method to enact the simulation-based analysis of healthcare processes. Blijleven et al. [39] proposed the process deliverable diagram consists of activity and class diagrams to represent healthcare processes. Braun et al. [40] extended BPMN based on clinical pathways to model hospital resources, activities and data objects.

However, these business process oriented approaches did not model business goals and technological devices. Gand [41] surveyed business model representations of the healthcare domain. He emphasized the importance to differentiate the business model concept from business processes. The business model consists of business goals that are realized by business processes. However, he did not mention the technology elements could be constituents of business models.

Fattah et al. [42] proposed a Conceptual Strategic Alignment Model for Organizational and Culture Adopter. However, the approach did not include business process models nor business goals. Although An et al. [43] showed the goal oriented approach is effective to elicit requirements for a healthcare system, they did not mention the business modeling using goal oriented approaches. Baslyman et al. [44] proposed using goals to represent indicators for the activity based process integration. They used goal models to quantitatively evaluate the performance of the healthcare process integration. They did not consider to design business models using goals but evaluate business processes.

Carroll et al. [45] introduced Design Thinking to align healthcare innovation and software requirements. Although they showed the connected health innovation

Table 20.2 Job analysis table elements using ArchiMate

JAT	ArchiMate
Customer	Actor
Concern	Value
Problematic situation	Driver
Cause analysis	Assessment
Progress	Goal
Solution	Requirements

framework based on Design Thinking process, the details of business models and processes were not clear.

Although business process modeling has been used in the e-Health domain so far, the key elements of business models are not clear. Moreover, a business model design method based on the key business elements has not been proposed. We are trying to propose the e-Health business model development method using EA modeling language ArchiMate to create e-Health services by extracting key e-Health business model elements.

20.3 Model Based Jobs Theory

20.3.1 JT Using ArchiMate

The JAT (Job Analysis Table) is designed based on the elements of JT except from the Hire and Job. Because the Hire and Job are related to the business processes. JAT contains the Cause analysis element that shows the reason of the problematic situation. JAT can be described by using ArchiMate [3–7]. Table 20.2 shows the mapping between elements of JAT and the corresponding elements of ArchiMate.

Concern, Problematic situation, Cause analysis, Progress and Solution of JAT are corresponding to value, driver, assessment, goal and requirements, of ArchiMate, respectively. By using the mapping, a generic form of JT is represented as showing in Fig. 20.1.

The diagram described by using ArchiMate. JT itself did not provide diagrams and tables to analyze and design innovations. By using JAT and ArchiMate, JT can be visualized.

20.3.2 Service Design Example Using MBJT

Schiltz and others proposed use cases for the business models of e-Health service [3]. For example, there was a use case of cloud based image archive service named PACS

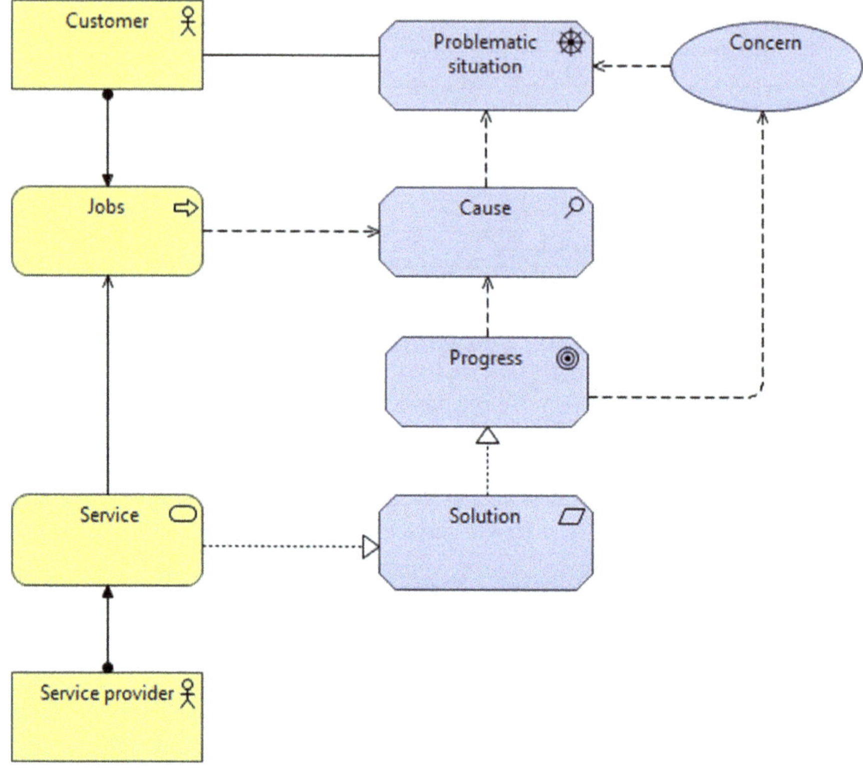

Fig. 20.1 Generic form of JT in ArchiMate

(Picture Archiving and Communication System) utilizing Radiology Information System (RIS).

The business model of PACS by using MBJT can be designed as follows. Actors of PACS are patient, medical service provider, and THS (Technology for Healthcare Service provider) service provides. JAT for medical service provider is analyzed as shown in Table 20.3.

Table 20.3 Job analysis table elements

JAT	Explanation
Customer	Medical service provider
Concern	Convenient use of medical image technology
Problematic situation	Medical image technology is hard to use
Cause analysis	There is no easy means to use medical image technology
Progress	To realize open use of medical image information
Solution	Medical image collaboration process to share and diagnose medical image remotely

The concern of medical service providers is convenient use of medical image technology. The problematic situation of medical service providers is that medical image technology is hard to use. The cause of the problem is that there is no easy means to use medical image technology. The progress to be achieved is to realize open use of medical image information. The solution is the medical image collaboration service to share and diagnose medical image remotely.

The MBJT model for the above scenario is shown in Fig. 20.2.

If the medical image collaboration service is realized by the medical technology service provider the following processes are achieved. First, the medical service provider takes X-ray diagnose images and stores them into the medical image remote collaboration service. Then, the medical service provider retrieves X-ray images on the service. The medical service provider also diagnoses based on the X-ray images.

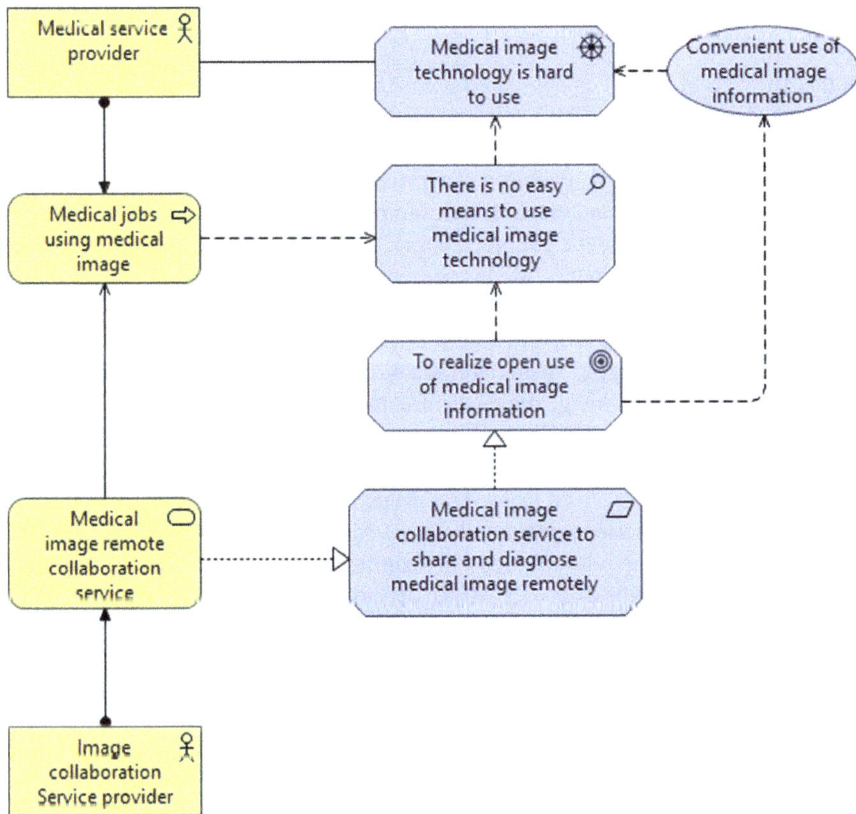

Fig. 20.2 MBJT model for PACS

Table 20.4 ASOMG analysis table

Actor	Service	Object	Means	Goal
Patient	X ray image capture	X ray image	X ray camera	Low overload, security
Medical service provider	X ray image capture/retrieval/diagnosis	X ray image	X ray camera/image retrieval SW	Efficiency, availability, security, reliability
Platform provider	Platform service	X ray image	X ray image server X ray image diagnosis	Rapid response availability

20.3.3 ASOMG Analysis

In this subsection, MBJT is compared with ASOMG (Actor Service Object Means Goal) approach [46]. ASOMG consists of the five key business model elements. The key elements are Business Actor, Business Service, Business Object,Business Means,and Business Goal. Patient and service providers are examples of business actors. Sensors, servers, and network are examples of business means. Actor, Service and Object are corresponded to Subject, Verve, and Object. SVO is the basic elements of natural language statements. Means and goals are also generic.

Actors of PACS are patient, medical service provider, and platform service provider. Services are X ray image capture, retrieval, diagnosis, and platform services. Objects are X ray images. Means are X ray camera, X ray image server, X ray image retrieval SW, and image diagnosis. Business goals are low overload, security, efficiency, availability, reliability, and rapid response.

The ASOMG analysis as described above is shown in Table 20.4. The PACS business model can be derived using ArchiMate is also shown in Fig. 20.3.

ASOMG table describes goals and means for each actor. Therefore, ASOMG table can clearly elicit relationships between actors, means and goals. Although MBJT clarifies problems of PACS, MBJT did not clearly define the detail of PACS service. In contrast, ASOMG can define objects, means and goals of the service solutions of PACS.

20.4 Discussion

20.4.1 Jobs Theory and Enterprise Architecture

In Jobs Theory, there is a “hire” relation between “Jobs to be done” and the product. In requirements engineering, there is a realization relation between requirements of

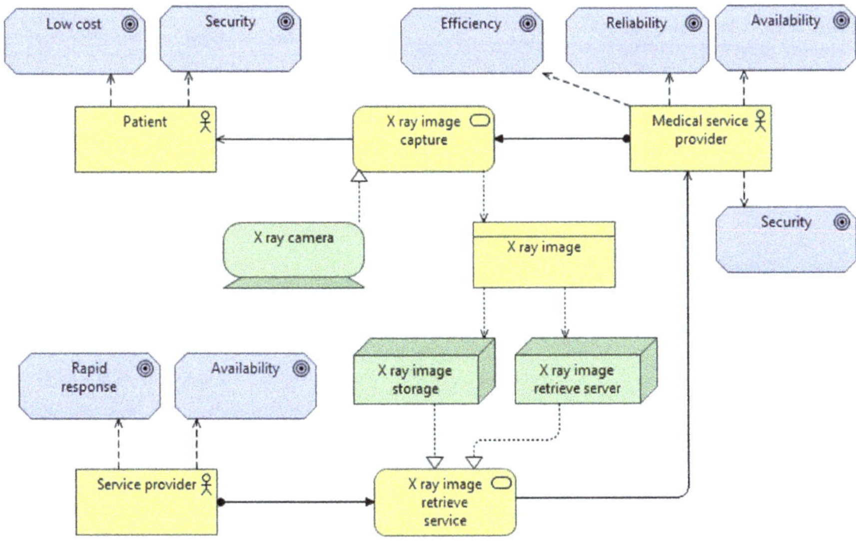

Fig. 20.3 ASOMG model for PACS

customers and system. In this way, there is a relationship between Jobs Theory and Requirements Engineering.

Traditional social science researches focus on the analysis and their goals are explanations of social phenomena including innovation. In the other hands, Engineering studies focus on the synthesis and their goals are designing artificial objects. Although traditional innovation theories only explained what kinds of innovations occurred, these theories could not design innovations.

Christensen spent 20 years for writing the book of Jobs Theory to design successful innovations [1]. The Jobs Theory provides the opportunity to connect with the innovation research to software engineering which aims to synthesis. The MBJT integrates innovation and enterprise architecture by using ArchiMate.

20.4.2 Applicability of MBJT

As described above, MBJT and ASOMG are able to visualize the business model of an e-Health services. In this subsection, we compare MBJT, ASOMG, and BMC (Business Model Canvas). Table 20.5 shows the comparison between elements of MBJT, ASOMG and BMC.

Except for the key resources and cost structure, MBJT can describe elements of BMC. In contrast, BMC cannot describe the problematic situation, cause analysis, and product of MBJT.

Table 20.5 Mapping among elements of MBJT, ASOMG, and BMC

MBJT	ASOMG	BMC
Customer	Actor	Customer segment, key partnerships
Concern, progress	Goal	Value propositions
Problematic situation		
Cause analysis		
Solution, hire	Service	Channels, revenue streams
	Object	Key resources
Jobs	Service	Key activities
Experience		Customer relationship
		Cost structure
Product	Means	

The customer relationship and cost structure of BMC are not also able to represent directly in ASOMG. In contrast, BMC cannot describe the means element of ASOMG.

Except for the object, MBJT can describe elements of ASOMG. In contrast, ASOMG cannot describe the problematic situation, cause analysis, and experience of MBJT.

If the cost structure of BMC can be mapped to the chain of value objects, it will be represented by ASOMG. Therefore, the integrated set of MBJT and ASOMG can cover BMC.

20.4.3 Limitations

This paper only showed an application of MBJT for a use case on the e-Healthcare service. More case studies are necessary to show the effectiveness of MBJT. Schiltz and others showed 15 use cases in [3]. The remaining use cases are candidates of MBJT application.

20.5 Conclusions

We have proposed the Model Based Jobs Theory using ArchiMate. Then we have shown the applicability of MBJT to e-Healthcare by using a simple business case. Moreover, we compared conceptual elements of MBJT, ASOMG, and BMC. The result showed the potential complemental relationship among the three approaches.

Future study includes applications of the proposed approach not only for e-Healthcare, but also e-Business and e-Government. The integration of MBJT and ASOMG is also expected.

Acknowledgements The authors express great thanks to Dr. Ishigure who introduced the reference paper on Business Model Analysis of eHealth Use Cases.

References

1. Christensen, C., Hall, R., Dillson, K., Duncan, D.: Competing Against Luck. HarperCollins Publishers LLC, USA (2016)
2. Commission of the European Communities: E-Health—making healthcare better for European citizens: An action plan for a European e-Health Area SEC. 539 (2004). Available at: http://www.ehealthimpact.org/download/documents/ehealthimpactsept2006.pdf. Accessed Aug 2018
3. Schiltz, A., Rouille, P., Zabir, S., Genestier, P., Ishigure, Y., Maeda, Y.: Business model analysis of EHEALTH use cases in Europe and in Japan. J. Int. Soc. Telemed. eHealth. 1.1: 30–43 (2013). http://journals.ukzn.ac.za/index.php/JISfTeH/article/view/26/121
4. IEEE Std 1362™-1998 (R2007) IEEE Guide for Information Technology—System Definition—Concept of Operations (ConOps) Document
5. Australian Government, National E-Health Transition Authority: Concept of Operations: Relating to the introduction of a Personally Controlled Electronic Health Record System. NEHTA Version Number: 0.14.18 (2011)
6. Osterwalder, A., Pigneur, Y.: Business Model Generation. Wiley (2010)
7. Brown, P., Kelly, J., Querusio, D.: Toward a healthcare business-process reference model IT Pro **13**(3), 38–47 (2011)
8. Braun, R., Burwitz, M., Schlieter, H., Benedict, M.: Clinical process from various angles—amplifying BPMN for integrated hospital management. In: International Conference on Bioinformatics and Biomedicine, pp. 837–845 (2015)
9. Baslyman, M., Almoaber, B., Amyot, D., Bouattane, E.: Using goals and indicators for activity-based process integration in healthcare. In: 7th International Conference on Current and Future Trends of Information and Communication Technologies in Healthcare, pp. 318–325 (2017)
10. Yu, E.: Towards modelling and reasoning support for early phase requirements engineering. In: 3rd IEEE Int. Symposium on Requirements Engineering, pp. 226–235 (1997)
11. ITU-T (2012) Recommendation Z.151 (10/12). User Requirements Notation (URN) Language Definition, Geneva, Switzerland
12. Yamamoto, S., KAIYA, H., Cox, K., Bleistein, S.: Goal Oriented Requirements Engineering—Trends and Issues. IEICE E-89D.11: pp. 2701–2711 (2006)
13. An, Y., Dalrymple, P., Rogers, M., Gerrity, P., Horkoff, J., Yu, E.: Collaborative social modeling for designing a patient wellness tracking system in a nurse-managed health care center. In: Proceeding of the 4th International Conference on Design Science Research in Information Systems and Technology (2009)
14. Hagglund, M., Henkel, M., Zdravkovic, J., Johannesson, P., Rising, I.: A New Approach for Goal-oriented Analysis of Healthcare Processes. MEDINFO2010 IOS Press, pp. 1251–1255 (2010)
15. Yamamoto, S., Olayan, N., Morisaki, S.: Another look at enterprise architecture framework. J. Bus. Theory Pract. **6**(2), 172–183 (2018)
16. The Open Group: TOGAF Version 9.1 Van Haren Publishing (2011)

17. Sharaf, A., Ammar, H., Dzielski, D.: Enterprise architecture of mobile healthcare for large crowd events. In: International Conference on Information and Communication Technology and Accessibility, pp. 1–6 (2017)
18. Ahsan, K., Shah, H., Kingston, P.: Healthcare modelling through enterprise architecture: a hospital case. In: Seventh International Conference on Information Technology, pp. 460–465 (2010)
19. The Open Group: ArchiMate 3.0 Specification C162, Van Haren (2016)
20. Wierda, G.: Mastering ArchiMate—A Serious Introduction to the ArchiMate Enterprise Architecture Modeling Language Edition II. The Netherlands Published by R&A (2014)
21. Lankhorst, M., et al.: Enterprise Architecture at Work—Modeling Communication and Analysis, 3rd edn. Springer (2013)
22. Archi: http://archi.cetis.ac.uk/
23. Meertens, L., Iacob, M., Jonkers, H., Quartel, D.: Mapping the Business Model Canvas to ArchiMate SAC'12, pp. 1694–1701 (2012)
24. Iacob, M., Meertens, L., Jonkers, H., Quartel, D., Nieuwenhuis, L., van Sinderen, M.: From enterprise architecture to business models and back. Softw. Syst. Model. **13**(3), 1059–1083 (2014)
25. Caetano A, Antunes G, Bakhshandeh M, Borbinha J, da Silva M (2015) Analysis of Federated business models—an application to the business model Canvas, ArchiMate, and e3value. IEEE 17th Conference on Business Informatics, pp. 1–8
26. Yamamoto, S.: Actor Collaboration Matrix for Modeling Business Values in ArchiMate. In: Asia Pacific Conference on Information Management 2016 Vietnam National University Press, pp. 369–378 (2016)
27. Yamamoto, S.: A survey on visualizing EA models toward consolidation. In: IIAI-AAI 2018, EAIS2018, pp. 775–780 (2018)
28. Yamamoto, S., Olayan, N., Fujieda, J.: e-Healthcare service design using model based jobs theory InMed2018. Procedia Comput. Sci. 198–207 (2018)
29. Muller, R., Rogge-Solti, A.: BPMN for Healthcare Processes ZEUS (2011)
30. Silva, N., da Silva, M., Sousa, P.: CO-EVOC: An Enterprise Architecture Model Co-Evolution Operations Catalog AMCIS2018 (2018)
31. Bork, D., Karagiannis, D., Pittl, B.: How are Metamodels Specified in Practice? Empirical Insights and Recommendations AMCIS2018 (2018)
32. Santana, A., Simon, D., de Moura, H., Fischbach, K., Kreimeyer, M.: Deriving Viewpoints for Enterprise Architecture Analysis AMCIS2018 (2018)
33. Smit, K., Zoet, M., Berkhout, M.: Functional Requirements for Business Rules Management Systems AMCIS2017 (2017)
34. Mertens, M., Gailly, F., Poels, G.: Supporting and assisting the execution of flexible healthcare processes. In: 9th International Conference on Pervasive Computing Technologies for Healthcare, pp. 329–332 (2015)
35. Behnam, S., Badreddin, O.: Toward a care process metamodel: for business intelligence healthcare monitoring solutions. In: 5th International Workshop on Software Engineering in Health Care, pp. 79–85 (2013)
36. Dogac, A., Kabak, Y., Namli, T., Okcan, A.: Collaborative business process support in eHealth: integrating IHE profiles through ebXML business process specification language. IEEE Trans. Inf. Technol. Biomed. **12**(6), 754–762 (2008)
37. Vilasdechanon, S., Sopadang, A.: Business process reengineering for the saline management in hospitals. In: 5th International Conference on Industrial Engineering and Applications, pp 84–88 (2018)
38. Antonacci, G., Calabrese, A., D'Ambrogio, A., Giglio, A., Intrigila, B., Levialdi, N.: A BPMN-based automated approach for the analysis of healthcare processes. In: IEEE 25th International Conference on Enabling Technologies: Infrastructure for Collaborative Enterprises, pp. 124–129 (2016)
39. Blijleven, V., Jaspers, M., Wetzels, M., Koelemeijer, K., Mehrsai, A.: A meta-modeling technique for analyzing, designing and adapting healthcare processes: a process-deliverable

perspective. In: IEEE-EMBS International Conference on Biomedical and Health Informatics, pp. 316–319 (2016)
40. Braun, R., Burwitz, M., Schlieter, H., Benedict, M.: Clinical process from various angles—amplifying BPMN for integrated hospital management. In: IEEE International Conference on Enabling Technologies: Infrastructure for Collaborative Enterprises, pp. 837–845 (2015)
41. Gand, K.: Investigating on requirements for business model representations: the case of information technology in healthcare. In: IEEE 19th Conference on Business Informatics, pp. 471–480 (2017)
42. Fattah, F., Arman, A.: Business-IT alignment: strategic alignment model for healthcare (case study in hospital bandung area). In: International Conference on ICT For Smart Society, pp. 256–259 (2014)
43. An, Y., Darlrymple, P., Rogers, M.: Collaborative social modeling for designing a patient wellness tracking system in a nurse-managed health care center. In: 4th International Conference on Design Science Research in Information Systems and Technology (2009)
44. Baslyman, M., Amyot, D.: A distance-based GRL approach to goal model refinement and alternative selection. In: IEEE 25th International Requirements Engineering Conference Workshops, pp. 16–20 (2017)
45. Carroll, N., Richardson, I.: Aligning healthcare innovation and software requirements through design thinking. In: IEEE/ACM International Workshop on Software Engineering in Healthcare Systems, pp. 1–7 (2016)
46. Yamamoto, S., Olayan, N., Morisaki, S.: Using ArchiMate to design e-Health business models. Acta Sci. Med. Sci. **2**(7), 18–26 (2018)

GPSR Compliance
The European Union's (EU) General Product Safety Regulation (GPSR) is a set of rules that requires consumer products to be safe and our obligations to ensure this.

If you have any concerns about our products, you can contact us on

ProductSafety@springernature.com

In case Publisher is established outside the EU, the EU authorized representative is:

Springer Nature Customer Service Center GmbH
Europaplatz 3
69115 Heidelberg, Germany

www.ingramcontent.com/pod-product-compliance
Ingram Content Group UK Ltd.
Pitfield, Milton Keynes, MK11 3LW, UK
UKHW021832270726
14058UKWH00001B/116

* 9 7 8 3 0 3 0 4 9 6 4 2 5 *